Praise for
*The Bond*

A *New York Times, Washington Post, Los Angeles Times*,
and Indie Bound Bestseller

"This is an important book, filled with information and inspired by a great compassion for our fellow animals—all animals. It is rich with fascinating, deeply moving, and sometimes deeply disturbing descriptions of our relationship with them. *The Bond* is thought provoking, and for some it will be thought changing. I hope that *The Bond*, written with such clarity and honesty by Wayne Pacelle, will be translated into many languages and become compulsory reading for students around the world. It will give new tools to those who are already working to help animals. But it must also be read by those who have not yet understood the true nature of animals and our relationship with them, that they too may help in the fight against cruelty and ignorance."    — Jane Goodall, founder of the Jane Goodall Institute

"A revolutionary book. . . . Wayne Pacelle is a fearless leader in today's modern animal protection movement."    — *Denver Post*

"An important read for anyone concerned about our responsibility as humans to all animals."    — *Christianity Today*

"Majestic in sweep and beautifully written, *The Bond* is a monumental achievement. . . . I can't think of any other book that is so valuable."    — Jeffrey Masson, author of *When Elephants Weep*

"Inspiring reading. . . . Pacelle presents the persuasive stance that by challenging cruelty to animals we are defending our own humanity."    — *Ventura County Star* (Ventura, California)

"Uplifting. . . . Pacelle makes a compelling case for more compassion toward animals."    — *St. Louis Post-Dispatch*

"One cannot read this book without coming away moved, disturbed, and compelled to reappraise one's own actions."
— Ted Kerasote, author of *Merle's Door: Lessons from a Freethinking Dog*

"A call to arms."  — *People*

"No animal escapes Wayne Pacelle's attention; nor should his book escape any human animal's attention."
— Alexandra Horowitz, author of *Inside of a Dog*

"Full of insight. . . . Covers the entire range of animal issues."
— *Winnipeg Free Press*

"*The Bond* stirs the heart and fuels emotions."
— *Deseret News* (Salt Lake City, Utah)

"The very best books change the way we see the world and the way we see ourselves. *The Bond* is that kind of work."
— John Mackey, CEO and Cofounder of Whole Foods Market

"The rallying cry for animal protection has grown louder."
— *Philadelphia Inquirer*

"A warrior for animals."  — *Santa Barbara News-Press*

"*The Bond* covers the full range of our interactions with animals."
— Peter Singer, author of *Animal Liberation* and Ira W. DeCamp Professor of Bioethics at Princeton University

"*The Bond* will resonate with many."  — *Washington Post Book World*

"Compelling. . . . Raises important points."  — *National Journal*

"Will hearten animal-lovers."  — *Washington Post*

"Very readable."  — *Sacramento Bee*

"A clear, profound statement."
                                        — AnnArbor.com

"Very much worth reading."
                                        — *Times Record* (Arkansas)

"A fascinating look at the lives we share with creatures."
                                        — *Ocean Drive* magazine

"Recommended reading."
                                        — *Savannah Morning News*

"An engaging and optimistic mediation on the future of human-animal co-existence . . . an inspiring read."
                                        — *Vegas* magazine

# About the Author

During his seventeen years with the Humane Society of the United States, including seven years as president and CEO, **Wayne Pacelle** has played a leading role in making HSUS, the nation's largest animal protection charity, into a dynamic public force and voice for animals. Taking a special interest in law reform, he has been the leading strategist in getting animal protection laws enacted by direct action of the electorate, designing winning campaigns in a dozen states for ballot initiatives that outlawed cockfighting, factory-farming practices, bear baiting, and a host of other inhumane practices. He has become the voice and face of the humane movement in this country. Pacelle was named one of *NonProfit Times*'s "Executives of the Year" for his leadership in responding to the Hurricane Katrina crisis. A graduate of Yale University, Pacelle lives in Washington, D.C.

Founded in 1954, the **Humane Society of the United States** is dedicated to confronting cruelty through public education, enforcement of humane laws, support of local humane organizations, investigations and litigation, and reform of public and corporate policies. The HSUS and its affiliates also provide veterinary services to remote areas, come to the aid of animals in natural disasters and incidents of large-scale cruelty, and operate a network of sanctuaries and rescue centers that directly care for thousands of animals each year. The HSUS family of organizations is supported by over eleven million people nationwide.

Learn more at www.humanesociety.org, and join the humane movement!

# THE BOND

# The
# BOND

*Our Kinship with Animals,*
*Our Call to Defend Them*

## WAYNE PACELLE

*wm*

WILLIAM MORROW
*An Imprint of* HarperCollins*Publishers*

A hardcover edition of this book was published in 2011 by William Morrow, an imprint of HarperCollins Publishers.

HarperCollins books may be purchased for educational, business, or sales promotional use. For information please write: Special Markets Department, HarperCollins Publishers, 10 East 53rd Street, New York, NY 10022.

First William Morrow trade paperback edition published 2012.

The Library of Congress has catalogued the hardcover edition as follows:
Pacelle, Wayne.
    The bond : our kinship with animals, our call to defend them / Wayne Pacelle. — 1st ed.
        p.  cm.
    Includes bibliographical references and index.
    ISBN 978-0-06-196978-2
1. Human-animal relationships.   2. Animal welfare.   I. Title.
    QL85.P33   2011
    179' .3—dc22

                                                    2010046177

ISBN 978-0-06-196980-5 (pbk.)

12  13  14  15  16  OV/RRD   10 9 8 7 6 5 4 3 2 1

# Contents

## Part II: The Betrayal of the Bond

## Part III: *Building a Humane World*

# Preface

W<small>E ALL HAVE OUR</small> own ideas about how to make the world a better place—and that's a good thing. Some are called to serve the poor, bringing food, shelter, medicine, and opportunity where the need is greatest. There are men and women devoted especially to the welfare of children, protecting them from violence and exploitation and finding homes for the orphans. Many dedicate themselves to preventing or curing disease, while others labor to protect the environment from pollution or careless development. And by the millions, men and women in America and beyond have set their hearts and minds to the work of preventing cruelty and alleviating the suffering of animals.

There's an endless division of labor in the good works of society. And though day to day all such worthy causes compete for our attention and support, in the end each is a part of the same fundamental enterprise of humanity. Imagine if we all focused on just one social concern, or even a handful of them. Where would that leave the other vital causes and needs that didn't make the cut? In a free and philanthropic society, it is for each of us to act and to give as our conscience asks, and in that pluralism of concerns, everybody is covered. In the famous phrase of Edmund Burke, each good cause and group is one more "little platoon" deployed in the work of building and defending a civil society.

My own little platoon is the Humane Society of the United States. And though I am a friend to many other causes, the cause of helping animals has always had a particular hold on me. I've always felt a bond with animals, and I have come to realize that so do people everywhere. At the same time, in more than twenty years of immersion in animal welfare, I've also seen incredible cruelty done to animals and heard ever more elaborate arguments offered to justify those abuses. This book is my attempt to confront these contradictions, to disentangle our sometimes conflicted attitudes toward animals, and to suggest a path forward in our own lives and in the life of our country. We all know that cruelty is wrong, but applying this principle in a consistent way can be awfully difficult when so many people and industries misuse animals so routinely and so blithely and often cannot even imagine doing things a different way. In each case, there is a different and better course, and our best guide is the bond with animals—that first impulse to do the decent thing for a fellow creature.

I've learned that in the animal-welfare movement no creature is quite forgotten, and there is no animal whose troubles do not matter to someone. Name any species and it has its defenders. It's not just the "charismatic" species, defended by such groups as the Mountain Lion Foundation, the Snow Leopard Trust, the Whale and Dolphin Conservation Society, the Gorilla Foundation, or Save the Elephants. Countless other groups have been formed to help farm animals, animals in laboratories, overworked animals like donkeys and camels, stray animals and feral cats, and other injured and needy creatures both domesticated and wild.

Some people are passionate about animals that most of us have never even heard of. After I completed the manuscript for this book, I came across a story by Kate Murphy in the *New York Times* about purple martins, the largest of North American swallows. Their numbers dropped in the twentieth century because of habitat changes and the introduction of exotic species. Today, all over America, you'll find nest boxes just for these birds, built by people who appreciate the martins for their beauty and want them to survive. Various blogs and YouTube videos are devoted to the birds, and there is even a Purple Martin Conservation Association, along with the Purple Martin Society of North America and the Purple Martin Preservation Alliance. Some might consider this preoccupation with a single species to be a little much, but I for one am glad for it. I love the idea that some people feel so connected to these creatures and are looking out for them.

I thought I'd heard about every category of animal rehabilitation until I read not long ago about the South Bay Wildlife Rehab, a group whose work includes saving injured and orphaned hummingbirds. Abby Sewell of the *Los Angeles Times* describes the work:

*"Only a crazy person would do this," said Terry Masear, 50, who has had as many as 60 hummingbirds at a time flitting around in the cages on her back patio in West Hollywood. . . .*

*During the summer, she takes a three-month hiatus from her job teaching English to foreign professionals at UCLA. Far from being a break, her summers with the hummingbirds entail 15- to 17-hour days of nonstop work. . . . From 5 a.m. to nightfall, every half hour, pre-fledgling hummingbirds must be fed with a syringe full of a special formula made in Germany. Masear guides a tube down the throats of the little birds, not much larger than bumblebees. Between feedings, she changes feeders for the older birds, cleans cages and monitors the birds' social interactions.*

*"You can't go out to dinner, you can't go out of town. You don't have a life," Masear said. . . .*

*When they are ready to live in the wild, Masear opens the*

*aviary door and watches the tiny birds spiral hundreds of feet into the air and disappear among the clouds. Even after seeing it hundreds of times, that moment still makes the long hours of drudgery worthwhile, she said: "When you release them, that's pure joy."*

We're told that not a sparrow falls without his Maker knowing, and millions of animal rescuers and rehabilitators like Terry Masear are paying close attention as well.

Of course, the flip side of all this benevolence is that such groups and their labors are needed in the first place. There is so much animal cruelty, homelessness, and suffering, and so much of it is a consequence of human action. In a rational world, the kinder people wouldn't be so busy dealing with the wreckage left by the cruel and careless.

As harsh as nature is for animals, cruelty comes only from human hands. We are the creature of conscience, aware of the wrongs we do and fully capable of making things right. Our best instincts will always tend in that direction, because a bond with animals is built into every one of us. That bond of kinship and fellow-feeling has been with us through the entire arc of human experience—from our first barefoot steps on the planet through the era of the domestication of animals and into the modern age. For all that sets humanity apart, animals remain "our companions in Creation," to borrow a phrase from Pope Benedict XVI, bound up with us in the story of life on earth. Every act of callousness toward an animal is a betrayal of that bond. In every act of kindness, we keep faith with the bond. And broadly speaking, the whole mission of the animal-welfare cause is to repair the bond—for their sake and for our own.

In our day, there are stresses and fractures of the human-animal bond, and some forces at work would sever it once and for all. They pull us in the wrong direction and away from the decent and honorable code that makes us care for creatures who are entirely at our mercy. Especially within the last two hundred years, we've come to apply an industrial mind-set to the use of animals, too often viewing them as if they were nothing but articles of commerce and the raw material of science, agriculture, and wildlife management. Here, as in other pursuits, human ingenuity has a way of outrunning human conscience, and some things we do only because we can—forgetting to ask whether we should.

Some object to the abuse of animals because they know that the habits of cruelty and selfishness easily carry over into how we treat one another. Yet in the end, the case for animals stands on its own merits. It needs no other concerns or connections to give it importance. Compassion for animals is a universal value, more so today than ever. Animals matter for their own sake, in their own right, and the wrongs in question are wrongs done to them.

Each of the chapters that follow expresses this truth in a different way. Chapter 1 reflects on the human-animal bond, its origins and its varying expressions across time. We are learning so much today about how animals think and feel, and chapter 2 examines that evidence along with the long history of denials among generations of scientists. In chapters 3 through 6, we'll survey some of the more systematic wrongs inflicted on animals—factory farming, animal fighting, and the abuse of pets and wildlife—and we'll see how some of these evils are being confronted and overcome through the power of democracy and the rule of law. From there, we'll venture into the world of the industries and interest groups that seek to hide or explain away the abuse of animals, and we'll listen carefully to their arguments and excuses. Finally, chapter 8 offers the best counterargument of all, by showing the great and growing possibilities of a humane economy—the new industries and practices that can thrive as we cast off old and cruel ways.

Over the years, I have rejoiced in the gains for animal welfare, and I've seen my share of setbacks, which you'll read about here too.

But the trajectory of progress is unmistakable and undeniable: by ever-larger majorities, the conscience of America is asserting itself. Animal protection has always been a noble cause. Now it's a winning cause too.

Today, more than ever, we hold all the cards in our relationship with animals. They have no say in their own fate, and it's up to us to speak and act on their behalf. International assemblies convene to decide which species will be protected and which will not—quarreling over terms and clauses that can either spare animals by the tens of thousands or destroy them on a similar scale. Humans control the births and deaths of billions of domesticated animals, and often the number of days or hours they are permitted to live. We even shape their very natures and temperament through selective breeding, genetic engineering, and now cloning—taking godlike powers upon ourselves, often with complete disregard for the original designs of God and nature.

When it comes to people and animals, power is asymmetrical, and all the advantages belong to us. Whether it's a subarctic nursery of newborn seals before the hunters come, or a herd of elephants about to be "culled," or dogs and cats at the end of their allotted time at a shelter and deemed too costly to keep alive, always their fate depends on our forbearance and our compassion. And one of the themes of human experience, since we first entered the picture ages ago, has been the expansion of that power and the moral test of how we use it—whether cruelly or kindly, selfishly or justly, pridefully or humbly. There have always been people and groups, in every time and place, who seek to dismiss and belittle the cause of protecting animals, as if the other creatures of the earth were just an obstacle to human progress that needs to be cleared away, subdued, or even wiped out as we decide. And there have always been those others who raised a clear voice in defense of animals, unafraid to question old assumptions, unworthy traditions, and practices and industries that can no longer hold up to reason or conscience.

Millions are carrying on in that same spirit of challenging, questioning, and calling cruelty by its name. The battle is unfolding on many fronts, as described in the pages to follow. In the end, whenever we humans find it in ourselves to help powerless and vulnerable creatures, we are both affirming their goodness and showing our own. In that way, their cause is also the cause of humanity, and this book is your invitation to join it.

# THE BOND

# INTRODUCTION

## *Sanctuary*

THE EAST TEXAS SUMMER sun was bearing down as I rumbled across the cattle guards at the front gate of America's largest animal sanctuary, the thirteen-hundred-acre Cleveland Amory Black Beauty Ranch. After a few familiar turns, I passed the chimp house and came to a stop at the elephant enclosure. That's where I knew I'd find Babe enjoying the shade of her barn or ambling about her spacious yard.

I was in luck. It was feeding time and she was inside. Arturo, Babe's caretaker and best friend, was serving the meal—no small task with a hungry, seventy-six-hundred-pound vegetarian. I was doubly lucky because Arturo let me get in on the act. He handed me a bundle of bananas, maybe fifteen in all—bigger than any bunch at a supermarket. Babe gave them a sniff with her trunk, then politely grabbed the entire clump, and guided it into her mouth. A few chomps later, and all those bananas were gone, peels and all.

Next on the menu was a watermelon. After Arturo handed me one of medium size, I grabbed the rounded ends and held it out in front of me with arms extended, as Babe's nimble and power-ful trunk—a natural wonder with forty thousand muscles—looped around it with impressive precision. Even though the fruit was big enough to feed a dozen people at a picnic, Babe wedged the whole thing into her mouth. She moved it around to position it just right, and there was a pop. You could hear the crunching for another thirty seconds, as the juice dribbled onto the ground, her eyes rolled back in her head, and an unmistakable smile swept across her face. It wasn't long before she was ready for more, which she indicated by sniffing me up and down, frisking me for another stash of fruit.

Babe allowed me a few pats on that amazing trunk—something I didn't take for granted considering what she had been through. During the 1980s, the South African government killed and removed entire herds of elephants—their idea of population control and the protection of habitat. It's euphemistically called "culling," and the practice is widely reviled because of the traumatic emotional effects it has on these highly intelligent animals. Babe's entire family was killed in a culling program just after she was born in 1984 in Kruger National Park. They spared her and a couple of other babies because they could be sold off for profit to animal dealers and used in circuses.

The U.S.-based Jorge Barreda Circus bought her, and she was shipped off in chains by boat to cross the Atlantic. After she was trained, almost certainly by forcible methods, the circus shuttled her from city to city to perform stunts before crowds that knew nothing of her sad travails. At some point, perhaps during transport to the United States or, later, in the circus, she sustained leg and foot injuries. Because the injuries were not properly attended to, they turned into a permanent disability. Babe was crippled—and no circus wants a performing elephant with a bum back leg. The circus cast her aside, and that's when Cleveland Amory and the Fund for Animals stepped in, acquiring her and retiring her to Black Beauty Ranch on Valentine's Day in 1996. Experts at the ranch knew that the injury might shorten her life span dramatically, but at least she'd pass that time in a loving and safe environment.

Even with such bitter stories as the backdrop to Black Beauty Ranch, every visit leaves me feeling hopeful, knowing that, in a world filled with so much trouble and misery for animals, at least these creatures are safe. And as it turned out, that visit in summer 2008 was my last chance to see Babe. We lost her just a couple of years later. She died at the age of twenty-six, as a result of heart problems that began with her abuse as a circus animal. That is a premature death for an elephant, but at least the second half of her life was spent in the care of people who respected and loved her. And somehow this gentle creature seemed to love everybody back, despite some very bad first impressions of humanity.

THE ANIMALS WHO FIND their way to Black Beauty have all experienced something frightening or hurtful in their lives, typically as a result of human neglect or cruelty. You can trace each animal back to a story of woe—to someone's shabby, heartless conduct—and taken together these rescued animals are a chronicle of the threats and sorrows visited upon animals in modern society. Creatures valued for commerce one day are treated as disposable the next. They are among the world's discards, long forgotten by the people who harmed them.

But this sad story also has a flip side. Our human instincts can run in the opposite direction. Every animal at the ranch is there because of some act of human kindness. For as much selfishness and callousness as they knew in their former lives, here they found a deep reservoir of human charity and goodness. This is true as well of the many thousands of animal shelters and sanctuaries across America and the world. And although these havens do not alone remove the causes of abuse and neglect, every rescued creature is a kind of messenger—showing the good that comes when we open our hearts to animals.

It would be so easy, in some ways, to "put down" such animals. It's done all the time, and often without a second thought. There's nothing in the law to prevent it, and strict economics do not argue for leniency. It's costly to rescue, transport, and, most of all, to provide lifetime care for animals. Those watermelons aren't free, and the price tag just for day-to-day operations at Black Beauty amounts to more than a million dollars a year. There is no end to the animals who need help, or to the help they need once they are rescued. From a certain perspective, all these works of charity are among the most impractical of human enterprises, and the only explanation for it is the bond of love and empathy that people feel for animals.

After the visit with Babe, I went back to the main gate and properly greeted the manager, Richard Farinato, and the other dedicated staff of Black Beauty Ranch. I'd been coming here on and off for almost twenty years, showing people around, or being introduced to the latest residents of America's largest and most diverse animal sanctuary. I walked a little ways from Richard's office to see Kitty, Lulu, and Midge—the second-chance chimps inhabiting the large rectangular enclosure just inside the front gate. Before they came here, these chimps had lived in laboratories, kept in metal cages not much larger than their bodies. Kitty was used for breeding purposes, and Lulu and Midge in invasive experiments. But at Black Beauty, they now had a chance to extend their long, sinewy arms. They had the run of an expansive field house nearly three stories high, with trees, giant tires, ropes, swings for them to climb and play on, and sun and fresh air streaming in. Once they saw that a visitor was there to greet them, they meandered over, one by one, to get a closer look. Lulu walked over on all fours from the sky bridge—which connects her sleeping quarters to the main outdoor enclosure and which spans above the dirt road I used to get to Babe's compound.

Years ago, in this same place, I had met Nim Chimpsky—the

chimp whose name was a play on the well-known linguist Noam Chomsky, perhaps the foremost theorist of human language structure. One of the most famous of the signing chimps, Nim had bounced around various research facilities, spending long stretches at the University of Oklahoma in Norman and then at Columbia University in New York, where for a number of years he was the subject of noninvasive language training under Dr. Herb Terrace. In his early life, Nim lived with people—watching television, sitting at the dinner table, and pulling snacks out of the refrigerator. He had many tutors through the years—a circumstance that undoubtedly had its emotional costs, as so many people passed in and out of his life. In spite of these challenges, he learned 125 American Sign Language (ASL) signs, and his nonhuman flirtation with human language was closely watched by academics from a variety of disciplines interested in whether the higher mammals could master one of the defining expressions of human culture. Whenever I came to Black Beauty, Nim Chimpsky typically greeted me and other visitors with a series of requests, usually for food, which always astounded me. The conclusions drawn from these language studies are, to this day, still debated—with the lead researcher, Dr. Terrace, concluding that Nim was unable to master syntax and the complex patterns of human speech.

But I always thought that Nim did a lot better with English than we did with the chimps' language and communication structures, and I was mightily impressed. His signing behavior provided plain evidence of meaningful thought and mental acumen, indicating that the differences in consciousness between humans and the higher mammals are ones of degree, and not of kind. In the last quarter century, scientists have decoded the genome of chimpanzees, revealing a 98 percent or so overlap with our own. When Nim threw things at a few of the people who came to see him at Black Beauty, or spit a stream of water at them with unbelievable precision from a considerable distance, and then ran around his enclosure with great glee, I wasn't quite sure if this was the 98 percent human or the 2 percent chimp expressing himself. It was one of those situations where you're not supposed to laugh, but you can't help it. I always told people not to wear their Sunday best when coming to Black Beauty.

Kitty, Lulu, and Midge are not signing chimps, and they are generally better behaved and less calculating than Nim, who died unexpectedly of a heart attack in 2000 at the relatively young age of twenty-eight. They communicate with their eyes. Once assembled on the other side of the metal fixture that separated us, they stared back at me, their eyes darting around in the same way that our eyes do. There's depth there, consciousness, something not far from personhood. And then there are the hands. They poked them out through the metal grating, seeking human contact in response. They possess the nimbleness of human hands, the gentle touch associated with our behavior. We talk with ours, and when they moved their hands in similar ways, it seemed so familiar.

Richard Farinato, who was a zookeeper before joining the Humane Society of the United States (HSUS) and then becoming manager of Black Beauty for some years, was with me on this tour, and he did me a favor by doing some chimp playacting. Richard began to run back and forth and to make other movements to excite Midge, the one male in the group. It didn't take much to get Midge going. He began walking on two legs, standing straight up, pumping his chest out with his hair standing on end all over his body, and making resonant hoots with his lips rounded and mouth open. He then made fast runs on all fours, including a lightning-fast scamper up a tree, before coming right back down. He picked up a giant truck tire, as if it were a normal car tire, and flipped it effortlessly. He was showing off, and it worked. Unless you see it, it's hard to comprehend the unbelievable speed and power of a chimpanzee— the males are five times as strong as a human adult, though their body weight is about the same. But the ladies, Kitty and Lulu, had seen it all before and displayed not the slightest interest in Midge's showmanship. Then just as quickly as it started, he settled down. He was done, and clearly pleased with his performance.

I wish these chimps didn't have to spend their lives in an enclosure, but there's no chance of releasing them successfully back into the wild. They've lived in captivity too long, their survival instincts have withered, and they are without family groups. But as captive settings go, this one is top-notch—and a dramatic upgrade from the significantly smaller enclosure that Nim and other chimps had once inhabited even at Black Beauty.

There is a world of difference between how we care for chimps at Black Beauty and how research laboratories treat the hundreds of chimps under their charge. We'd looked into the latter problem, exposing in March 2009 the abuses of the animals at the New Iberia Research Center (NIRC), just about three hours southeast of the ranch, near Lafayette, Louisiana. There, working as a lab employee for nine months and recording images on a tiny, hidden camera, an HSUS undercover investigator documented how 325 chimps live at the facility, with many languishing in small cages for years on end. Karen, the eldest chimp at NIRC, was captured in Africa in 1958, and she's been living in laboratories for more than half a century— since the end of the Eisenhower years. Karen and hundreds of other chimps at NIRC are used primarily for hepatitis research and are subjected to regular blood removal and invasive procedures, after being shot with a tranquilizer gun or stick. They live with constant anxiety—fearful of the people wielding guns and needles, whose presence signifies that something unpleasant or traumatic is about to happen. Several chimps, including Sterling, a twenty-two-year-old male, have gone mad as a result of their lifelong, unrelentingly boring, often solitary confinement—a situation that is particularly harsh for social animals who live in extended family groups in the wild. As a result, Sterling is a self-mutilator—or, as the technicians at NIRC say, capable of "self-injurious behavior"—severely

biting himself and throwing himself against the walls of his enclosure.

About a thousand chimps are kept in research facilities in the
United States, and I am relieved that Kitty, Lulu, and Midge have
been delivered from the wretched existence of a laboratory. Yes,
they are confined at Black Beauty, but in an open-air enclosure.
They are now a kind of family, and the humans who approach them
are carrying only offerings of food and toys. I often wonder how
these creatures could reconcile, if they were able to set their minds
to it, the different lives they have known—the before-and-after circumstances that they've experienced, and the two faces of humanity
they have seen.

Animals are so intuitive, and it does not take long for any of
them to realize they are in a safe place at Black Beauty—be they
chimps, buffalo, or horses. But I still wish that the incoming animal
residents at Black Beauty could read the sign hung from its front gate
and have an added measure of assurance. It reads: "I have nothing to
fear, and here my story ends. My troubles are over, and I am home."


THE GREAT MAN WHO gave them this home was Cleveland Amory.
This author and humanitarian cofounded, with Marian Probst, the
Fund for Animals. He was an animal rescuer in a category all his
own, and he had dreamed for years of creating a wide-open place
for homeless, abused, or neglected animals. Cleveland didn't like it
one bit when in 1979, the National Park Service announced plans to
shoot burros at Grand Canyon National Park, on grounds that they
were an invasive species and a threat to native vegetation. Cleveland
and the Fund intervened, orchestrating a dramatic capture and air-
lift of the animals that many had thought impossible in the canyon,
with its treacherous terrain and steep paths. He defied the skeptics
and in the end, he and his hired cowboys net-gunned and then air-
lifted 577 burros.

But this led to a second problem: he needed a place to put them. Fortuitously, the Fund for Animals had received a $600,000 bequest from a donor earlier that year, and Cleveland and Marian invested the money in a property in the small east Texas town of Murchison to fulfill their dream. Cleveland named the new sanctuary in honor of the horse whose story so captivated him as a child. The words that greet visitors to the ranch are drawn from Anna Sewell's *Black Beauty,* which she wrote in the 1870s "to induce kindness, sympathy, and an understanding treatment of horses." The book turned out to have even broader appeal and became one of the best-selling children's novels of all time and an enduring plea for kindness toward all animals.

Cleveland Amory envisioned the Black Beauty Ranch as mainly a sanctuary for horses and burros, but he was not a man to turn away any animal in need. And when he learned of the plight of Babe and Nim and others desperate for a safe home, he opened wide the front gates at Black Beauty, like a modern-day Noah. By the time Cleveland passed away in 1998—nearly two decades after he established the ranch—this haven created for burros was home to more than a thousand animals of all sizes and species. During my visit, I checked in on the mountain lions and bobcats who were formerly kept as pets but given up once they became too difficult or dangerous. I saw a small herd of exotic deer from Africa and Asia, saved before they ended up as live targets at a Texas trophy hunting ranch. And—in keeping with Cleveland's conviction that the ranch should always take in the downtrodden and sometimes the surprising—I was especially glad to see twenty-one prairie dogs, who faced gassing and poisoning by ranchers before they were plucked from their burrows and given new "digs."

When the prairie dogs rise from their dens here, they not only see burros, but also a wide range of other creatures not native to America's Great Plains. Among the most unusual sights for them must be the more than a half-dozen ostriches. Some years ago, a boom in raising ostriches for meat and feathers in Oklahoma and Texas went bust, and Black Beauty ended up taking some of the castoffs from bankrupted businesses. The prairie dogs may find some comfort in looking out at the American bison, also native to the Great Plains. But they must feel befuddled when they see a water buffalo together with a "zony" (a mix between a zebra and a pony) and a one-armed kangaroo named Roo-Roo, who lost a limb because of injuries suffered while being forced to box. In all, there are thirteen hundred animals at Black Beauty. And Cleveland Amory is still in their midst: he asked to be buried at Black Beauty with his beloved Polar Bear, the scruffy, hungry stray cat he rescued on a frozen winter night in New York City and made famous in three of his best-selling books.

After seeing the prairie dogs, Richard and I hopped in a truck and headed to some of the horse pastures. I had new residents to meet—two horses with quite a backstory. About half of all of Black Beauty's residents are equines—Belgians and other big-bodied draft horses, athletic thoroughbreds, stout quarter horses, and even miniature ponies. The hoofed animals have room to roam, constrained only by a perimeter fence. Here the pastures rise and fall in the distance, the grass is thick, and landscape is sprinkled with trees. There's plenty of rain in east Texas, and during much of the year the fields are rich and green.

On our way to find the two new equine arrivals, we passed by the exotic deer, more than a dozen bison, and Omar the dromedary camel, who had formed an improbable friendship with Babe the elephant. They played together over the fence that separated them, with Babe often reaching over with her trunk to wrap Omar's long neck in a hug—a sight you won't see anywhere else and certainly won't forget. Before long, we wended our way into the back pastures, driving off the dirt road and onto the grass. The horses cleared a path for us as Richard looked left and right to find Mari Mariah and Josie Sahara, two Arabians. "There they are," he said.

He steered in their direction and drove slowly toward the two gray-coated horses with black markings. We hopped out of the vehicle.

Trucks and people at the ranch signify feeding and friendship, so nobody scattered—not even the wild mustangs or the wild burros rescued from public lands in the West and previously unaccustomed to human contact. I gave a few gentle pats to horses on the way, and then approached the two girls, who let me get right up next to them. I stroked the back of Mari's neck first, and then gave equal attention to Josie, rubbing the coarse hair on her left side. They are mother and daughter, and they had perhaps the closest brush with death of any animals at Black Beauty—no small statement, considering nearly all of the animals here had some dangerous encounters with people.

The pair of thoroughbreds had been purchased somewhere by "a killer buyer"—the people who collect horses for the meat trade—and had been shipped in a cattle truck to Cavel International, a Belgian-owned horse slaughter plant in DeKalb, Illinois. Mari and Josie were on the slaughterhouse floor with about two dozen other horses, organized in single file to be shot with a brain-penetrating captive-bolt gun, then hoisted upside down by a chain wrapped around a rear leg, and finally bled out after a sharp cut to the neck with a long knife. With their blood drained, they'd then move down the disassembly line, where their legs and head would be severed, their skin peeled away, and then their remaining body parts processed, shrink-wrapped, frozen, and flown to France, Belgium, or Japan. There, the meat would be carefully prepared before being served to high-end patrons. But just before that final, fatal series of events took place, a federal judge ordered the plant to suspend operations. Mari and Josie had seen the machinery of death, but then were turned around and marched alive out of that facility.

A few years before, HSUS had launched a national campaign to stop this slaughter of horses. We knew it was a despicable business, with killer buyers purchasing horses at auction and often misrepresenting their intentions to sellers who assumed that their horses would end up in good hands. Once acquired, the horses would be loaded up on cattle trucks and sent straight to slaughter. The trip, sometimes more than a thousand miles, was a harrowing experience, in which stallions, mares, and foals were all bunched together in trucks too small for the adults even to stand. If they survived the transport, they would meet their end at the plant, with eyes bulging as these fright-and-flight animals witnessed other horses dying before them.

We made our case against this cruelty and the betrayal of these animals to Congress, which in 2005 voted to bar the U.S. Department of Agriculture (USDA) from funding inspections at horse slaughter plants—effectively shutting down the plants because a slaughterhouse cannot operate without federal meat inspectors. We won the vote, but the USDA, which often acts as an agent of agribusiness and the slaughter industry, came up with a scheme to allow the slaughter plants to pay the federal government's inspectors, in a fee-for-service arrangement. With the USDA now looking the other way, the plants kept operating, and HSUS sued to stop this clear disregard of the will of Congress. U.S. District Judge Colleen Kollar-Kotelly ruled in our favor and ordered the plants closed. It was this merciful intervention that had spared Mari and Josie and the others. Just weeks later, the Illinois legislature banned horse slaughter in the state, and the plant was finally and permanently shuttered. And just two months earlier, we secured a separate victory in federal court, upholding a Texas law to ban slaughter in that state—an action that forced the closure of the only two other plants in the nation. As it happened, one of the two Texas plants was in the town of Kaufman, just a half hour from our Black Beauty facility.

Many of the animals at Black Beauty have been through harrowing ordeals, but I don't think any of the others have been marched onto a slick, bloody concrete floor at a slaughterhouse and somehow emerged alive into the light of day. Mari and Josie were very lucky girls. In the pasture, on a cool but sunny day, I whispered to them that we were happy they were here, and that they needn't worry anymore about their safety. This was their permanent home, I said, and all of the people here were their friends.

ANIMALS HAVE ALWAYS TUGGED at my pant leg, in one way or another. I've had an instinct to draw closer to them, watch them, learn about them, and sketch them. An interest in animals, and a concern for them, has been a big part of my emotional outlook for as long as I can remember. My childhood dogs—Pericles, Brandy, and finally Randi—were among my best friends. On the athletic fields across the street from our home, I'd throw the ball for hours on end to Brandy, a retriever mix. She'd have kept at it for the entire day if I'd kept throwing, and I could have watched her run the ball down and drop it at my feet for just as long.

I had a protective instinct toward animals—all animals, not just the ones I knew well and loved—before anyone gave me any moral guidance on the subject. The books I read as a boy about animals did not spark my interest in them, but rather fed an interest that seemed to be there from the start. Children's publishers, writers, and illustrators have for decades understood the connection that kids have with animals, and animals are everywhere in their works. Dr. Stephen Kellert, a Yale professor who has studied human attitudes toward animals and nature, found that "animals constitute more than 90 percent of the characters employed in language acquisition and counting in children's preschool books." Along with books, TV shows featuring animals were my favorite diversions. I never missed an episode of *Mutual of Omaha's Wild Kingdom,* the first in a long line of nature programming that's become so much more popular, widespread, and authentic today.

I never had a gift for drawing or art, but I filled sketchbooks with my renderings of animals—deer, moose, wolves, and any others that caught my interest. I had all our encyclopedias dog-eared to the entries for animals, and I had much of the information memorized. I knew the major mammals and birds of every continent, as well as the marine mammals, such as seals and walruses, who lived on the margins of the land masses. I'd anxiously await the arrival of the latest *National Geographic;* I wasted no time in scanning the subjects teased on the cover and turned to the spread featuring an animal species. Almost at once, my imagination would take me to a distant place.

I knew animals were different from humans—but different in good ways. They had surprising and beautiful forms: the lowered, shaggy enormous heads of American buffalo; the bizarre elongated noses of anteaters; the powerful, muscular frame of the grizzly. They were either fast like pronghorn antelopes, or slow like the three-toed sloths of South America. They were fragile-looking like white-tailed deer and Dama gazelles, standing on champagne-glass legs, or they were sturdy and stout like rhinos and hippos. They were small but possessed with a strong internal motor, like hummingbirds, who are the only birds able to fly backward. Or they were prehistorically large like African elephants and blue whales, whose languid movements in the water belied the long distances they traveled in a day. Classes of animals like monkeys came in so many novel and distinct forms and species—from the lemurs of Madagascar, to the gibbons of Asia, to the New World monkeys with their prehensile tails.

Truly, I was dazzled. The world was a kaleidoscope of unbelievably interesting creatures—depending on your worldview, the gifts of the Creator, or the finely honed works of natural selection. The baleen whales feed on krill, the tiniest of sea creatures, yet these marine mammals grow larger than any other animals ever to inhabit the planet. Wolverines are solitary and no larger than a medium-sized dog, but they have been known to drive larger predators, even a bear or a pack of wolves, from their kill. The small Arctic tern makes an annual, winding, nonlinear trip from the Arctic to the

Antarctic and then back again, a staggering forty-four thousand miles—by far the longest migration of any animal on the planet.

That the animals were different from us was no reason to diminish their worth, to deny them comfort, freedom, or the respect due a fellow creature. We have so many remarkable and distinguishing attributes as a species, including our language and our ingenuity. But animals, sometimes humbly and sometimes grandly, have distinctive attributes of their own that command our appreciation. At least in their physical characteristics, some animals made us look plodding, cumbersome, and a bit inelegant, with our less impressive musculature and patchy hair. I loved them all the more for the strange, unique, and often beautiful traits that set them apart from us. This is a feeling common to most of humanity, across time and culture, and captured in the enduring words from the Old Testament that all creatures sing their Maker's praises and are dear to Him for their own sake.

Growing up in Connecticut, my parents took me a couple of times each year to the state's modest, unaccredited zoo—the Beardsley Park Zoo in Bridgeport. There, I made the rounds to see all of the creatures, but I spent the most time gazing at the wolves I had studied so much in books from the National Geographic Society. Occasionally, we'd head to the Bronx Zoo in New York City, and even though I was a die-hard baseball fan I'd always choose a visit to the animals over a game at Yankee Stadium. A couple of times we even trekked several hours farther to the Catskill Game Farm—a drive-through safari park I especially liked because the animals were not in cages and freely roamed the grounds. I later learned that this facility was an awful place that mistreated its animals, procured them from disreputable sources, and sent "surplus" animals to exotic animal dealers and even captive-hunting operations. I also learned as an adult that the animal scenes on *Mutual of Omaha's Wild Kingdom* were often staged: for instance, a bear would be dropped in a rushing river and then saved in an exciting made-for-TV rescue.

There's often a hidden story to the places and people who trade on the public's fascination with animals, and I've learned that skepticism about motives and methods is usually in order.

I felt a bond with animals in an intense way, and it had all sorts of expressions in my childhood—my love of pets, reading, drawing, gazing at their exotic forms. I guess that's not surprising given where I've ended up professionally—as president of the Humane Society of the United States, and fully engaged in the cause of protecting animals. But even as a teenager, I also realized that my passion for animals was not a peculiar or unusual interest. So many others I knew, including family members and friends, shared my instincts. I rarely ever came across anyone who said he or she disliked animals, and most everybody I knew had a pet, along with a general benevolence toward other creatures.

And today, as I scan with a wide lens the many expressions of the human-animal bond, I see it everywhere in our culture. Some 170 million dogs and cats are living in American homes, compared with 65 million just thirty-five years ago. The pet product and services industry generates more than $45 billion in annual sales. There are nine million horses in the United States. More than seventy million wildlife watchers take to the forests, fields, and waterways each year to see animals in their native habitats, and they spend billions in the process. There are more than ten thousand organizations devoted to helping animals in the United States alone. We now have a television network dedicated entirely to animals—Animal Planet, which does for animals what MTV did for pop artists, giving them a regular presence in our homes and in our daily consciousness. Animal books like *Marley and Me* often sit on the best-seller lists alongside political memoirs and self-help blockbusters. There is something universal about the bond we have with animals—an instinct to have them in our lives, to be near them and to care for them.

But this bond is just part of the story in our relationship with animals. There have always been countervailing forces that trump or negate the positive effects of the human-animal bond. Even today, despite organized efforts to protect animals and the environment, human activity inflicts harm and death on animals to a degree that was once impossible, much less conceivable. We slaughter upwards of fifty billion domesticated animals for food every year in the world, with an ever-growing share confined for their entire lives on factory farms.

Other human actions pose a threat to thousands of species and billions of individual wild animals, including habitat destruction and fragmentation; increasing volumes of carbon dioxide, other greenhouse gases, toxins, and animal waste released into the atmosphere and environment; overfishing and floating plastic and tangled fishing lines in the oceans; the introduction or release of nonnative species on land and in freshwater lakes and rivers; and direct killing or trapping of wild animals for a variety of trade-related purposes. Some frogs and other amphibians are now born without limbs or with other deformities, and scientists blame chemicals and endocrine disrupters introduced into the environment. In some areas of the world, such as Angola, Mozambique, and parts of the Congo, once-rich forests are now nearly empty of mammals, reptiles, amphibians, and even birds, as a result of poverty, civil war, and the bush-meat trade. The animals of Southeast Asia are being collected in extraordinary numbers to meet the demand for animals at live-animal markets and for trinkets and other wildlife products in China. Because of these human actions, some experts believe we are now moving toward the greatest extinction crisis since the last Ice Age—the sixth extinction crisis in our planet's history, but the first one with unmistakable human fingerprints on it.

Our contradictory impulses toward animals are a central theme of this book. And it's fairly clear that these opposing impulses—our concern and fellow-feeling for animals, and our capacity for complete disregard—have always been at work for good and ill. The contradiction is sharpest in our own time, and becomes more untenable as our power over the animals and their world reaches the outer boundaries. Never before have we known or cared so much about animals, and never before have we been so callous or ruthless in the things we do to them. Humanity cannot go much farther along both paths at once. We have to choose one way over the other, and it's my hope that this book will help to move us all in the direction of kindness, mercy, and life.

# Part I

## A Special Bond

# CHAPTER ONE

# The Ties That Bond

I'LL NEVER FORGET AN encounter I witnessed, as a teenager, between my mother and a mouse. I was in the dining room, Mom was in the kitchen a few steps away, and we were chatting about nothing in particular. I was doing most of the jabbering, and as I looked toward the kitchen doorway I was the first to spot the tiny intruder. I was surprised but gave no audible reaction. But when Mom caught a glimpse of our visitor, she let out a sound none of us kids had ever quite heard before—a gasp followed by a piercing shriek. I guess the mouse had never heard such a scream either, because in an instant the creature scurried across the kitchen floor and disappeared into the tight space beneath the stove.

By then Mom was standing on a chair, just like in a cartoon. Not yet able to exhale, she whimpered, "There's a mouse!" I ran into the kitchen just as she stepped down from the chair and rushed out. I can still recall her wide-eyed expression of fear as she dashed by me and then ran up the stairs to her bedroom. I got my bearings, and then followed to calm her, as well as any thirteen-year-old could manage.

Mom's reaction was not exactly proportional to the threat. The mere sight of the mouse triggered some deep-seated fear, planted not by any incident in her own life but ages earlier by the experiences of our human ancestors. I imagine the little guy was pretty scared himself, running for his life once he found himself out in the open.

Rodents, snakes, and certain other species have this effect on us. Many people have a conditioned fear of snakes in particular, even in areas without poisonous species. Younger children rarely exhibit such fear, but often by age five or so become more wary. Primates react much as we do. They don't have chairs to jump on, but they get up those trees in a hurry.

Large predators, of course, elicit a similar instinctive reaction in people. An experience of my friend and colleague Katherine Bragdon has stayed with me because it still seems so completely irrational. Katherine and I have run a number of political campaigns for animals, including a ballot initiative in Oregon in 1994 to ban the use of hounds in hunting bears and mountain lions. Critics of the ballot initiative—mostly leaders of sport hunting organizations—shamelessly played off people's fears, arguing that if the initiative passed, mountain lions soon would prowl the suburbs and stalk children in school yards. In debates with our opponents, Katherine and I countered that hunting with packs of dogs was unsporting and inhumane. We also pointed out that killing two hundred or three hundred mountain lions annually, out of a statewide population of twenty-five hundred, did little to further reduce the very slight risk of an attack—in fact, there was no record of any mauling in the state's history. Statistically, a person in North America had a far greater chance of being struck by lightning or a falling tree, or killed by a bee sting, than of tangling with a mountain lion.

Even so, just a few weeks before the big vote (which we won), Katherine confided that she and her boyfriend had been hiking in the Cascade Mountains when she was suddenly overcome by the fear of a lion attacking her. She had been swimming in a natural pool in a forest and suddenly felt terrified. She had not spotted a lion, or seen or heard any evidence of one nearby, but nonetheless believed that a big cat was on her trail. For a time, Katherine couldn't move. Eventually, she shook off the paralysis but not the fear, and she led her boyfriend on a panicked run back to their cabin. All the while, she banged sticks and yelled loud enough to ward off any lion in the vicinity or, for that matter, probably every other living creature within earshot of her clattering. When she made it back safely, relief washed over her. Even though she knew as well as anyone that the chance of such an attack was remote, and had herself made that case a hundred times before, a primal fear had trumped all reason.

Maybe Katherine was channeling the fears early hominids had of saber-toothed tigers or other large predators. These may be the same kind of fears that today lead people in some areas to exaggerate the threats posed by wolves or grizzly bears, and to urge their eradication. In the case of my dear mother, her rodent sighting had triggered an innate fear response. After all, seven hundred years ago the Black Death decimated Europe, setting off a series of political, cultural, and perhaps even genetic aftershocks still being felt today. Likely the greatest pandemic ever, it claimed the lives of some 150 million human beings—one-third of the entire population of the continent. Rats had a lot to do with its spread, and in such an environment, avoiding rodents was a distinct evolutionary advantage. We humans are programmed not to forget it.

None of this is to say that these coded, fear-based reactions must dictate our behavior. Although some people want to kill wolves and mountain lions, there is also a powerful movement to protect such creatures. The federal government has reintroduced predators in parts of the West, and there have been well-organized and often successful efforts, like our 1994 Oregon campaign, to restrict the hunting of wolves, mountain lions, and bears.

In other words, our fascination with wild creatures has two sides. If fear is one face, kinship is the other, and our instinct to draw more closely to animals is the basis of the human-animal bond, inspiring our attachment to pets and our preoccupation with wildlife. This bond is not, however, merely a contemporary concern or the product of an affluent society, though it's certainly true that the humane movement was born in the midst of staggering exploitation of animals in the industrial era. Long before that, the same feeling of connection that inspired the humane movement was at work in the world. It can be traced across human history and detected even in such ancient practices as hunting and animal sacrifice—as antithetical as these might seem to any sense of kinship. Our bond with animals has taken many forms, both violent and benevolent, but it's always been there. Animals are not just in the backdrop of our own story, but at the center of the whole drama, and how we treat them is one of the great themes of the human story.

## The Biochemistry of the Bond

IN OUR DEALINGS WITH animals, we are a blend of nature and nurture. Many positive responses to animals come naturally to us, and these are the building blocks of the human-animal bond. This instinctive benevolence, just like our instinctive fears, can be encouraged or suppressed by culture. But either way it is forged of strong stuff and explains much of the fascination with animals common to every society.

Some compelling scientific theories and recent studies even point to evidence of a bond in the very biochemistry of mammals. The idea was most recently advanced by Meg Olmert, who has spent twenty years studying the human-animal relationship. Her work brings together the latest scientific evidence that the mammalian hormone oxytocin provides a crucial explanation for the presence of the human-animal bond.

It's hardly radical to suggest that biochemistry has a strong influence on human and animal behavior. The fear we feel upon encountering a human intruder or a threatening animal derives from a chain reaction of responses, ultimately releasing the hormones adrenaline and cortisol. These, in turn, influence our emotions and trigger glucose production, which prepares the body for a fight-or-flight response. Biochemical changes that are now well understood help explain aggression, panic, and sexual attraction, among many other behaviors. Advances in neurobiology have led to the development of medicines that better control depression and other mental disorders. If biochemical reactions drive aversive responses, it seems plausible that hormones must also affect social behavior, especially for a gregarious species like ours.

Oxytocin is a neuropeptide long known to promote maternal care in animals. It plays a major role in human childbirth, inducing labor contractions and also stimulating lactation. Olmert argues that oxytocin also calms new mothers and intensifies their fascination with their babies. In rodents, oxytocin triggers a mother's protective aggression against intruders.

But oxytocin does not just circulate in the bloodstream during childbirth and the nursing phase of child rearing. Present in all men and women, it has a broader prosocial, bonding function in human-to-human interactions. It is a social recognition hormone, helping us to identify faces and to exude warmth toward others. It even appears to have stress-relieving effects, such as lowering blood pressure and anxiety. Several researchers have found that administering oxytocin to people encourages trust and empathy, while facilitating better reading of emotions.

Sue Carter, codirector of the Brain-Body Center at the University of Illinois, Chicago and one of the leading oxytocin experts, says much of what she learned about the bonding powers of oxytocin came from studying prairie voles, small burrowing rodents who mate for life. Monogamy is highly unusual for mammals, and rare even among other species of voles. It seems that the reason prairie voles are able to form their powerful social bonds is because their brains make more oxytocin and vasopressin (a closely related hormone) in the regions that release dopamine and serotonin. Together, these pleasure-inducing neurochemicals both initiate and reward social behavior, weaving bonds that last a lifetime.

What holds for prairie voles is also true for us and many other mammals. Researchers are recognizing that common social behaviors trigger the release of these pleasure-inducing hormones. So a smile, a hug, or the smell or laughter of a baby trip the oxytocin switch and stimulate this biochemical cascade, making us want to reach out and approach others and to build more cooperative and nurturing relationships.

Maia Szalavitz and Dr. Bruce Perry, in their book *Born for Love,* go further, arguing that oxytocin is the basis for familiarity and empathy in mammals. They speculate that its work begins with the mother-child bond but continues throughout life, allowing "mammals to make the connection between a particular individual and pleasure. Without it, many animals can't even tell each other apart."

There's little doubt now that oxytocin plays a critical role in bonding for humans, too. But what about bonding between humans and other species? Two South African researchers, Johannes Odendaal and Roy Meintjes, measured the blood pressure and blood chemistry of eighteen humans and dogs before, during, and after they had friendly interactions. They found that the interactions almost doubled oxytocin concentrations in both the people and the dogs, and that the humans' blood pressure also dropped dramatically. This finding and others suggest that pets may be "one of the most potent triggers of oxytocin production in humans."

In 2009, the U.S. Army began pairing service dogs with veterans suffering posttraumatic stress disorder. Just weeks after Iraq war veteran Chris Goehner got a service dog to help him deal with PTSD, he reduced his doses of anxiety and sleep medication by half. Aaron Ellis, another Iraq veteran, ceased his medication and was able to set foot in a grocery store for the first time in three years. Nobody has done the blood work to prove that oxytocin aided the rehabilitation of these veterans, but it's a decent bet.

A quarter century before Meg Olmert's synthesis of oxytocin research, the eminent Harvard biologist E. O. Wilson offered his own novel theory to explain the human-animal bond. He argued that humans have "an innate tendency to focus on life and lifelike processes." "From infancy," he explained, "we concentrate happily on ourselves and other organisms. We learn to distinguish life from the inanimate and move toward it like moths to a porch light." This bond between humans and nature is "biologically encoded," built into us as a result of hundreds of thousands of years of experiences with other living beings. Wilson called his theory "biophilia," and although the idea was deeply influenced by his love for the natural world, it was grounded in years of study and observation.

To support his conclusions, Wilson drew upon the behavior of the few hunting-and-gathering cultures that remain. For them, the exploration of nature is not a form of recreation or retreat, as modern societies regard it, but still the central activity of their lives. And a working knowledge of nature, such as the ability to distinguish between poisonous or edible plants, or to understand the habits and behaviors of animals, is a matter of survival. Their whole existence is built around the lives of other creatures and the rhythms of nature, and this outlook affects everything from those individuals' behavior to their most basic beliefs.

Living in small nomadic bands, early hominids knew how to hunt, but probably ate mostly nuts, plants, and the leftovers from the kills of larger and better equipped predators. For much of prehistory, they maintained stable and relatively low population levels. Population geneticists at the University of Utah guess that one million years ago, there were just fifty-five thousand people "spread across the entire Old World"—half the population of my hometown of New Haven, Connecticut. There is evidence that hominids have been around in various forms for at least two million years, and a few clues trace them back another five million years before that. It was during these innumerable generations that we gained our basic first impressions of animals, impressions that left a permanent mark on our genetic coding.

It was not until the arrival of modern humans, perhaps eighty thousand years ago, with their larger brains, more lethal weapons, and control of fire that our species became a much more dominating predator, able to exert its will over other species and paving the way for steady growth. The human population, scattered across the Old World, expanded from tens of thousands to several million in the years to come. And it was not until about twelve thousand years ago that humans wandered across a Bering Sea ice bridge and first left footprints in the New World.

A few tentative conclusions can be drawn about these diverse hunting-and-gathering peoples. Along with archaeological evidence, we have the records of European explorers who first encountered the natives of the Americas, Australia, and other regions centuries ago. A limited number of these communities still survive and have been studied by anthropologists and ethnographers.

There's a tendency to think of these preagricultural peoples as "noble savages," existing in harmony with nature and all its creatures. But this romantic notion does not always square with the evidence. Many archaeologists speculate that about eleven thousand years ago, Clovis people, the first makers of sharp-pointed tools, wiped out many species of large mammals throughout much of the world, including mammoths, mastodons, and giant beavers in the Americas. This "overkill" hypothesis is a contested theory, and other scientists believe that climate change may explain the demise of so many species. But whatever the true causes, there is little doubt that Clovis people were very successful hunters who killed massive numbers of animals and hardly had an ecologically benign impact.

Yet it is still plausible to describe many hunting-and-gathering people as the original naturalists, who depended "on an exact learned knowledge of crucial aspects of natural history," as E. O. Wilson puts it. Biological anthropologist Jared Diamond of UCLA observes that the hunting-and-gathering societies he's encountered in Papua New Guinea are "walking encyclopedias of natural history, with individual names (in their local language) for as many as a thousand or more plant and animal species, and with detailed knowledge of those species' biological characteristics, distribution and potential uses." Recounting an exchange with a man named Teu, a native of Kulambangra, one of the Solomon Islands in the Pacific, Diamond writes: "For every one of Kulambangra's eight resident bird species, Teu dictated to me an account consisting of its name . . . its song, preferred habitat, abundance, size of the group in which it usually foraged, diet, nest construction, clutch size, breeding season, seasonal altitudinal movements, and frequency and group size for over-water dispersal."

Many tribal communities not only understand other life-forms, but also hold beliefs that acknowledge their complete dependence on other creatures for survival. In their cosmology, "people can turn into animals, and animals turn into people," according to Dr. James Serpell, an expert on the history of the human-animal bond. The Inuit worry about offending animal spirits and are careful about killing animals, perhaps because in the high Arctic they are so dependent on animal protein for survival. Other communities recognize totem animals—creatures especially revered, and species that are not to be killed. "They are apologetic when they kill animals," Serpell observes. "It is an interesting moral dilemma. They have to eat animals, but they feel bad about it. They have all sorts of rituals and belief systems to absolve them of the guilt." Social anthropologist Tim Ingold adds, "The hunter hopes that by being good to animals, they in turn will be good to him. But by the same token, the animals have the power to withhold if any attempt is made to coerce what they are not, of their own volition, prepared to provide. . . . Animals thus maltreated will desert the hunter, or even cause him to fail."

It may be that our modern-day appreciation of animals is, in part, an inheritance from tribal peoples who needed these observational skills to survive. Stephen Kellert of the Yale School of Forestry and Environmental Studies says that we are drawn to vistas because that's the way our ancestors behaved, surveying the landscape for threats or hunting opportunities. And still today we are drawn to the sight and sound of running water because it is clean and drinkable. We especially like to look at waterfalls and find the sound of rushing water soothing. In short, many modern-day behaviors may be partly attributable to the survival benefits they provided to prehistoric hominids.

## Pet Keeping Through the Ages

ALMOST EVERYONE WITH A dog or cat notices the daily displays of instinctive action—a cat stalking a toy mouse; a dog chasing a squirrel or rolling on his back in play to expose an open stomach. Dogs or cats who are separated from their mothers at birth behave the same way. It's in their "wiring," and while parental teaching or play with others of their kind can perfect it, the core behavior is innate. We domesticated these animals thousands of years ago—the dog at least fifteen thousand years ago, perhaps considerably earlier, and the cat some four thousand years ago—and they have been selectively bred over many hundreds of generations. Yet these predatory or playful behaviors are still there.

It's probably not so different for us—the connections to animals and to the physical features of the natural world are still with us, ready to be expressed with the proper cues. We have not shed these inclinations or instincts in the last ten thousand years. As Kellert notes,

> The various expressions of biophilia hardly constitute "hard-wired" instincts like breathing or eating. In fact, the human affinity for nature represents a collection of relatively weak biological tendencies. All the various strains of biophilia depend on adequate learning and experience.

Even in our era, with meat at supermarkets in every neighborhood, hunting still preoccupies a portion of the male population. It is practiced throughout the world, and thirteen million Americans go hunting every year. Some reasonably argue that hunting behavior is still encoded within us. There is something atavistic in men. Yet it is also clear that hunting is a learned behavior, with fathers or grandfathers passing on the tradition to the young—the sort of "weak" tendency that without cultural support would fall away. But once triggered, the thrill of stalking and killing a living creature is addictive for some hunters, providing an adrenaline rush. Many hunters speak of "buck fever"—the physical reaction that occurs when an animal comes into shooting range. As one hunting blogger writes,

> It is the severe case of the jitters. It's the adrenaline rush associated with the sight of your target. It's the nervous system on overdrive. Crazy things such as the shakes, sweats, increased blood pressure and just plain poor shooting can occur when Buck Fever overtakes you.

Hunting animals can become a mania. Some well-heeled trophy hunters travel the world in search of big game and have killed three hundred different species and subspecies of animals, making it almost a full-time avocation. They will tell you they consider their rapacious pursuits a "God-given right."

It's also true that more than 90 percent of men in industrialized societies choose not to hunt at all, and demographic trends show the practice is in a gradual but steady decline. Social scientists speculate that other activities now provide an outlet for this biological impulse. "Humans are not hardwired to hunt, but they are hardwired to engage in physical activities that resemble hunting," argues James Serpell. "Most sports have to do with chasing and catching and hitting small objects that are traveling fast."

Although hunting is a long-standing, biologically based human behavior, other, gentler behaviors better demonstrate the human-animal bond. Generally, we think of pet keeping as a modern Western practice. But people have been drawn to pets throughout human history, and even prehistory. Prior to the last two centuries, it was more a custom of elites than the underclass. The British monarchs, for instance, displayed a particular affection for small dogs. The ancient Egyptians had an incredible affinity for animals, even mummifying them as a demonstration of respect. And in ancient China, Serpell notes, "the ancestors of the modern Pekingese enjoyed a privileged status unrivaled by any other variety of pet before or since. They were given the titles of princes and princesses, and huge personal stipends were set apart for their benefit."

Serpell's most surprising revelation, however, is that pet keeping was also a social activity among tribal societies. "The practice of capturing and taming wild animals is widespread and incredibly common, especially in South America and Southeast Asia," he writes. This seems counterintuitive, given the popular perception that hunting-and-gathering peoples engaged in an unyielding struggle for survival, and that pet keeping would be an unaffordable luxury for people living on the edge. Nevertheless, it's true that our remote ancestors found a place in their lives for the care of animals.

It may be that some of these pets were the orphaned young of animals killed in the hunt. The same people who killed the mothers of these creatures felt pity at their bleats and cries—an early expression of our conflicting impulses toward animals.

Early Native Americans, for instance, kept tame raccoons, moose, bison, wolves, and even bears as pets, and they treated their dogs affectionately. Serpell cites the Norwegian explorer Carl Lumholtz who observed that Australian Aborigines treated their pet dingoes "with greater care than they bestow on their children," kissing the animal on the snout and even grooming them. Occasionally, tamed wild animals were used for food, but often they were just companions. "In the New Guinea villages where I work," Jared Diamond relates, "I often see people with pet kangaroos, possums, and birds ranging from flycatchers to ospreys." According to Serpell, Spanish explorers reported that South American Indian women kept wild animals in their huts. The women would never eat or sell them; when strangers killed one, the women would shriek, dissolve into tears and wring their hands "as if it had been an only son."

There is also evidence that native peoples took things a step further than pet keeping. The cries of baby animals may have drawn empathetic responses from women in native communities. Women suckled animals, whether wolf puppies or other baby animals, as a sign of kinship and affection, in one of the first steps toward domestication. These experiences remind us that the human-animal bond is not a modern-day invention, but just the latest expression of it.

## Domesticated Animals: "Bonded to the House"

AT SOME POINT, PROBABLY in the latter half of the twentieth century, we crossed a divide in the animal kingdom: the planet became home to more large domesticated mammals than wild ones. Cows, pigs, horses, goats, and others of their kind outnumbered the great herds of caribou in the Arctic, the wildebeest and zebra and Cape buffalo who migrate across the plains of East Africa, and all the bison, bighorn sheep, deer, elk, and other large mammals in the world.

China alone, for example, has an estimated seven hundred million pigs—nearly seven times as many as the United States. India has nearly three hundred million cattle. And if the human population grows by two to three billion in the next fifty years, as expected, we're likely to add billions more domesticated animals to today's totals, causing further habitat loss and declines in populations of wild mammals. The domestic-to-wild ratio has generally been a zero-sum game, with increases in domesticated animals resulting in greater loss of the wild ones. There still exists an amazing diversity of life on the planet, and major efforts are under way to preserve what remains. But the most abundant animals are a relative handful of domesticated breeds. If you count chickens and rabbits, there may be close to a hundred billion domesticated mammals and birds in the world. The Mesozoic era was the age of dinosaurs; the Cenozoic era was the age of mammals; and today, we live in the age of domesticated animals.

Among large animals, most are here because we encourage or allow their presence. With domesticated animals, we create them, kill most of them for food, and then create more of them. Our involvement in their lives is direct, complete, and intimate, as we typically control everything from the location and timing of reproduction, to their diet, to the means of death. Although that's not the case for wild animals—unless they are captive—even they survive largely because of a conscious decision on our part to protect them, or at least to not eradicate them. Every other year, for example, representatives from more than 150 nations gather to set rules on the commercial exploitation and trade in wild animal populations, with life-and-death debates and votes on the fate of elephants, whales, turtles, and hundreds of other species.

Domestication may seem like a historical inevitability, perhaps because cattle, sheep, and others are so central to our way of life. Christian tradition places the newborn Jesus in a manger with domesticated animals gathered round. What could better demonstrate the rootedness of these animals in Western culture than their presence at the Savior's birth? But it's important to remember that for most of history, humans killed animals without ever assuming complete control over them, except at the point of taking their lives. The animals were free—reproducing naturally, using their wits and physical skills to find food, and living day to day without any human influence. Domestication is among the most far-reaching innovations of humankind, as significant as the development of human language, the control of fire, or the harnessing of steam and fossil fuels for energy.

Early hominids could not possibly have foreseen how domestication would play out. It may have been driven by raw need, but much came of what they started, and the human-animal bond traces back to them—including one of the most enduring friendships on earth. Our special regard for dogs likely began when prehistoric people scavenged the remains of wolf kills, or wolves scavenged the remains of human kills, instilling a sense of familiarity. With wolves acting as sentries, and benefiting by feeding on meat scraps, humans may have slept more soundly, providing what must have been a very prized benefit. Maybe humans studied the way wolves hunted in groups and learned from it. Quite possibly our ancestors also took in the orphaned young of wolves and raised the pups among humans. Women and children may have played with wolf pups, and nursing mothers may have even allowed them to suckle. With the barriers broken down, humans and juvenile wolves may have kept each other warm by sleeping side by side. However it all happened, it is indisputable that Paleolithic wolves not uncommonly came under human control, and an everlasting bond was formed.

Over the millennia, dogs have become remarkably specialized, with widely different sizes and functions, through countless generations of selective breeding. Some dogs have been bred to protect people and their property, whereas others have been bred to herd farm animals. There are a variety of breeds with specialized hunting abilities. Some chase and corner prey; others point or flush upland game birds such as pheasants or grouse; and still others retrieve waterfowl. In the Arctic, huskies pull sleds, allowing people to cover vast terrain; though snowmobiles have diminished the need for such work, today's mushing events are competitive celebrations of this legacy. Dogs also work for us in other ways, helping police to chase down criminals and the military to detect explosives. They serve as rescuers, sniffing out people trapped in the rubble of buildings brought down by earthquakes or other calamities. They guide the blind, and they act as therapy animals. If you have a dog and think of him or her as part of the family, you are in good company with people everywhere on earth.

Sheep, goats, pigs, and cattle (probably in that order) came after the domestication of the dog—all about ten thousand to thirteen thousand years ago. Humans enlisted them to produce meat, milk products, and clothing, and, in the case of cattle, labor. Humans domesticated plants about the same time, allowing them to reduce their dependence on hunting. Domestication allowed humans to live in fixed settlements, which grew as people accumulated food. These expanded communities required more complex social, political, and economic organization. And this affected human health, religious practices, literature, art, science, culture, and just about everything else. Take any ancient civilization, in China, Egypt, or the Indus Valley, and progress began with the domestication of plants and animals.

Since that defining point cattle have been present in almost every culture. Today, the U.S. cattle industry is worth $76 billion annually, while the Brazilian and European industries generate billions in commerce too. Along with goats and sheep, cattle are kept by the thirty million to forty million nomadic pastoralists in Asia and Africa, and these communities still have a wide range of rituals associated with the care of the animals. In India, which has more cows than any other nation, the animals are venerated in Vedic scriptures and are treated like "one's mother" because they give milk. They are generally not killed for food. There are cow shelters throughout India, which were some of the first animal shelters anywhere in the world. The followers of Jainism, an ancient religion in India, were the first to set up animal hospitals.

Of the one hundred fifty or so species of large, herbivorous wild terrestrial mammals in the world above seventy-five pounds, only fourteen have been successfully domesticated. Zebras, wisents (European buffalo), and Cape buffalo are just a few of the species that did not submit to the demand to live and breed in close quarters with people. Candidates for domestication needed the right temperament and behavioral traits to be tamed. Humans have had more luck with smaller species, from the fox to the mink, in the last two centuries. But for those larger animals who were trusting and malleable enough, the new relationship changed everything.

Horses were among the most difficult of animals to domesticate. They are high-strung, fright-and-flight animals, and early humans were their unyielding predators. Horses were probably headed toward extinction as a result of human overhunting in the Pleistocene era before domestication changed their fortunes. It's a sharp turn to go from prey to partner, but horses managed to make it.

Once it happened, first about six thousand years ago, the domestication of horses had an enormous impact. Originally used for meat and milk, and then for labor, their greatest utility came in transport and battle, aiding the ascendancy of empires throughout the world. Horses enabled the extension of Roman power, tilling fields and transporting cargoes to the far corners of the empire. Genghis Khan and his army galloped on horseback from the steppes of eastern Asia to establish the largest contiguous empire in history in the thirteenth century. With the aid of horses, Spanish conquistador Francisco Pizarro vanquished the much larger force of the Incan empire in the Andes Mountains in the sixteenth century. Without horses, these and many other momentous turns in history would never have come to pass, and of course the same is true in American history. Manifest Destiny advanced across the continent with the gallop of a horse.

Yet despite their service to humanity, many horses were overworked, whipped, and exposed to extreme weather—abuses so common that they gave rise to the humane movement in the nineteenth century. By the twentieth century, railroads and later the internal combustion engine diminished the need for horses, who assumed a largely recreational role in America and other advanced countries. In much of the developing world, though, horses and mules are still widely used for labor. Historian David Anthony summarizes, "Horses changed the way people hunted and made war, altered concepts of distance, extended interregional trade, brought previously isolated cultures into contact, provided new standards of wealth, opened the world's grasslands to efficient human exploitation, and redefined the cultural identities of those societies that became equestrian." Even today, "horsepower" is the measure of engine might in high-speed transportation.

We can hardly begin to measure the extent to which domestication changed the way humans lived, and it necessarily altered the terms of the human-animal bond. "From being independent co-equals or even superiors," as James Serpell notes, "animals became servants and subordinates, increasingly dependent on people for care and protection." Instead of simple dependence, the human-animal relationship became one of dominance.

More broadly, agriculture changed not only communities' means of gathering food, but also their basic economic order. Now people could accumulate goods and exchange them. Bartering turned into trade, and money became the medium of exchange.

Where hunting-and-gathering cultures practiced communal living and egalitarianism, agricultural peoples established social hierarchy and class. Property owners brought in laborers to work the land, and before long feudalism and serfdom took hold in some places, and slavery in others.

But it's simplistic to view domestication as a new form of hierarchy, with animals at the bottom and all people above them. The mere act of raising and tending plants and animals promoted a new intimate connection with other beings. Historian Lewis Mumford argues that pastoralism, by involving humans in the process of producing and maintaining animals, created a new interdependence between living organisms.

This was especially true for domesticated animals. Although people ultimately killed the animals, they first helped give them life, and then nourished and cared for them. The term animal *husbandry,* literally meaning "bonded to the house," nicely expresses a sense of kinship and responsibility that once characterized this relationship. There was some level of mutuality, even more so than among nonagricultural people who killed animals but gave nothing back to the animals in return—no nourishment, shelter, or protection. Like Mumford, the contemporary social commentator Jeremy Rifkin believes the practice of nurturing other creatures spilled out and had broader implications for the functioning of these societies. The garden civilizations in the early days of domestication, according to Rifkin, were some of the most peaceful, stable, and self-contained societies.

From the beginning of domestication, to be sure, the human-animal bond was conflicted—with exploitation and compassion living in close quarters. But just as the killing of wild animals posed a moral conflict to the hunter, it was perhaps a greater dilemma for the farmer. This was an era of agrarian living, when by far most people lived on farms tending animals. Farming was modest and personal, with the shepherd and his family knowing the flock and nourishing them—a bond beautifully captured in the biblical story of the poor man and his cherished ewe lamb, who "drank of his own cup, and lay in his bosom, and was unto him as a daughter."

## Animal Sacrifice: "The Perfect Gift"

EVEN THE ANIMAL SACRIFICES we read of from ancient times signify a highly personal moral tension—a recognition that living creatures belong to God and not to man, and that killing animals is a shedding of innocent blood. Animal sacrifice—which today seems needlessly cruel—was humanity's way of showing respect, to the creature and the Creator alike, and of washing away the taint of violence.

For thousands of years, animal sacrifice was common to agricultural societies. It was practiced in the ancient civilizations in Egypt, Mesopotamia, Assyria, the Indus Valley, and Greece. And for a time it was a ritual in the major monotheistic religions.

There were rules that guided much animal sacrifice, including biblical codes. Often the killing was done just before eating the meat, to ensure that the animal's carcass was not wasted. Religious scholar Laura Hobgood-Oster says that animal sacrifice "may have also been a way to distribute food," allowing not just the wealthy to consume meat. Sacrificial rituals mirrored absolution rituals in hunting communities, where shaman conducted rites of atonement after the killing of wild animals. Indeed, priests typically conducted animal sacrifice, with laypersons forbidden from killing domesticated animals.

Only unblemished animals could be sacrificed, and that meant they had to be cared for before slaughter. Jonathan Klawans writes of sacrifice in ancient Israel: "Before any animal can be sacrificed, it must first be protected when born, fed, and then finally guided to its place of slaughter." The whole practice is predicated on a decent treatment of animals to ensure that they are fit offerings for sacrifice to the Almighty. If they were not well treated, Klawans adds, "there would be nothing left to offer."

A well-established order prevailed, with the Creator above all things, the people below Him, and the animals below them. The people of Israel begged God for mercy and understood that they had to be merciful to their flock as well—Moses himself, we are told, was chosen because of an act of mercy to a stray lamb. Death was the inevitable result for the sacrificial animal, but there was much more to the exercise before the killing. The general idea was to demonstrate devotion, care for animals, and a humble spirit—virtues today entirely missing from the world of industrialized agriculture.

In the Book of Leviticus, killing maimed animals is strictly forbidden. Leviticus also prohibits killing an animal and her offspring on the same day, partly in recognition of the bond between the animals. Historian Kimberly Patton notes that the animal is a "perfect gift or offering to God . . . expressing through its own form God's perfection."

Patton also argues that ritual slaughter elevates the place of the animal among the flock, though it could hardly have felt that way to the chosen creature: "The sacrificial animal is special, even unique; it is perfect; it is ritually adorned and beautified for its death. It has a special relationship to God and in sacrifice is given back to Him."

As with the killing of wild animals in many primitive communities, a sense of sorrow is expressed in animal sacrifice, reflecting an understanding of the moral implications of the action. The feeling was also observed by the famed ethnographer Bronislaw Malinowski, who examined animal sacrifice among Trobriand Islanders in the Pacific islands near New Guinea in the early 1900s:

> For the act they were about to commit elaborate excuses were offered; they shuddered at the prospect of the sheep's death, they wept over it as though they were its parents. Before the blow was struck, they implored the beast's forgiveness. They then addressed themselves to

*the species to which the beast belonged, as if addressing a large family*
*clan, beseeching it not to seek vengeance for the act that was about to*
*be inflicted on one of its members.*

Typically, there is a tradition that sacrificial animals voluntarily submit to their fate. Similarly, hunting-and-gathering people prefer to think that animals offer themselves to the hunter and will not do so again if the pact is violated.

Of course, in the Christian tradition, ages of animal sacrifice ultimately gave way to the willing sacrifice of the Lord, who went to the cross as a lamb to the slaughter. With Jesus having taken upon Himself the sins of man, the rationale for animal sacrifice had vanished. It faded away in Christian practice, even while surviving elsewhere.

Some readers may be surprised to learn that animal sacrifice continues even in our time, though whatever air of piety it might once have conveyed has long since given way to mindless, wholesale butchery. Humanity has learned a few things in two thousand years, and one of those lessons is that real respect is shown in sparing animals from death instead of meting it out. We don't have the excuse of ignorance anymore, and we don't operate under the same harsh necessities of earlier times. Without those justifications, all that's left are violent and self-serving spectacles with the hollow pretext of religious devotion.

In November 2009, at the Gadhimai Festival in southern Nepal, followers of a Hindu sect killed a staggering five hundred thousand animals in an orgy of animal sacrifice—doubling the number of animals killed at the previous festival five years earlier. The "panch-bali," or five offerings, involves slicing the throats of five kinds of animals (buffalo, goats, pigs, roosters, and rats). One participant in the slaughter commented, "The more animals I kill, the more satisfied I feel. I am helping an ancient tradition to survive." A priest declared, "The goddess needs blood. Then that person can make his wishes come true." In reality, the exercise has no point but to excuse and even encourage bloodlust, and religious and secular leaders in Nepal and India have called for an end to this frenzy of killing. In many Indian states bordering Nepal, it's seen for what it is and is strictly forbidden.

In the United States, the city of Hialeah, Florida, outlawed animal sacrifice in the 1980s, and authorities arrested a follower of the Afro-Caribbean religion of Santeria for practicing animal sacrifice within city limits. The defendants cast the case as a test of religious freedom, and eventually it was heard in the U.S. Supreme Court. The justices ruled in favor of the defendants on grounds of religious freedom, deciding the legal question but leaving the moral ones unresolved. Animal advocates, including the Humane Society of the United States, sided with the city of Hialeah and against the practitioners of ritual killing, arguing that it is unnecessary and wrong. Although animal sacrifice is still practiced throughout much of the world, and in particular the Islamic world, it does not find much favor in Western societies and is no longer a part of rituals for Buddhists, Christians, Jews, or other major religions. And no one who takes those great religions seriously feels that anything has been lost.

## Animal Welfare and the "Great Republic of the Future"

THIS HISTORY IS THE backdrop to our modern debates about the care and treatment of the other creatures with whom we share the earth. Critics of today's animal-protection movement often argue that animal welfare is ultimately a trivial matter—the product of effete modern sensibilities. But the truth is that our relationship with animals has always been profoundly important. How we treat the world's animals—whether coldly or compassionately, selfishly or justly—is a measure of who we are. It defines our character, our moral progress, and our ability to look beyond self-interest. There's a reason why the decent treatment of animals commands ownership of the word *humane*.

Far from a trivial concern, the care of animals and the fight against cruelty can bring out the best in the human heart—all the more so because it inspires pure altruism, with no angle or payoff. For the earliest reformers in the humane movement, religious conviction was no hindrance to an active concern for the treatment of animals. It was their greatest inspiration, and it gave them the courage to call cruelty by its name. Theirs was the spirit of Britain's William Wilberforce, the great nineteenth-century Christian champion against both human slavery and cruelty to animals. Then as now, such reformist zeal was often dismissed as extreme and subversive, but Wilberforce had his answer: "If to be feeling alive to the suffering of my fellow-creatures is to be a fanatic, I am one of the most incurable fanatics ever permitted to be at large."

The campaigners of Wilberforce's time brought new passion to the cause, and added the first major animal-welfare reforms to Western law, but in fact they were carrying on a debate and a mission that had begun much earlier. For millennia, the Christian, Jewish, and Islamic traditions have called followers to be compassionate toward animals, and Hinduism, Buddhism, and Jainism are even more solidly grounded on the principles of kindness toward all creatures. During the Middle Ages, numerous stories of saints and their solicitude to animals provided a rich lore of tradition that continues today in the timeless example of St. Francis of Assisi—a man honored by Pope John Paul II for "his solicitous care, not only towards men, but also towards animals." Even St. Thomas Aquinas, sometimes cited to support the view that we have no direct duty to animals, recognized animal cruelty as a moral issue, because he feared it fostered human wickedness.

In the early modern era, questions about cruelty and kindness were given even more serious attention. Even in 1641, while the

French philosopher René Descartes was arguing that animals were automata with no souls, mind, or conscious experience of pain, New England Puritans were approving the first legal code to protect animals—the Body of Liberties, which prohibited "tyranny or cruelty towards any brute creatures which are usually kept for the use of man."

Only a few years later, in *Some Thoughts Concerning Education*, John Locke advised parents to chastise thoughtless cruelty by their children, and within a few decades the kindness-to-animals ethic became a common theme in children's literature. By the late eighteenth century, a few influential thinkers, including Jeremy Bentham and John Lawrence, were arguing that the law should extend its protection to animals. Bentham famously argued that "the question is not, Can they *reason*? nor, Can they *talk*? but, Can they *suffer*?," while Lawrence plaintively asked his readers, "Can there be one kind of justice for men and another for brutes? Or is feeling in them a different thing to what it is in ourselves?" In 1751, the English satirist William Hogarth painted a series of engravings, *The Four Stages of Cruelty,* showing how cruelty to animals preceded violence against people. The authors and poets of the Romantic movement weighed in with admonitions against cruelty and a call to respect nature—Percy Bysshe Shelley looked forward to the day when "All things are void of terror," and "man has lost his terrible prerogative."

By the early nineteenth century, these moral convictions were widely shared. Although the German philosopher Immanuel Kant believed with Aquinas and Locke that humans had no direct duty to animals, he agreed that cruelty was wrong because it debased humans and hardened their hearts. Several decades later, Kant's countryman Arthur Schopenhauer directed his own serious challenge to human presumption in treating animals as nothing. "The assumption that animals are without rights," wrote Schopenhauer, "and the illusion that our treatment of them has no moral significance, is a positively outrageous example of Western crudity and barbarity. Universal compassion is the only guarantee of morality." Before the end of the century, in 1892, the reformer Henry Salt laid the groundwork for the modern approach to animal rights with a powerful work that makes remarkable reading even today, *Animals' Rights Considered in Relation to Social Progress*. Salt encouraged people to "recognize the common bond of humanity that unites all living beings in one universal brotherhood," asserting with conviction that the "great republic of the future will not confine its benefi-cence to man."

America's first anticruelty laws actually predate any animal-protection organization. Maine passed the first such law in 1821, and more than a dozen states followed suit by 1860. The states also began to prohibit staged animal fighting—more than half of them doing so before 1900. Bull baiting, ratting, gander pulling, dog-fighting, cockfighting, and other such spectacles were often viewed as attracting gambling and other vices. There was a commonsense understanding that cruelty to animals went against the better in-stincts of humanity and was an affront to any notion of Christian stewardship.

Organized concern for animals first emerged in the nineteenth century. Just two years after Britain enacted the world's first na-tional anticruelty law in 1822, several dozen men met in a London coffeehouse to form the Society for the Prevention of Cruelty to Animals—the first of its kind in the world. Queen Victoria lent her authority to the group some years later, and in 1840 it took on the noble name it carries to this day, the Royal SPCA. It is still one of the best-known charities in the United Kingdom, with an annual budget exceeding 110 million pounds.

In 1866, New York socialite Henry Bergh founded this nation's first animal-welfare charity, the American Society for the Preven-tion of Cruelty to Animals (ASPCA), just three days after the New York legislature passed an ambitious anticruelty law. Inspired by

Bergh, and concerned for the plight of Philadelphia's animals, Caroline Earle White and her husband formed the Pennsylvania SPCA and a women's branch of the organization about a year later. And after learning of a grueling forty-mile race involving two horses each carrying two riders over rough roads, an ordeal that killed both animals, George Angell, also a man of wealth and stature, founded the Massachusetts SPCA in Boston in 1868. Historian Bernard Unti calls Bergh, White, and Angell "the founding triumvirate" of the nineteenth-century American animal-protection movement. Inspired by these leaders in the Northeast, humanitarians throughout the United States formed SPCAs and humane societies, from Portland, Maine, to Portland, Oregon.

In *Pets in America,* historian Katherine Grier notes, "As families abandoned keeping livestock, another change that took place gradually into the 1900s, pet keeping became the only way many could directly express kindness to animals." For the emerging urban class, pets may have fulfilled a basic impulse to be close to other creatures—what some have since described as "nature-deficit disorder."

The new middle-class interest in pet keeping inevitably produced a new expression of the human-animal bond, but also stray cats and dogs, especially in cities. The newly formed humane organizations responded by taking on the tasks of animal control, sheltering, and often the grim business of euthanasia. Bergh and Angell, in particular, focused on all cruelties, from inhumane transportation and slaughter of livestock, to pigeon shoots, animal fighting, and vivisection. Congress's first animal-protection law, enacted in 1873, actually protected farm animals: the Twenty-Eight-Hour Law required the off-loading, feeding, and watering of livestock from railcars on the long journey from the West to the East. In the United Kingdom, Parliament regulated vivisection in 1876, but despite parallel efforts by American advocates, the U.S. Congress did not follow, and animal experimentation went largely unregulated until Congress enacted the federal Animal Welfare Act nearly a century later. The result of these long-ago failures by Congress was to leave a free-for-all in animal experimentation and agriculture—a long period of unrestrained abuse that we are just now beginning to correct in law. So many of the animal-welfare controversies of our day could have been avoided had there only been some basic moral framework for the conduct of these industries, put in place before new and ever more severe cruelties were allowed to become the norm.

Darwin published *On the Origin of Species* in 1859, and it's tempting to think that his argument about the continuity of life kick-started the formation of the first animal-welfare charities in America just a few years later. But, as Bernard Unti notes, humane reformers "never cited such influences, preferring to ground their arguments in religious conviction, concern for suffering, and animal individuality." Indeed, it's closer to the truth that Darwin's arguments made a handy case against animal protection. Critics of the humane movement invoked Darwin (in ways he would not always approve) to rationalize interspecies competition and violence, excusing human brutality as "survival of the fittest." If life is a ruthless struggle for advantage over other creatures, then why on earth should anyone bother to care for a weaker, more vulnerable species? It was not until the 1890s that some animal advocates, Henry Salt among them, began to find support in Darwin's writings. At the heart of Darwin's work, after all, was a recognition of commonality among the "higher" mammals, and a shared experience of suffering in a world filled with struggle.

## Self-Evident Truths

THE BIRTH OF THE organized humane movement, however, required more than philosophers, preachers, and sympathetic scientists. Like most social movements, animal protection was born from both ne-

cessity and crisis—as a reaction to the intensifying, seemingly .
less exploitation of animals in the industrial era. In the nineteen.
century, market hunters stalked the vast forests and plains of America and slaughtered wildlife with abandon to supply meat, furs, and
feathers to the boomtowns of Chicago, New York, and Philadelphia.
In the old agrarian economy, food had often been produced right
on the farm, connecting people intimately to its production. But in
the new urban-rural economy, many city dwellers were completely
disconnected from the wasteful killing of wildlife and the clear-
cutting of forests. It dawned on some people—the would-be leaders
of the conservation movement—that all wildlife would be lost if
boundaries and standards were not established.

Around the same time, city dwellers became all too familiar
with cruelty, since it occurred right before their eyes—and not just
the gruesome abattoirs exposed by Upton Sinclair in *The Jungle*.
More than twelve million horses lived in America, most of them in
cities, and they were the primary means of transport for people and
goods. Carriage drivers often overworked and mercilessly whipped
the animals. They denied them adequate shelter, or forced them to
work in punishing heat or cold. Henry Bergh and others decided
to confront this abuse, and the years afterward would see a sudden
rise in prosecutions for the abuse of horses—in fact they accounted
for 70 percent of the 8,256 prosecutions for animal cruelty between
1868 and 1880. With those actions in New York to protect horses
began the animal-protection movement we know today.

Katherine Grier argues that a variety of other social factors prepared the way for the humane movement, including the emerging
ethic of gentility in domestic households. The new urban middle
class adopted fresh social rules, among them an ethic of kindness
and of pet keeping. In the Second Great Awakening, a period of
religious revival in the 1820s and 1830s, there was a rapid growth
in liberal evangelical Protestantism, and with it a belief in salvation
through benevolence. Even before then, other religious-minded re-

kers and Methodists, advocated a Christian ethic
imals. In churches and meetinghouses across the
e was any formal humane movement to speak
hear stirring sermons on the theme. The great
ley himself, the founder of Methodism, offered a sermon
entitled "The General Deliverance," which would inspire a lot of
soul-searching if it were heard in any Christian church today. Why,
Wesley asked, should we consider the plight of mistreated animals?
"It may enlarge our hearts toward these poor creatures to reflect
that, vile as they may appear in our eyes, not a one of them is forgotten in the sight of our Father which is in heaven."

Such direct appeals to conscience had a power of their own,
but in an earlier age would have had no possibility of ever gaining
the force of law. By the eighteenth century, of course, humanity
was ready to question a good many old assumptions and to upend
entrenched institutions. This was the time of a great shift in political thought, which changed forever our understanding of the rights
of the individual, and over time would establish the moral rationale
for extending legal protections to animals against the particular tyranny directed at them.

Political philosophers of the Enlightenment asserted the claims
of the individual against the capricious power of monarchies and
other undemocratic institutions. Suddenly the world had a new and
unstoppable idea—that rights were inherent and that principles of
justice were universal. With the ideals that inspired the French and
American revolutions, Western societies had a new creed that held
laws and governments alike to "self-evident truths" and to the con-
sent of the governed. Laws that perpetuated any wrongs or abuses of
power were unjust laws and stood in need of democratic reform to
meet the demands of impartial justice.

With these new principles ringingly affirmed in the Declaration
of the Rights of Man and above all in the American Declaration of

Independence, there could be no going back, and old and obvious evils could no longer be excused or explained away as the natural order. Of course, the most glaring of wrong in that era was human slavery, which would take generations of agitation and conflict, and in America all the carnage of a civil war, before it was finally abolished. For good reason, the cause of abolition would command the devotion and energy of reformers for generations, and the first successes of the organized animal-protection movement would have to wait until slavery was overcome. But once that fundamental advance had been made, the humane movement suddenly found its footing, and it's not by coincidence that its ranks included many of the same men and women who had labored against slavery.

In the United States, the very first humane organization—the ASPCA—was founded in the year after the Civil War. And the same woman who had helped awaken the conscience of America about slavery was also a stirring voice in defense of animals. "The care of the defenseless animal creation," wrote Harriet Beecher Stowe, "is to be an evidence of the complete triumph of Christianity."

A half century earlier, a similar sequence had played out in England, with the conviction and fervor of abolitionists quickly turning to the abolition of cruelty to animals. The United Kingdom banned the slave trade in 1807, and little more than a decade later enacted national laws against cruelty that were the first of their kind anywhere. William Wilberforce is rightly remembered as a determined opponent of both slavery and animal cruelty, and often for the same reasons. This was a man who moved comfortably between the antislavery cause and the SPCA and saw both efforts as part of the same humane enterprise—defending different victims from the same kind of arrogant and merciless abuse of power. The spirit was expressed by a young compatriot of his in both causes, Anthony Ashley Cooper, the seventh Earl of Shaftesbury. "I was convinced," Shaftesbury wrote, "that God had called me to devote whatever advantages He might have bestowed upon me to the cause of the weak, the helpless, both man and beast, and those who had none to help them."

Across the years, new generations of animal advocates have made the case in their own way, but the words of Shaftbury express the fundamental moral impulse that has guided us from the very beginning. In every time, there have always been people who abuse power and are blind to the misery they cause. No matter who the victims are or how ruthless the practice in question, they always find some pretext, excuse, or high-sounding justification for the harm they inflict. And in every time there have also been people who see the wrong and try to right it—who confront injustice no matter how acceptable, routine, or customary it may seem to others. They are the ones who rid the civilized world of slavery, at a time when settled opinion still held that slaves were the rightful property of a few and an economic necessity for society. They are the ones who dared to challenge powerful industries and to assert the right of children not to be exploited for cheap labor. They're the ones who refused to live in a country that treated women as second-class citizens and secured reforms that protected women from domestic violence, opened doors of opportunity, and finally granted them the right to vote. And for a hundred and fifty years, they are the kind of people who have seen cruelty and couldn't abide it, and often at risk to themselves have come to the aid of afflicted animals.

Look around today and you'll find thousands of groups carrying on this work, often against abuses of a kind and scale that men like Shaftsbury could not have imagined. Though critics try to cast the animal-protection movement as something foreign, eccentric, and subversive, this cause has long been a worthy and natural expression of the great Western moral tradition.

And the cause runs even deeper than that. It calls us back to a bond of kinship and respect for animals that humans have felt since before we had words to describe it. It calls us to be faithful shepherds of Creation, alert to our duties and alive to the suffering of our fellow creatures. It is a cause that speaks to moral aspiration, asking us to live up to our own highest ideals of justice and mercy. It writes common moral standards into law, so that liberty does not give way to license, and cruel people are not left as judges of their own conduct. Like all the best moral causes, in the end animal protection reminds us of what we know already—that to mistreat an animal is low, dishonorable, and an abuse of power that diminishes man and animal alike.

# CHAPTER TWO

# The Mismeasure of Animals

IT WAS A HOT, mid-August day at the Brookfield Zoo in Chicago. As they do each day, visitors by the thousands, young and old, passed through the turnstiles for a close-up look at wild animals, many of them rare and most from faraway corners of the earth. For many visitors, this was the closest they'd come to seeing these creatures, outside of Animal Planet. And on this summer day in 1996, many visitors would leave with a story to tell.

Like even the best-managed zoos, Brookfield cannot possibly re-create the fullness of an animal's life in the wild. Today's curators may pay more attention to the behavioral and psychological needs of the animals than in other times, but the settings are still nothing like their native habitats, where the lives of countless creatures are all bound up together. In captivity, the animals are normally housed separately by species—the hippos here, the polar bears there, and the jaguars somewhere else. Even if the zoo architects plant some familiar vegetation, each animal's quarters bear little resemblance to plains, forests, or wetlands that the creature would know in the wild.

With the zookeepers providing food and water, the animals have little need of their foraging, hunting, or evasive abilities. There's not a lot of problem solving going on for the animals—just eating, a lot of lying around, and occasionally some playing. On rare occasions, we'll see something different and exciting—a lion or tiger makes a ten-foot leap within the enclosure; baboons chase, play, and climb; or the hippos spar. Usually, though, patrons get just a hint of the full range of behaviors these creatures normally display in the wild. What the visitor does take away, even in this manufactured setting, is an appreciation of the physical majesty and beauty of the animals, and that's what keeps the crowds coming back.

On this day, too, it all began with a flash of action. A woman and her three-year-old son had come to the zoo for all of the obvious reasons. But the mother's plan went awry when her son suddenly scampered ahead and out of view. She made a mad dash to find him, but it was too late. Somehow the boy had climbed over the railing of the gorilla exhibit and tumbled to the bottom of the pit—a fall of eighteen feet. According to visitors who saw it, he landed with a thud, striking his head.

Witnesses gasped, not only at the boy's fall but at where he had landed. A crowd soon gathered, with those nearest the railing peering down to see the motionless child, facedown and splayed out. Some of the spectators must have felt an urge to climb over the fence and save the little boy. But for anyone who tried, getting into the pit would be the least of the dangers. Once down there, they would be confronting seven adult lowland gorillas, who had the run of the exhibit. They are the biggest and strongest of the world's great apes and many times more powerful than a grown man. (The largest males of the species exceed four hundred pounds.) If the long fall hadn't killed the boy, then surely the gorillas would do it themselves.

It didn't take long for the gorillas to notice. They moved in on the boy, a little startled themselves and clearly curious. As onlookers screamed, zoo staff ran to the scene and emergency personnel were on the way.

A mother gorilla, Binti Jua, with a baby on her back, was the first to reach the boy. She seemed to have a purpose in mind. As the males came close, she stopped them with a straight-arm gesture that clearly translated to "Back off!" Then, to the astonishment of everyone watching, she gently picked up the child and cradled him.

With her own baby still clinging to her, Binti Jua carried the boy over to the sliding door that zookeepers use to access the exhibit. She set him down right beside it and stepped away, so that keepers could get to him. They snatched him out and rescuers immediately began working on him before the boy was rushed to the nearest emergency room.

The boy would survive and recover, and images of the drama would be broadcast all over the world. What the footage showed was an understanding, caring animal coming to the aid of a vulnerable child. At that moment, we glimpsed a side of our fellow creatures that we often overlook and are sometimes told is not there at all.

## Thinking and Feeling: Shared Capacities

IN DECEMBER 2005, IN the waters off the Farallon Islands, near San Francisco, boaters spotted a female humpback whale so entangled in the ropes of crab traps that she was unable to free herself. They radioed authorities, and a team of divers quickly answered the call. She was at the surface and thus able to breathe, but she wouldn't last long without assistance.

A century ago, when men in boats set their sights on a whale, they typically had harpoons in hand. These divers also carried sharp weapons, but for a different purpose. Their large, heavy knives were meant to cut through rope, not flesh. To do so, however, they'd have to bump right up against a frightened and frustrated fifty-ton animal, and there was no telling how she would react. With the rope also caught in the whale's mouth, the rescuers had to get eye to eye with her to accomplish their mission. As they cut the rope piece by piece, she remained calm and quiet, not thrashing her tail or using her mass to harm the divers or even to signal that she feared them. She seemed to know they were there to help, the divers said, and she put her trust in them.

"When I was cutting the line going through the mouth, its eye was there winking at me, watching me," said one of the divers. Once she had been set free, she did not swim away, but went up to the divers, one by one, and nuzzled them.

In West Africa, there was an animal rescue of a different and less dramatic kind. German field researchers studying forest chimpanzees in the Ivory Coast have discovered that unrelated adult chimpanzees, both males and females, adopt and provide care for orphaned chimps. One male chimp they observed shared his nest with the baby every night, carried him on his back for long travels, and shared Coula nuts he expertly opened. In all, the researchers witnessed eighteen adoptions of orphaned chimps. With no benefit to themselves or their group, and indeed despite the sacrifice required to help the needy juveniles, these adoptions have every appearance of charity.

Had such observations of chimp adoptions been recorded fifty years earlier, they would very likely have stirred controversy and perhaps not even been published. The authors would have been accused of attaching human attributes to chimps. Adoptions of the orphans might have been explained away as an evolutionary strategy of some kind, rather than an act of kindness or sympathy. The scientists would have been accused of straying from their lane and ascribing consciousness to animals—and the whole study dismissed as another case of "anthropomorphism."

Indeed, just such charges were made in 1960, when Jane Goodall first began sharing her findings about chimpanzees in Gombe National Park in Tanzania. Her fellow scientists quickly brushed off her findings about the lives and emotions of chimps as a woman's sentimentality. Goodall wasn't even formally trained, they noted. What could *she* possibly know about primates?

In her dispatches from the field, which would later gain an enormous popular following in *National Geographic* magazine and on television, Jane Goodall named the chimpanzees and treated them as individuals. There was David Greybeard, a striking adult male who first accepted Goodall into the group, and his friend Goliath, the daring alpha male. She produced rich accounts of the chimps' activities and displays of intelligence and emotion, describing their individual personalities. Over time, Goodall was even accepted as if a member of the troop, living among them for twenty-two months. To the strict mid-twentieth-century behaviorists, observations of this kind were heresy. Humans, as dogma held, were the only animals capable of deliberate action, conscious thought, and emotion. No animal other than man was to be considered a *who*, with personality and awareness, but merely an *it* to be studied much as one might observe a food-gathering, reproductive machine.

Louis Leakey, the legendary Kenyan paleontologist, had seen too much of animals in the wild to accept such orthodox thinking. It was Leakey who urged Goodall to go into the field and document chimpanzee society, just as he did in later years with Dian Fossey and Birute Galdikas in their study of the other great apes. Equipped only with a pen and paper, the twenty-six-year-old Goodall began recording the hidden lives of chimps: their games, family squabbles, power struggles, interpersonal violence, and affection.

One day she watched as David Greybeard used a stick to catch food from a termite mound, and soon after she noticed Goliath stripping leaves off a stalk to fashion it into tool for catching termites. The implication was tremendous: toolmaking, long considered a defining attribute of humans, was shared by other animals. When Goodall shared her observation with Leakey, he replied, "Now we must redefine 'tool,' redefine 'man,' or accept chimpanzees as human."

Today Goodall's discovery seems less startling, and Leakey's conclusions overstated. We now have learned that all sorts of animals—from New Caledonian crows to bottlenose dolphins—can make and use rudimentary tools. And we also know that animals can do much more—that dogs can feel empathy, elephants can suffer emotional trauma, and birds can reason and solve problems.

But at the time, Goodall's methods, vocabulary, and findings were unheard of. Decades of academic papers, layered over centuries of philosophical dogma, had sealed our species off from all others, somehow completely separate from them. We occupied our own special citadel, claiming tool use, cognition, any capacity for language, and all the elements of reason as ours alone.

It would take a long time, starting with Goodall's reports, for the reality that animals think and feel to sink in. So many people didn't see it, or simply denied what they saw. All along though, it was right there—playing out in the behavior of our own dogs and cats, in elephant and chimpanzee societies, and in the lives of so many other creatures. In their humble ways, animals do think and feel—and this should awaken our empathy and command our respect.

## Denying Animal Intelligence

DENYING OR DIMINISHING THE intelligence of others has always been a strategy to justify oppression, cruelty, or indifference, or at the very least to divert moral responsibility. We've seen it used for centuries to downgrade the moral worth of animals, but it's also been invoked to justify plenty of human wrongs—whether perpetuating slavery, denying rights to women or minorities, or locking away the mentally ill.

In *The Mismeasure of Man,* the late Harvard paleontologist Stephen Jay Gould looked back at the so-called science of "biological determinists," who argued, perhaps in part unconsciously, that certain races, notably blacks, had a lower intelligence quotient than whites. Through craniometry, or the measuring of the size of the brain, and certain psychological testing methods, proponents argued that there was a biological basis for the supposed inferiority of blacks. With cool reason, step by step, Gould stripped away the pretense of objective science and demonstrated that their data were cooked and their conclusions preordained. His work stands as a reminder that scientists sometimes labor under prejudices of their own. They may simply shore up the reigning falsehoods of the day, instead of knocking them down as we expect good science to do.

Of course, Gould had exposed a form of pseudoscience that was employed to invent racial differences. It's not quite so easy to argue that similar methods have been used to deny the intelligence of nonhuman animals, since there are so many actual differences of brain size and intelligence between humans and other species. That humans possess special cognitive abilities, powers of thought and creativity that long ago set us apart, is among the most obvious facts in nature.

The question is not whether animals possess a level of intelligence equal to ours, but instead whether they possess some meaningful level of intelligence and awareness. Do they think and feel and suffer in ways that command our moral attention? Uniquely in the world, human beings are the creature of conscience, and that very capacity allows us to perceive the worth of other beings and to care about how they're treated. Perhaps alone among animals, we know what is right and fair, and it falls to us to give animals their due. The nineteenth-century abolitionist and women's-rights advocate Sojourner Truth, in speaking up for dispossessed people of her time, said it well: "If my cup won't hold but a pint, and yourn holds a quart, wouldn't ye be mean not to let me have my little half-measure full?"

René Descartes, a stern man in his measuring of cups and pints, had confidently declared that animals were "mere automatons"—objects incapable of feeling conscious sensations, and hence incapable of suffering. Language, Descartes argued, is essential to thought, and since animals couldn't speak, therefore they couldn't think. Emboldened by this belief, vivisectionists in the succeeding decades cut open live dogs without anesthetic to provide anatomical lessons, and they dismissed the dogs' cries as mere stimulus response. The absence of reason, or even sentience, meant that the dogs' tormentors had no moral obligations to other creatures and could do with them as they wished—which they did, to the point of staging vivisections in public for the enjoyment of crowds.

Whatever the theories used to justify such practices, there have always been those who knew better. Voltaire, writing in the eighteenth century, went to the heart of the matter by challenging the assumption that animals don't feel: "Answer me, machinist, has nature arranged all the means of feeling in this animal, so that it may not feel?" He reasoned that since animals share anatomies strikingly similar to our own, it was highly likely that they feel pain like we do too. In time, these biological similarities would only become more evident, and they seemed undeniable to the man who would become the most famous scientist of the age.

In 1859, in *On the Origin of Species*, Charles Darwin called into question much more than man's unique place in Creation. He argued that the differences between humans and animals were differences of degree, and that animals had complex emotional lives. Summarizing his studies of animals in the wild, he wrote that mammals "experience (to greater or lesser degrees) anxiety, grief, dejection, despair, joy, love, 'tender feelings,' devotion, ill-temper, sulkiness, determination, hatred, anger, disdain, contempt, disgust, guilt, pride, helplessness, patience, surprise, astonishment, fear, horror, shame, shyness, and modesty."

In 1872, Darwin published *The Expression of the Emotions in Man and Animals,* which documented how people and animals share similar facial expressions and body language to express the same state of mind. He observed how monkeys laugh and dance when happy and dogs stand taller when angry, much as people do. He pointed out that animal emotions make evolutionary sense too; without fear, a gazelle wouldn't know when to run, and without aggression, a dog wouldn't feel the same urgency in defending his territory. It was equally obvious that animals took pleasure in the basics of life such as eating and mating.

Still, the denials of intelligence and rationality by Descartes (and earlier but similar denials going back to Aristotle) shaped ages of thinking on the subject. Although the human-animal bond predates all known philosophy, the dominant view among scientists in the fields of psychology and animal behavior was that animals operate by instinct, somehow mimicking consciousness or intelligence without truly possessing them.

By the middle of the twentieth century, most scientists rejected the Cartesian notion that animals could not feel pain, but they still dismissed the idea of animal cognition. They embraced Darwin's theory of evolution, but rejected or ignored altogether his findings on animal intelligence and emotions. The leading voices of the new school of behaviorism, including American psychologist John B. Watson, decreed that scientists should reserve comment for the external behavior of animals and avoid assumptions as to internal states such as thought and feeling—if animals had any feelings at all. This was consistent with what animal behaviorist Lloyd Morgan called his law of parsimony: if a behavior can be attributed to a lower cause (that is, a direct stimulatory response), it should not be attributed to a higher one, and least of all to intelligence. In 1913, Watson wrote,

*Psychology as the behaviorist views it is a purely objective experimental branch of natural science. Its theoretical goal is the prediction and control of behavior. Introspection forms no essential part of its methods, nor is the scientific value of its data dependent upon the readiness with which they lend themselves to interpretation in terms of consciousness.*

The most prominent behaviorist of the era, B. F. Skinner, did not discount the possibility that animals can reason, but theorized that it would be too difficult to know. We should just observe behavior, he argued, and describe it in a scientific manner free of "anthropomorphism."

In our own day, the author Jeffrey Masson sees the imprint of the Skinnerian model of behaviorism in the work of scientists throughout the twentieth century. Masson wrote:

*From the belief that anthropomorphism is a desperate error, a sin or a disease, flow further research taboos, including rules that dictate use of language. A monkey cannot be angry; it exhibits aggression. A crane does not feel affection; it displays courtship or parental behavior. A cheetah is not frightened by a lion; it shows flight behavior.*

These are subtle descriptive distinctions, but loaded with implications for the way we view animals. The animals did not think, but simply demonstrated certain programmed behaviors. They operated by instinct, devoid of emotion or consciousness.

The scientists' belief in evolution did not solve this faulty perception. They had little difficulty adapting evolutionary theory to fit their worldview, placing animals in some sort of behaviorist straitjacket. In their view, natural selection programmed animals to gather food, reproduce, promote the survival of their offspring, and, in the end, to pass on their genes. Animals were locked in an unyielding and competitive struggle for survival. All of their energy was devoted to achieving reproductive success, and all of their behavior could be understood according to this single, blind, all-explaining force.

In this view, animals in the wild, or in your own house, never play just for the fun of it; if play occurs, it's always about honing predatory skills or other training for survival. Animals never feel concern or alarm for us or for another creature; any act of assistance to humans or to other animals must have had a survival benefit for the social group or the herd. And reproduction is, well, about reproduction only; whether the animals exhibit monogamy, polygamy, or polygyny, it is, in a word, workmanlike—mechanical copulation to make sure the genes were passed on to succeeding generations.

Descartes viewed animals as machines, and though behaviorists rejected his formulations, they held the view that animals were machines of a different sort—biological machines, shaped by natural selection, and operating through instinct alone.

We know as human beings that there is more to life than reproduction. That instinct to reproduce is certainly a strong biological impulse. But not all of our behavior is a run-up to sexual reproduction or the raising of offspring. We like to do all kinds of things that have no particular reproductive purpose or advantage, like playing, laughing, or just lying around. There may be some evolutionary explanation for such behaviors, or maybe it's just that we like to do these things. They're enjoyable, and no grand scientific theories are needed to explain why.

And that's one of the flaws in the reductionist thinking of the behaviorists, whether it's applied to human action or to all the things that animals do. Animals also like to have fun and enjoy life. When a dog sticks his head out of the car window, he's having a great time. When cats are swatting at each other, they are playful. And as baby calves scamper and chase each other, it is part of the fun in their lives. Animal behaviorist Jonathan Balcombe, who has challenged the views of his more conventional peers, says that too many scientists treat nature like a "constant, joyless struggle," bereft of happiness, or even an idle moment.

There are theories and even entire books that try to explain all of canine behavior as an elaborate evolutionary strategy to get food and shelter from us, as if dogs were just clever parasites on the earth's dominant species. What we see as love and loyalty in dogs, the behaviorists insist is all just an evolutionary game playing out in our homes—and boy, have they put one over on us.

Darwin's observations about animal emotions and thought got the science of animal behavior off to a great start. But because many of the same animal scientists who accepted his doctrine of evolution also ignored his views on animal minds, it would fall to later researchers to reawaken the field of study. In the 1950s, the famed zoologist Konrad Lorenz provided a counterweight to the dominant school of behaviorism, arguing that animals were conscious and emotional. But it took the growth of the new field of cognitive ethology in the 1970s to begin to dismantle in a more comprehensive way the falsehoods left by Skinner and others.

In 1976, Harvard zoologist Donald Griffin provided a jolt to the discipline with *The Question of Animal Awareness,* in which he argued that animals had conscious minds much like humans. Griffin was an innovative scientist; as an undergraduate in the 1930s, he had strung guitar strings across his dorm room and noticed that bats still navigated effortlessly through them in the dark—a finding that led to the discovery of echolocation.

For most of his career Griffin had belonged to the orthodox behaviorist school, and he rightly earned the respect of his peers as a rigorous scientist. But starting in the 1970s, he pioneered new techniques to study animals in their natural habitats, demonstrating that animals do not operate merely by instinct. "Nature," Griffin wrote, "might find it more efficient to endow life-forms with a bit of awareness rather than attempting to hardwire every animal for every conceivable eventuality." Griffin argued that intelligence itself began as an adaptive characteristic. Animal behaviorist Marc Bekoff told me that when Griffin first began discussing animal intelligence at conferences, the idea was so rattling that people thought he had lost his mind.

Thanks in large part to Griffin, the discipline has seen great change for the better in the last three decades. Gone are the days when talking about the emotional lives of animals will get you laughed out of an academic conference. Cognitive ethology is now a reputable and growing field, and every year more studies are published on animal intelligence and emotions. Marc Bekoff notes that in 2000, when he sought essays from respected scientists on animal emotions, more than fifty signed up. The resulting book was called *The Smile of a Dolphin,* with a foreword by Stephen Jay Gould, and it signaled a long-awaited shift away from the mechanistic and miserly views of the behaviorists.

## Signs of Change

ON THE NIGHT OF September 6, 2007, Alex, a thirty-one-year-old African grey parrot, wished his trainer, Dr. Irene Pepperberg, goodnight as usual.

"You be good, I love you," said Alex.

"I love you too," Dr. Pepperberg replied.

"You'll be in tomorrow?"

"Yes, I'll be in tomorrow."

The next morning, Dr. Pepperberg arrived to find Alex dead. In the ensuing weeks, Alex would receive more public attention than perhaps any bird in history. Television shows from *Good Morning America* to *The Tonight Show with Jay Leno* ran segments on his death, while the *New York Times* published three articles on his life and feats of intelligence. Thousands of people from around the world wrote heartfelt letters to tell Dr. Pepperberg how Alex had opened their eyes to the incredible minds of animals. *The Economist,* which usually devotes its weekly obituary to statesmen and celebrities, devoted it instead to Alex, noting, "by the end, Alex had the intelligence of a five-year-old child and had not reached his full potential."

When Dr. Pepperberg bought Alex at random at a pet store in 1977, there was nothing to suggest he would bring about immense change in animal science. To no avail, researchers had been trying to teach chimpanzees to talk, and few thought that parrots, with their walnut-sized brains, would do any better. (It was only later that researchers turned to teaching chimps sign language, with far more success.)

But evolutionary explanations for behavior were taking a new turn. British scientist Nicholas Humphrey, for example, argued that intelligence evolves in response to the social environment rather than the natural one. According to Humphrey, the reason chimps are so smart is that they live in complex societies, in which they must consistently reason, learn, and negotiate to survive. That got Dr. Pepperberg thinking. Since parrots also live in complex societies in the wild (and, like chimps, have long enough life spans to make investing time in learning worthwhile), perhaps they had also evolved advanced intelligence. And as parrots have vocal cords more suited to talking than chimps do, perhaps Alex really could be taught human language.

Alex's education began slowly. Dr. Pepperberg and her colleagues at the University of Arizona (and later Harvard, MIT, and Brandeis) would hold objects in front of Alex and discuss them in detail. Conventional wisdom held that parrots could simply repeat words—"parroting"—but Alex soon began to express thoughts that seemed awfully similar to human ones. He'd ask to be taken to his play area, and then complain when he was taken to the wrong place. He understood and could discuss concepts like "bigger," "smaller," "same," and "different." By the end of this life, he knew fifty objects by name, and he could describe the colors and shapes of objects he'd never seen before. He even knew when and how to apologize when he antagonized Dr. Pepperberg. Before he died, Alex had been learning about optical illusions—which he perceived with an incredible likeness to the way humans see them.

As Dr. Pepperberg would later write, "Scientifically speaking, the greatest lesson Alex taught me, taught all of us, is that animal minds are a great deal more like human minds than the vast majority of behavioral scientists believed—or, more importantly, were even prepared to concede might be possible."

Nor are parrots the only birds to show such amazing abilities. In 2003, Cambridge University researchers camped out for months in the jungles of New Caledonia to study crows. By attaching miniature video cameras to the crows' wings (the cameras fall off when the birds molt, leaving them unharmed), they observed the birds making advanced use of tools. Some used sticks to burrow into rotten trees for larvae, while others adapted dry grass to fish for ants. One intrepid crow even hauled a large stick hundreds of yards because it made such a useful tool. Researchers have since found that even crows raised by people, without the benefit of learning by experience, are capable of working out how to make and adapt tools.

More recently, scrub jays have gone even further toward upending our assumptions about avian intelligence. Scrub jays not only have episodic memories of the past that mimic our own, but also plan for the future—a trait that even infant humans lack. The researchers tested the jays' memories by having them hide perishable wax worms and nonperishable peanuts. If they could access their cache after just a couple of days, they chose the wax worms, but if they knew they wouldn't get back to the cache for weeks, they'd go for the peanuts— knowing the wax worms would have decayed in the interval.

To test their "future-planning" capacity, the researchers rotated the birds between different locations and gave them the opportunity to hide food in each one. They found that the birds not only stored food in locations where they anticipated they would be hungry the next morning, but also stored preferred foods where they anticipated only nonpreferred foods would be available. In other words, the birds were stocking up on their favorite breakfast foods, much as we might fill up the fridge before a busy weekend.

Although the countless mental feats of dogs are still routinely dismissed by behaviorists as mere "anecdotal" evidence, they, too, have often surprised researchers in formal tests. Just how smart are dogs? Rico, a border collie, astounded scientists in 2004 by demonstrating the ability not only to recognize the meaning of more than two hundred words, but also to infer the names of items she hadn't seen previously. In a test described in *Science,* researchers placed objects in another room and then had Rico retrieve them by name. When asked to retrieve an object with a foreign name, Rico used what the researchers called "simple logic" to infer it must be the one object whose name she didn't know, and she retrieved that one.

Stanley Coren, a psychology professor at the University of British Columbia, doesn't think these results are exceptional. He administers language and cognition tests devised for human infants to dogs and says that the most intelligent breeds square up well. Poodles, retrievers, Labradors, and shepherds can learn around 250 words, signs, and signals, according to Coren, and some even "get the idea of being a dog"—they are able to differentiate photos with dogs in them from photos without dogs.

As for the great apes, recent studies have turned up evidence of mental acuity that would impress even Jane Goodall. At Kyoto University, Ayuma, a five-year-old chimp, gave humans a humbling lesson in memory recall. He was placed in front of a computer screen and a sequence of numbers from one to nine flashed in front of him for a second before being replaced by white squares. He was able to remember the sequence almost perfectly, while humans could barely remember four or five numbers.

They even put Ayuma up against British memory champion Ben Pridmore—who can remember the order of a shuffled deck of cards in thirty seconds—and the chimp performed three times better. As Jonathan Balcombe relates, "When the numbers flashed for just a fifth of a second, Ayuma correctly recalled all nine digits 90 percent of the time, compared to 33 percent for Pridmore." Balcombe concludes that chimps probably have a "botanist's memory" for up to two hundred plant species, and that may have aided in their development of this amazingly rapid recall. In a head-to-head matchup on memory recall, the chimp won.

With language experiments, of course, great apes and parrots are no match for us. But what they can do is pretty impressive. The lowland gorilla Koko (whose niece, Binti Jua, was the one who rescued the fallen three-year-old boy at the zoo) has famously mastered more than a thousand signs in American Sign Language. Here we have a case of a gorilla learning human language—so how good are we at learning gorilla language? We probably wouldn't do all that well. Balcombe frames the matter another way:

> But how would a lowland gorilla do if she were tested on things that are important to her, such as the ability to recognize native plants and to distinguish edible from inedible ones? Or to predict weather changes in the lush African jungles where lowland gorillas live? Or to gauge the moods of other gorillas based on facial expressions, body postures, or their scent? By such measures, she would be a genius among humans. Gorillas' evolutionary history and survival depend on these mental skills. Gorillas are intelligent at being gorillas. Similarly, a rat is probably no less intelligent than a rhino, but more to the point, a rat is intelligent at being a rat, as is a rhino at being a rhino.

One animal commonly praised for its intelligence is the dolphin, but even dolphins keep surprising us with all that they can do. They have brains remarkably similar to the basic architecture of our own, with folds in the neocortex that are thought to enable many of our most impressive mental feats. Dolphins off Western Australia, confronted with jagged coral reefs and spiny fish, have learned to hold sponges over their mouths as they trawl the ocean floor. And in one test, a captive dolphin named Akeakamai was able to grasp and act upon commands given in a form of sign language. To do so required not just receiving elementary commands like "fetch," but to understand a series of thoughts that she was able to interpret and act upon. In a separate study some years ago, dolphins' keepers asked them, by hand signals, to devise a trick of their own. Beneath the water, they communicated among themselves and then performed a synchronized trick—having never done it together before.

The tricks of trained dolphins have even found their way into the wild. In captivity, bottlenose dolphins will bide their time by creating underwater rings and helices of air to swim through, and then teach the trick to other dolphins. In the wild, they will surf in waves and play fight with other dolphins. In 2009, Australian wildlife rehabilitators taught a rescued dolphin to tail walk during a three-week recovery period, and then released the animal to the wild. Weeks later, they were astounded to see other dolphins performing the trick. The rescued dolphin had passed along the lesson, and before you knew it a new craze had swept the entire pod.

## Lifesaving Dolphins and Dogs, Altruistic Elephants, and Other Remarkable Animals

IF ANYTHING HUMBLES US in what we learn about dolphins and other creatures, it's not just their intelligence but how they've been known to use it. Stories abound of dolphins rescuing sailors and swimmers, even to the point of risking their own safety, and you'd be hard-pressed to find any adaptive advantage in that behavior. In 2004, for instance, four lifeguards in New Zealand found themselves suddenly encircled by a pod of dolphins, which seemed at first to spell big trouble. As it turned out, the real lifeguards that day had fins and could swim at thirty knots:

> *Lifeguard Rob Howes said he and three female lifeguards were on a training swim about 100 metres off Ocean Beach near Whangarei on the North Island.*
>
> *About halfway through the swim, a pod of dolphins "came steaming at us" and started circling, startling the swimmers, he said.*
>
> *Howes said he was unnerved by the speed of the approach, thinking perhaps it was a group of aggressive males or dolphins protecting their baby.*
>
> *The dolphins bunched the four swimmers together by circling about 4–8 centimetres from them, and slapping the water with their tails for about 40 minutes.*
>
> *Howes said he drifted away from the main group when an opening occurred. One large dolphin became agitated and submerged toward Howes, who turned to see where it would surface.*
>
> *That, he says, is when he saw a great white shark about two metres away in the beach's crystal clear waters.*
>
> *"The form came and travelled in an arc around me. I knew instinctively what it was," he said.*
>
> *When the shark started moving toward the women, including his 15-year-old daughter, the dolphins "went into hyperdrive," said Howes.*
>
> *"I would suggest they were creating a confusion screen around the girls. It was just a mass of fins, backs and . . . human heads."*
>
> *The shark left as a rescue boat neared, but the dolphins remained close by as the group swam back to shore. At no point did the shark break the surface of the water, remaining near the bottom, he said.*

Humans have been hearing stories like this since the days of the ancient mariners, and with the frequency of these rescues in our own day we know it's not just lore. Off the coast of California in 2007, it was a surfer in trouble, a man named Todd Endris, and the great white had already gotten a piece of him: "The shark—a monster great white that came out of nowhere—had hit him three times, peeling the skin off his back and mauling his right leg to the bone. That's when a pod of bottlenose dolphins intervened, forming a protective ring around Endris, allowing him to get to shore, where quick first aid provided by a friend saved his life."

If ever a man needed a miracle, it was Todd Endris. And it came in the form of a group of strangers who happened to be passing through, acting on motives that, if they were human, we would instantly recognize as altruism and incredible courage. Humans who do things like that get the keys to the city, but all the dolphins receive in return, as in all such cases of animal heroics, is more labored theorizing from the skeptics, more attempts to explain it all away as so much preprogrammed behavior.

Anyone watching a service dog on the job, guiding a blind person through the perils of streets and sidewalks, is left with the distinct impression of an awareness, attentiveness, and commitment going beyond anything that training can instill. The animals really care, and they know that a lot depends on their jobs. The singular devotion of service dogs—their special bond with the people who depend on them—is much more than the right mix of positive and negative reinforcement. And it's not just the professionally trained ones who show such qualities. Dogs rise to the occasion all the time, perhaps especially shelter pets. Most anyone who has rescued a homeless dog will tell you that it's a favor they don't forget.

In recent years, the HSUS has run a "Dogs of Valor" contest, recognizing canines whose alertness, loyalty, and bravery saved lives. The 2010 winner was Kenai, a fourteen-year-old Bernese mountain dog mix, from Erie, Colorado. Kenai's owners, Todd Smarr and his wife, Michelle Sewald, were staying in the basement of a vacation house during a weekend getaway with family and friends, when

Kenai awakened them at 4:00 A.M. by whining and barking. When Todd got up to comfort Kenai, another friend sleeping in the basement, Karen Hull, woke up and said she wasn't feeling well. Minutes later, Michelle collapsed into Todd's arms, unconscious. Todd rushed upstairs, roused their friends, and soon realized there was a carbon monoxide leak in the house. Saved by Kenai's warning, the seven adults, two children, and four dogs in the house escaped into the fresh air to breathe another day.

The runner-up in the HSUS contest, Calamity Jane, was still recovering from having her leg amputated from an old gunshot wound and from giving birth to seven puppies when her foster mother, Shar Pauley, took her for a walk on a cold January night in Aledo, Texas. As Shar walked past the neighbors' house, Calamity Jane suddenly bolted into the neighbors' front yard, barking furiously, in spite of her somewhat feeble condition. A few moments later, Shar heard a car door slam and then saw a car speed out of the driveway. She quickly returned home, and minutes later the neighbor arrived, telling her to call 911. He said that his family and guests, including two children, had been held at gunpoint for close to an hour during a violent home invasion. But when the intruders heard Calamity Jane, they yelled to one another that there were people outside and fled.

Then there's Jack, a terrier mix rescued from a trash bin. His best friend was little Maya Pieters, who as a three-year-old had been diagnosed with congenital bilateral perisylvian syndrome (CBPS), an extremely rare neurological condition that mainly affects the oral motor functions. One morning, as Jack slept downstairs in his open crate, he suddenly darted upstairs to Maya's room where he began clawing and barking at the door. Maya's parents heard the commotion and realized something was wrong. It turns out Maya was having her first grand mal seizure in her sleep. Her parents scooped her up and rushed to the emergency room. After Maya recovered and the family returned home, Jack dutifully stayed by the little girl's side, knowing that he had a special ability to help her. Now, each time Maya has an epileptic event at home, Jack seems to sense it in advance. Once, as she suffered a seizure, he stepped in to cushion her with his body.

Like Jack, many dogs have an uncanny ability to detect oncoming seizures—whether it's by smell or observing behavioral changes is unclear. Hungarian researchers trained a dog to warn a blind and epileptic man with barks and licks minutes before a seizure, allowing the man to call for help. Other dogs have been trained to turn on the lights for trauma victims afraid of the dark, to remind their sick owners to take medication, and to aid severely depressed patients under suicide watch.

SUCH SKILLS AMONG DOGS take both smarts and empathy, and it's seen in their wild cousins as well. When Marc Bekoff first began studying coyotes, he noticed a code of moral behavior among members of the pack. At the time, any capacity for a moral sensibility was considered a strictly human virtue, one of the scientific barriers that conveniently separated *us* from *them*. But Bekoff, leaving behind the constraints of a laboratory setting that can warp natural social behaviors, gathered his evidence in Grand Teton National Park near Jackson, Wyoming. There he observed coyote dogs in the wild, discovering complex societies with surprising rules of their own.

Video camera in hand, he found that coyotes and other canids begin their play fights with a "bow," in which the animal crouches on her forelimbs and often barks or wags her tail vigorously. Seeing this gesture again and again, he established that this was a signal understood by one and all in the pack—saying, in effect, Let's spar but keep it playful. Coyotes will often repeat the bow during play fights, especially before biting their opponents or doing anything else that could be confused with real fighting. And when they overstep the bounds, for instance by biting an opponent too hard, the coyote will bow to apologize. Coyotes who break the rules of play are duly punished: those who bow and then attack are less likely to be chosen as play partners, and more likely to be shunned.

Bekoff tells other stories of what he calls "wild justice." In one, a hormone-crazed male elephant had knocked over a female elephant suffering from a leg injury. A third elephant rushed to her aid, touching her trunk to the sore leg as if to soothe the pain. In another case, an elephant matriarch set free a group of captive antelopes, using her trunk to undo the latches on the gate of their enclosure. Scientist Jonathan Balcombe notes that vampire bats even appear to exhibit reciprocal altruism—individuals will share food with ill and nursing bats, helping them out when they really need it.

Does all of this prove that animals possess morality, at least in the very basic sense of being able to care for one another and for us? There is certainly evidence of empathy among some animals, and empathy is often the starting point of moral action. Primatologist Frans de Waal has tested the extent of empathy in capuchin monkeys. He put the monkeys in pairs and gave each one an option: the monkey could choose a token that dispensed treats only to himself, or one that produced treats for both. He found that the monkeys consistently chose to share as long as their partner was familiar, visible, and receiving equal rewards. De Waal concluded that "they seem to care for the welfare of those they know."

If acts of kindness are one expression of empathy, then grief is another. Researcher Cynthia Moss has documented elephant burials in which elephant families cover their dead relatives in branches and then stand vigil over the body for the night. Bekoff notes that such grief is common in the wild. Animals who lose a mate, family member, or friend may withdraw from their group and seek seclusion. They might try to retrieve the dead animal or stay with the remains for days. Some distraught animals even give up eating and mating—a response utterly at odds with evolutionary self-interest.

In 2009, chimpanzees at a West African sanctuary amazed volunteers by forming a kind of funeral procession to mourn the loss of an elder named Dorothy. As caregivers at the Sanaga-Yong Chimpanzee Rescue Center in eastern Cameroon bore Dorothy's body by wheelbarrow, the normally boisterous chimps rushed to the edge of their wire enclosure and fell silent. They stood there—wrapping arms around each other, some leaning on the shoulders of troupe mates—as Dorothy was prepared for burial and lowered into the ground.

Along with sorrow, animals also can fall into depression just like us. It's especially common among captive zoo animals, who are given medications to mitigate their condition—usually the very same antidepressants given to people. Rats separated from social contact will often choose morphine-laced foods over regular foods—a preference that ceases when the rats are returned to a more enriching environment, and not in the solitary confinement of a laboratory cage. When they're together, in fact, and have a chance to play, rats show increased levels of dopamine—biochemical evidence of happiness.

That animals experience emotional trauma when they're isolated, mistreated, or bereaved should not surprise us. Neuroscientist Jaak Panksepp notes that most fellow mammals show neural reactions in the brain similar to ours, and that human pain-relief drugs typically have the same effects on them. Mammals are attracted, likewise, to the same environmental rewards as people—food, social contact, and so on. In 2008, researchers at the University of Sussex even found that gorillas' facial expressions are controlled by the same processes in the left side of the brain as they are in us—suggesting that when a gorilla winks, grimaces, or smirks, he really means it.

The hurts and losses that animals experience are more than skin deep, as in the especially poignant case of elephant calves who have seen their parents killed by poachers. Well into their lives, if not forever, they show signs of posttraumatic stress disorder. Scientists in Uganda were the first to document this after investigating a series of violent attacks by elephants on villagers over the last decade. Traditional wisdom held that elephant attacks were motivated by competition for scarce food, but at the time of the attacks, food supplies had never been so abundant and elephant numbers had never been lower. That led scientists to consider a different explanation: rampant poaching over three decades had wiped out some 90 percent of Uganda's elephants, leaving just four hundred elephants, many of them orphans.

Scientists wondered whether these orphans, as they grew to adult size, were still acting out the effects of their childhood traumas. And they couldn't figure out why the elephants were knocking down and stabbing rhinos with their tusks—a bizarre kind of violence almost never seen among these herbivorous animals who often share the same habitat. In South Africa, delinquent elephants were responsible for killing as many as thirty-nine rhinos—10 percent of the population—in Pilanesberg Park. It turns out that the parents and grandparents of these elephants had been culled by sharpshooters in Kruger National Park. The babies were spared, but apparently never forgot what they saw and had grown up without adults to teach them how to behave.

Sure enough, researchers found neurological signs of posttraumatic stress. In particular, orphaned elephants had nightmares and also had trouble forming emotional ties with other elephants— all clearly linked to having seen their mothers and families slain. Some researchers even suggested the possibility that these elephants might now be avenging the terrible things they witnessed. As Joyce Poole, research director at the Amboseli Elephant Research Project, says, "They are certainly intelligent enough and have good enough memories to take revenge."

Considering all that has befallen elephants, it would be understandable if they held it against us. Yet such cases are so very rare, and the remarkable thing is that so many animals, when they come to know humans a bit, still see the best in us. They seem to feel the bond even when we have fallen short.

## The Emotional Lives of Animals

WITH OUR NEW KNOWLEDGE of how animals think and feel come new obligations in how we care for them. Many challenges in animal welfare today are an attempt to apply our understanding in consistent ways, and a good example of the difficulties involves the care of captive wild animals in zoos.

I think of my friend Ron Kagan, who for thirty-five years has worked in the world of zoos. He's still going strong despite years of resistance from the leadership of his profession. He's challenged many of the set ways and myths within the field—lobbying for animal-welfare legislation, using the zoo as a sanctuary for many abandoned and confiscated exotics, and starting an animal-welfare teaching center at the zoo.

What really got Ron into trouble, however, was his decision to send the elephants living at the Detroit Zoo to an elephant sanctuary—essentially conceding that even the best-intentioned zoo, with financial resources to hire the best staff and to improve and expand the elephant enclosure, could not match the conditions of a dedicated elephant sanctuary in a warm climate with room to roam.

It was not an impulsive decision for Ron. In fact, it came about through a series of experiences that moved him gradually to the conviction that elephants are awfully difficult to maintain properly in a zoo.

In 1974, he was a new keeper, caring for elephants and rhinos at the Boston Zoo. On his first encounter with an elephant, the animal pinned him against the wall with his head—a message of just who was in charge and of the dangers to zoo staff dealing with enormous and highly intelligent animals. Several years later, that same elephant killed a worker at the facility, and the elephant himself was destroyed.

When Ron was the general curator of the Dallas Zoo a decade later, he helped to rescue a showbiz orangutan, a gorilla from a shopping mall in Washington State, and a rhino from a Florida circus. These experiences only heightened his concerns about the quality of life for certain captive mammals, such as great apes and elephants.

In 1990, while still working at the Dallas facility, Ron went to Kenya and met elephant researcher Joyce Poole at Amboseli National Park. Poole and her colleague Cynthia Moss have studied elephants in the wild for years. Their work has revealed that the animals live in matriarchal family groups, with elders living up to seventy years, and traveling as far as forty miles in a day. Kagan noticed that almost all of the captive elephants he knew about suffered from chronic and extremely painful foot and skin problems. He asked Poole whether the wild elephants at Amboseli had these same problems. "In the most polite way, she told me what a foolish question this was," Kagan told me. She explained to him that they simply do not get this condition in the wild—it is an affliction that besets only captive elephants.

Just two years later, Kagan was selected for the top zoo job in Detroit, which, with its cold winters, can hardly be considered an ideal environment for a species built for savannahs and tropical forests. Kagan had two elephants in Detroit—Winkie came from a zoo in Sacramento, and Wanda from San Antonio. As the truck carrying Wanda approached the city, Kagan was excited at the thought that he'd have a new showcase animal for his exhibit, despite his growing doubts about keeping elephants in captivity. But his excitement turned to anguish when he heard her screaming and wailing when she arrived. "She didn't know what was ahead of her," he told me. "It was heartbreaking." With everything else he had learned about elephants, he knew too much. "That experience was unnerving, and it contributed to the discussion of whether certain animals should be in captivity at all."

So in 2003, Kagan and his colleagues decided to give up elephants at Detroit and send these two animals to the Performing Animal Welfare Society sanctuary in central California, where they would have dozens of acres, a mild climate, the company of other elephants, and the daily attention of people who understood them. It took a year and a half to complete the public discussion process, with some members of the city council and the community initially objecting. Eventually, Kagan won over the locals, but he couldn't persuade the leaders of the other major zoos in America to rethink their own policies. They didn't like the outcome in Detroit, and they sure didn't care for the drift of the discussion over this whole matter. To the zoo establishment, this was more evidence that Kagan was trouble, and now he was opening up a larger debate that few were willing to have.

It didn't help ease their discomfort that the elephants seemed to adapt quite nicely to the sun and wide open spaces of their new digs in California. After years on display, shipped around from one small place to the next, this was a kind of deliverance. They would never again be in the wild, but this was surely the next best thing.

Who takes in a homeless elephant, or any other captive wild creature in need of space and care? To his great credit, Kagan is among the rare zoo administrators who keeps an eye out for captive animals in need—the discards of roadside menageries and exotic pet fanciers who get in over their heads. For that gorilla whom Ron helped rescue in 1995—a thirty-year-old creature born in Africa, purchased by a shopping mall, and forced to live in a concrete-and-steel cage as a curiosity for shoppers—moving to an accredited zoo was a big upgrade. But giving sanctuary to the many thousands of animal castoffs and refugees is largely the work of a rather remarkable network of private charities and volunteer rescuers—smaller versions of Black Beauty Ranch. Name just about any animal or any kind of abuse, and somewhere there are people devoted to bringing shelter and relief. And different though they are, in their focus and their means, they are each called to the same kind and noble work, which in practice usually involves cleaning up after careless, selfish, or malicious people.

It's not just good work, it's hard work. And it takes a special kind of person to do it—guys like Matt Smith of the Central Virginia Parrot Sanctuary. It was a beautiful spring day when I stopped by to see Matt in May 2009. When I climbed the steps of a white-fronted colonial house, I knew I had the right place because I could hear the ear-piercing screeches of parrots inside. Before long, Matt, a clean-cut thirty-two-year-old with a winning manner, greeted me. Matt runs one of a handful of sanctuaries devoted to rescuing, rehabilitating, and if possible adopting out, these strikingly intelligent, demanding, and long-lived birds.

Matt's youth is an asset in his chosen field. It's a young man's job to race around to care for dozens of parrots, build the enclosures they need, and give them the attention they deserve. If Matt lives into his eighties, many of his birds will still be with him. And the younger birds in his care will likely outlive him.

Matt told me people rarely know what they are getting into when they acquire a parrot. The data support that claim. Most birds have five to ten homes before they die, meaning that most of them experience a never-ending cycle of loss and separation from their owners.

As we walked out back, into the open-air aviaries, Matt asked me to hold out my left arm while a cockatoo named Callie stepped onto it. She was a large bird with a beautiful gorgeous head, thick with colorful feathers that would make any mother bird proud. But Callie had no feathers below the neck. She had plucked out every last one—leaving only little bumps across her bare, sickly-looking body. Animals always look so small and frail without their fur or feathers. All that was left of this poor creature's glorious plumage were the feathers she couldn't reach and pull out.

Callie's previous owner had trouble coping with the noise and the persistence of the cockatoo, and his solution was to sequester her in a room where she was alone almost all of the time. In the wild, the birds live in flocks, fly for miles every day, and spend time breaking open nuts and other food with their powerful beaks. She had none of that stimulation in this man's home—she was effectively in solitary confinement. With the downturn in the economy in 2008, he needed extra income and had to rent out his spare room—and that spelled eviction for Callie. It came to a better end than many other such stories, and at least her owner sought out a good sanctuary and found Matt Smith. Callie did not come with a dowry, only with a lot of problems, leaving Matt with the responsibility of indefinite care for a troubled creature.

"Some birds can come from great homes and pluck," Matt told me. "Other birds come out of outright abuse and do not pluck. But what we do know is that feather plucking does not exist in the wild." In captivity, 30 to 40 percent of parrots pluck their own feathers.

Matt said the best he can do is try to replicate the birds' wild habitat as much as possible to minimize self-destructive behavior. Branches, ropes, flocks, and flight can improve the situation and provide some needed stimulation. But sometimes the birds are just traumatized. And he says that the dreadful disorder just beyond plucking is self-mutilation. "They tear into themselves. The muscle is exposed and bloody. We can use a collar, and then try to bring them a better life."

The numbers are depressing. Matt estimates that between parrots and the passerines, there may be fifty million birds in captivity in the United States, a much higher number than industry surveys indicate. Mira Tweti provides a similar estimate in *Parrots and People,* her indictment of the captive bird trade. She notes that many of the

older birds in people's homes were captured from the wild and im-
ported to the United States for the pet trade.

Young birds are typically captive-bred, because of the restrictions of the Wild Bird Conservation Act of 1992, which forbade
the importation of wild-caught birds. This law unintentionally gave
rise to an enormous industry of "bird mills"—the equivalent of
high-volume breeding operations for dogs, or puppy mills—where
birds are kept by the hundreds in overcrowded and permanently
dark sheds, caged until they are removed for sale. It is pathetic, Matt
told me, to see these extraordinarily intelligent birds locked in such
squalid environments.

"There is no escaping the intelligence," as Matt put it. "Rescuers and sanctuary folks can use the research on their intelligence as
a call to protect them, but the pet trade uses that same research to
promote their ownership as pets. To realize how smart a bird is, all
you have to do is live with one. Eventually, most people will realize how wrong it is to keep them in captivity. A lot of people feel
guilty—it's a recurring theme."

As I was about to leave, Matt showed me the flight cages he'd
designed. "Flight is the most important component to a bird's well-
being," he said. In these cages, the birds are together, and can stretch
and use their wings, assuming their wings have not been clipped.

I just wish that potential buyers of exotic birds could see his
sanctuary—a large bird colony filled with examples of how difficult
it is to keep and maintain creatures made for the sky. Most captive
birds eventually become the responsibility of someone else. Matt
and others to follow, including some good-hearted souls yet to be
born, will spend years of hard work cleaning up after the foolish decisions of others, and trying to make things right for these creatures.

I also feel so grateful to this young man for devoting his life to
these birds. He is one in the growing ranks of people who see the
need and answer the call, and whose unselfish efforts make a mark

in the world every day. It is the mark of respect and appreciation, of understanding and empathy for creatures great and small who have the same spark of life that we do, and who so often deserve better than they receive at our hands.

Parrots yearn for the sky, like all the fowl of the air. Elephants are called to roam, unbounded by the designs of man. Chimps want to climb and swing and dance, rejoicing in the lives intended for them. They all have their own minds and desires; they all have a place and purpose of their own. They have their own dignity and their own destinies to fulfill, in a plan ultimately beyond any man's power to know. Sparing or rescuing them from cruelty is a picture of humankind at our best. And so often in our dealings with animals, the greatest power we have is to stand back and let them be.

There's so much to the mental and emotional lives of animals, and though the research affirming that fact is fairly recent, it all suddenly seems so obvious. Outside the professional journals of the behaviorist school of thought, or the animal-science departments subsidized by animal-use industries, very few people will tell you anymore that animals neither think nor feel in any meaningful way. All the evidence and every ounce of common sense tell us otherwise, even if we still do not put that understanding into everyday practice.

Darwin himself, in the 1870s, recognized and captured the rich emotional lives of animals. Yet by a strange selectivity, his theory of evolution had a profound impact on how we humans see ourselves, while his evidence about the emotional lives of animals had almost no impact on how we treat them. And for all the pathbreaking work of Donald Griffin, the Harvard ethologist who a century later picked up where Darwin left off, somehow we've still had a hard time getting past the mechanistic dogmas of the behaviorists. In part because of the falsehoods they have spread, with their way of fitting every animal they study with the same scientific straitjacket, we've been in denial. Whatever common sense or our own good instincts tell us, we are still reluctant to ascribe sadness where we see tears and suffering where we hear cries, and to act accordingly.

It all reminds me a little of the scene of *Terminator 2: Judgment Day* when the Terminator, played by Arnold Schwarzenegger, notices human tears and asks, "Why do you cry?" He has developed an unlikely bond with the boy he's been sent to save, and the boy explains to the Terminator that people cry when they are sad or experience loss, or occasionally even when they are happy. The Terminator looks on skeptically, but the boy is providing a lesson on what it means to be human—seeming, for just a moment, to stretch the Terminator's understanding beyond his programmed knowledge.

For a long time, in our dealings with animals, we have conducted ourselves like beings from another world, unfeeling, all-powerful, and strangely disconnected from the realities of animal consciousness and emotion. But it shouldn't be such a stretch to recognize how much we have in common with creatures made of the same flesh and blood, or to imagine ourselves in their place— helpless, vulnerable, and afraid. It's been slow to come to us, since in some manner we seem to have been programmed to see the world a certain way. As we gain greater understanding about animals, we cannot help but begin to develop a closer bond with them and to open our hearts to their plight.

Refusing to believe that animals have intelligence, or even conscious life, is not only counterintuitive, but also a little too convenient. Leaders of animal-use industries have hired their own veterinarians and other scientists to deny the emotional intelligence of animals or to say the industry's treatment of other creatures is just fine—much like the biological determinists of another time who twisted reality to prove that certain races were inferior. When pseudoscientists tell us that animals are not conscious or aware, and are instead driven by mechanical, unfeeling instinct, the moral path has been cleared for economic interests to do as they please.

Mechanical and unfeeling, moreover, better describe humanity in action when we permit the boundless cruelties that modern industries inflict. In the treatment of animals, there's a vast gap between what we know and what we allow, what objective science affirms and what the laws permit. And much of the modern animal-welfare movement is working to close that gap—to bring consistency into the moral equation. We know too much, and what might have been excused in other times can no longer stand up to reason. We know that pigs and other animals are intelligent, social creatures, and with that understanding comes moral responsibility. It is wrong to condemn them to the dark, wretched existence of the factory farm. In the same way, we now know better than to treat primates as if they are the raw and disposable material for experimentation, or to drown dolphins by the thousands as if they were just acceptable bycatch for fishing fleets, or to consider companion animals as expendable surplus when there are no companions to claim them.

In matters of animal welfare, as in everything else, stubborn denial only makes things worse. And the better outlook is to view these questions not just as moral problems, but as moral opportunities. Thinking for yourself always takes a little extra effort, and shaking off old ways always requires that extra measure of courage. But it sure beats having to go through life making excuses for harsh and unpleasant things, and there are plenty of brave and good-hearted people to show us the way.

# PART II

## The Betrayal of the Bond

# CHAPTER THREE

# A Message from Hallmark: Exposing Factory Farming

W E HAD A FULL day ahead of us, and I told Maggie Jones, a staff writer for the *New York Times Magazine,* that I'd scout out a good place for lunch. The plan was, I'd pick her up midmorning at Los Angeles International Airport and from there we would drive into the heart of the Inland Empire—the sprawling desert counties of Riverside and San Bernardino that extend all the way to the borders of Arizona and Nevada.

This was our first meeting, although Maggie and I had already covered a lot of ground on the phone. She was a mother and a Bostonian, perhaps in her late thirties, and pleasant as can be. I could tell from our conversations that she was well informed on animal-cruelty issues, and not afraid to face the details. Even so, I was on my guard. We've all seen the occasional public figure become too relaxed during interviews with reporters who tag along, only to produce some gaffe or silly statement that defines the whole story. I didn't want that to happen to me or to the case I was making for Proposition 2, a California ballot initiative addressing some of the severest confinement methods in animal agriculture. On a long drive with a *New York Times* reporter, I knew I had to stay focused.

Maggie was doing a story about the rising political strength of HSUS, with an emphasis on our campaign confronting the mistreatment of animals in industrial agriculture. I wanted her to see some of the places where animals raised for food live and die. Only from afar can one be indifferent; there's nothing like a firsthand experience to awaken the detached skeptic. I knew well enough what she would see, and how people typically react when they haven't seen it before. And reacquainting myself with the sights and sounds of industrial agriculture could only renew my energy for the big fight we had taken on in California. She was here to cover that unfolding battle. And I wanted to give her all the background on a campaign that would be either an epic failure for the cause, or one of the biggest victories ever for animal welfare in America.

It was the last Saturday in July 2008, and the vote on Proposition 2 was then just three months away. I was as keyed up as I always get before one of our ballot initiatives. The election seemed at once around the corner and an eternity away. After Labor Day, thirty-eight million Californians, including the state's nearly eighteen million registered voters, would pay more attention to the candidates and issues in November's general election. We'd been building this campaign from the ground up, starting just weeks after we achieved resounding wins on our slate of ballot measures in the 2006 midterm elections. There was plenty of reason to be optimistic, but we knew that the polls don't count for much until the electorate is actually tuned in to the campaign.

As in any competitive statewide race, the battle would largely be won or lost in the final sixty days. During this stretch, the press, too, would pay closer attention, and its coverage would frame the debate for voters. Most important, both sides would spend the bulk of their money—ultimately, close to $20 million combined—on television advertisements to sway the undecided, to respond to perceived or actual misrepresentations, and then to make their closing arguments.

The main event was, of course, the presidential race between Barack Obama and John McCain. It would not be close in California; the state had last gone Republican in the 1988 presidential election. But there were lots of interesting down-ballot races, and some were highly competitive. California is the nation's prime staging ground for initiatives and referenda, and the debates that play out there can quickly become national debates. The political contest that drew the most coverage and campaign cash was Proposition 8, the attempt to ban gay marriage that attracted $80 million in spending from rival campaigns. On this same ballot were measures requiring parental notification for minors seeking abortions and alternative energy development.

In this crowded field, HSUS and other animal-protection groups would be fighting for Proposition 2, a reform to ban the extreme confinement of twenty million animals on concentrated animal feeding operations, known as CAFOs and more descriptively referred to as factory farms. Prop. 2 was the latest and biggest political clash between animal advocates and Big Agribusiness. Voters in other states had approved similar restrictions on factory farms—Florida in 2002 banned the confinement of breeding sows in small confinement stalls called gestation crates, and Arizona in 2006 began to phase out both gestation crates and veal crates. Now California was considering a proposal to outlaw those crates, and to take it a step further by adding a ban on the egg industry's confinement of chickens in battery cages.

## The Slaughterhouse Next Door:
## Inside a California Meat Plant

AS A WELL-INFORMED REPORTER, Maggie knew this recent history, and she knew as well that the agribusiness lobby was determined to stop our momentum. The way they figured it, if they could hold the line in California—a largely Democratic state, with its major population centers closer to sand than soil—then that would likely discourage us from taking the fight to heartland states. But if they lost a vote of significance in California, it could well prepare the way for major reform of factory farming across the country.

Laying all this out for Maggie, I explained that today HSUS is not your grandmother's humane society, helping only stray dogs and cats while other forms of neglect and cruelty went unanswered. In fact, this image of HSUS has never done us justice. From its founding in the 1950s, our organization has never limited its everyday work and moral concern to companion animals. We have major programs to protect pets, but our concern reaches to all animals, and I have sought as CEO to assemble the whole range of assets that any modern animal advocacy group needs—lobbyists, lawyers, scientists, undercover investigators, tech-savvy writers and editors, grassroots organizers, and more. That is news to some people—but not to Big Agribusiness, the furriers, and other industries that try to portray our broader reform efforts as a neglect of traditional sheltering of needy pets. In fact, we help animals of every kind wherever help is needed—and the need is greater than ever. These industries regard us as a threat to business as usual; if they didn't, I'd start to worry.

When I became president of HSUS in 2004, I pledged that America's largest animal-protection group would not fear to confront America's largest animal-welfare problem. I promised investments in programs to challenge the systematic mistreatment of animals in industrial agriculture. Nearly ten billion animals are caught up in the food production system in the United States. How could we not try in earnest to reform this industry, when even modest and incremental changes in production, transportation, and slaughter practices could reduce suffering enormously?

I had suggested we take this day trip outside of Los Angeles because there were factory farms and slaughter plants to see, and also because this was a crucial swing voting area in the state. Riverside and San Bernardino counties are also an area where middle-class families are striving to find affordable housing and their piece of the California dream. It has its share of white-collar professionals, but it is still largely working class, with a fast-growing Latino population that will soon become the largest ethnic group in the region. More important, it is where two powerful trends in California collide. There's the vast and enormously productive agricultural economy, dominant in the state's interior, and then there's the sprawling, suburban new economy typified by Los Angeles, Orange County, and other coastal counties in the Southland. This area is a bellwether of California, with lots of votes to be won or lost.

The Inland Empire is due south of the state's Central Valley—the expansive floodplain between the Coastal Range and the Sierra Nevada that is one of the most fertile regions in the world. Thanks to its year-round warm climate and rich, deep topsoil, California has long been the nation's top farm state in revenues generated, with its wide array of vegetable crops, fruits, almonds, avocados, and more. It's not the breadbasket of America, but rather its salad bowl. And, for our purposes, it is also a major animal agriculture producer. It is the biggest dairy state in the nation, with two thousand farmers managing about 1.8 million cows, yielding 41.2 billion pounds of milk in 2008 and $6.9 billion in revenue.

Not too long ago, it was also the nation's leading egg producer, though it recently slipped to fifth place. The state still has twenty million laying hens in cages, producing an astounding five billion eggs a year. (Recently Iowa, Ohio, and a couple of other midwestern corn-growing states have surpassed California in egg production. Corn and other crops used as animal feed are heavy, whereas eggs are light and easier to ship than corn. Thus, egg production is shifting to areas where chicken feed requires less cost in transport.)

The Inland Empire still houses much of California's egg and dairy industries, including an egg factory farm in Riverside County that confines eight million birds. Within this same community, however, suburban sprawl and family-oriented neighborhoods are beginning to overtake the old agricultural landmarks. Egg farms sit alongside prefabricated housing tracts, slaughterhouses alongside shopping malls, dairy farms alongside office park complexes. In most regions, there is a separation between the agricultural economy and bedroom communities, but here they sit together uneasily.

On the ride with Maggie, I was the designated driver so that she could take notes. I gave her an overview of HSUS and our mission, as well as the contours of the debate over Prop. 2, as we passed through Pomona and the rest of eastern Los Angeles County. It was about an hour's drive to our first stop—the Hallmark/Westland Meat Packing Company in Chino, which straddles the border between San Bernardino and Riverside counties.

We pulled up to the plant from the south and parked on a side street just behind the building. I had talked to the press many times about what went on here, but this would be my first visit to the scene. It is a workmanlike facility—a standard-issue aluminum and concrete factory, with air vents, pipes, and air circulation fixtures protruding from box structures. The main building looked to be about three stories tall, but smaller, ramshackle buildings were connected to it. There were parking areas, and holding areas for the animals, with overhanging structures, pens, and chutes—together covering much of a city block.

Hallmark looks like a cow in a pig herd, oddly out of place, as though dropped from above into a modern business district. There is an open field across the street on the north side of the building, but the south face has a modern office complex bumping right up against it. There is a busy thoroughfare on the west end, on a neatly paved road with the yellow stripes freshly painted. Across the street from this slaughter plant are more office complexes. What did the accountants, businesspeople, and other office workers think of their unsightly neighbor? As they settled into their cubicles and sipped their morning coffee, did they somehow block out the cries and bellows of the cows, or had they just become accustomed to it?

Prop. 2 wasn't about slaughter practices, but it did in a sense begin here, with reports of scenes that unfolded at the Hallmark/Westland plant every day. The plant remained a backstory to Prop. 2, and lest anyone forget, we kept reminding voters about the connections throughout the campaign for reform.

In the fall of 2007, HSUS sent a seasoned investigator to apply for a job at an egg production facility in the Inland Empire, to get an inside view of what was happening there and to capture it on film. This was a relatively new strategy for us—an employment-based investigation deploying an undercover operative for weeks or months at a factory farm or slaughter plant. Through the eyes and camera of that one person, people across the state and nation could see for themselves what was done to animals in industrial agriculture and who was doing it.

Our undercover guy didn't get off to a great start. He applied for jobs at a number of egg facilities but had no luck in getting hired. On one of his drives through the area, however, he saw the Hallmark plant. He decided to put in for the job, while waiting for any opening at an egg facility. It was purely opportunistic and not based on knowledge of questionable practices at the plant; in fact, we'd never even heard of this place before. It was a default option, but he took it because he knew the work might turn up something valuable. If it didn't seem like the right fit, he'd abandon the effort and resume his original search for work behind the lines at a battery cage facility.

Hallmark, it turned out, wanted him right away. This wasn't entirely surprising since many slaughter plants have up to a 100 percent annual turnover rate in personnel. From year to year, there's typically an entirely new crew of workers because the labor is so demoralizing and dangerous. Imagine killing and dismembering animals for every hour of every workday. Not much to look forward to in the morning, nor much to feel good about in the evening.

Slaughterhouse workers wield sharp tools all day long. The floors are slippery with blood. Large animal carcasses pass by swiftly on a mechanized pulley system, and the plants have continued to increase line speeds, placing ever greater strain on the workers. Accidents are common, and some workers develop carpal tunnel syndrome from the repetitive cutting. It is very tough, physically demanding, low-paying, deeply dispiriting work, and it's nobody's idea of a dream job.

Hallmark specialized in the slaughter of older, "spent" dairy cows, and was known in the industry as a "cull cow" slaughter plant. Through selective breeding, we've engineered Holsteins to produce enormous volumes of milk. But they can only endure this level of production for about three years, and when they are used up and no longer commercially productive, they are hauled to places like Hallmark.

Upton Sinclair wrote a hundred years ago that the meat industry uses everything but the squeal of the pig. That's true today for virtually all farm animals. There are no retirement homes for spent cows, and the dairy farmer can make additional revenue by selling cows for ground beef, rather than paying to dispose of the animals himself. In all, 17 percent of U.S.-produced beef comes from spent dairy cows.

The cows coming from the factory dairies often look emaciated—the result of years of being treated as nothing more than milk-producing machines. The average cow produces more than twenty thousand pounds of milk a year—five times more than cows of past generations. More than 90 percent of America's dairy herds are afflicted with mastitis, an inflammation of the udder, which in the modern cow is large and freakish-looking, often hanging so low that it drags along the ground. Many also have foot rot from standing all day in their own manure.

OUR MAN AT HALLMARK was a tall, lanky, dark-haired fellow who went by the name "Adam." After being hired at Hallmark, he was assigned the job of off-loading the cows upon arrival, sorting them into various pens, and then sending them single file up a chute into the "knock box"—the term at the plant for the slaughter area—where another group of workers took over. No HSUS investigators are allowed to harm animals, so his assignment was consistent with our policies.

On his very first day, Adam spotted problems at Hallmark. The cows looked very sickly and were typically lame. The animal handlers were shocking the cows with electric prods, or hot shots, to keep them moving toward the knock box. (Though not strictly forbidding the practice, federal regulations require that electric shocks be used as little as possible.) Adam said the cows' eyes would roll back in their heads in response to the pain inflicted by the electrical charges. He decided to stay on the job and documented all that he could.

At times, he had a clear view of the slaughter process. In the knock box, one worker would shoot the cow in the head with a captive-bolt gun—a device that looks like a large handgun and discharges a thick metal bolt. The gun is aimed at the cranial area, and the bolt is designed to penetrate the skull, stunning the animal and rendering her unconscious. Once an animal is stunned and then hung up by a rear leg, another worker slits her throat to kill her. She is then attached to a mechanical pulley system on a rail, which moves her body past various workers, who are each tasked with removing a specific body part until nothing is left. It has the efficiency of an assembly line, but is in fact the opposite—a disassembly line.

Adam worked ten hours a day, five or six days a week, and got paid a little more than the minimum wage. He was the only non-Latino among half a dozen or so workers who did animal handling and slaughter. That was no surprise, since plants throughout the country now use almost exclusively immigrant labor. During the last years of the Bush administration—and with the public clamoring for action on illegal immigration—U.S. Immigration and Customs Enforcement raided several major slaughter plants, arresting hundreds of undocumented workers. That was just a glimpse of an industry that now relies on the most desperate and easily exploited workers, people without options or legal standing to complain about what they are forced to see and do.

Adam had to report to work at 6:30 A.M., well before the office buildings next door opened for business. In the early hours of the day, the plant's workers would slaughter about 350 cows. Hallmark's daily total was about 500 cows, or just over 2,500 a week—making it a small slaughter plant by federal standards, but a relatively large one among cull-cow plants. By contrast, Smithfield Foods runs a pig-processing plant in Tarheel, North Carolina, that slaughters 32,000 pigs a day and employs 4,650 workers.

A U.S. Department of Agriculture (USDA) veterinarian was assigned to the Hallmark plant to enforce the Humane Methods of Slaughter Act—his job was to inspect each cow and ensure humane handling and slaughter procedures were observed. This, though, was not your standard veterinary examination. He didn't bend down on one knee and take a look at each animal and evaluate her condition. There just wasn't time for that, given that there were hundreds of cows. At HSUS, we had been working for years to get more money appropriated for the inspections program, and we had been steadily building support for it. Still, the funding was far from enough to enforce the humane laws that are supposed to protect the 33 million cows and 113 million pigs sent to slaughter annually, along with millions of goats, sheep, and other mammals. And the USDA doesn't apply the humane slaughter laws to poultry, even though birds are more than 95 percent of all animals slaughtered (about 9 billion chickens and 250 million turkeys are killed each year).

Adam and the other workers would herd dozens of cows past the government inspector, and as long as the vet didn't see anything unusual in the animals' gaits, he'd clear the entire group for slaughter. If he saw an animal clearly unable to stand, he was supposed to "condemn" that animal and not let her into the slaughter area. After he completed his morning inspection at 6:30 A.M., he'd leave the unloading area and holding pens to perform other duties. The unsupervised plant workers would then start slaughtering the cows, one by one. Like clockwork, around 12:30 P.M., the vet would come back and approve another 150 or so for slaughter. The vet admitted to federal investigators that he took "shortcuts" during these inspections to save time.

Before work each morning, Adam strapped a miniature video camera to himself, the device hidden beneath his clothes. Because the camera was so small, the battery life was very limited and he could record only about an hour of footage per day. At night, he'd head back to his motel, undress, unstrap the camera, and then do his best to scrub the filth of Hallmark off his body. His evening routine was to make notes, catalogue the tape, recharge the batteries, insert a new tape, and get ready for the next day.

At work, Adam generally kept to himself, and he couldn't converse much with the other workers since he didn't speak Spanish. They had no idea he was a vegan, and that the "meat" in his sandwich was actually a soy riblet he'd bring to work every day. Every week or two, he checked in with us so we would know he was okay and to update us on the investigation. He didn't have time to review his tapes in any detail; he was just amassing as much evidence as he could. He had to perform his animal-handling job well enough to prevent undue scrutiny, and to make sure he didn't get fired—all of which limited his opportunities for filming.

Along with the federal veterinarian, USDA had four other inspectors at Hallmark stationed inside the plant where the meat was processed. Their primary charge was food safety. They would test carcasses for salmonella, *E. coli*, and other pathogens, and certify that the meat was fit for human consumption.

The USDA presence at slaughter plants had always been weighted toward the food safety functions, with inspectors focused on postmortem examinations. The animal handling and antemortem inspections seemed like a nuisance to the department, and this lack of commitment showed. When President Dwight Eisenhower signed the Humane Methods of Slaughter Act in 1958, he remarked to a group of congressmen, "If I relied on my mail, I would think that the country is concerned only about humane slaughter." Unfortunately, the USDA never quite got the memo, and its oversight has been poor, or, in the case of birds, nonexistent.

In 1997, Gail Eisnitz, an animal-welfare investigator, wrote a book called *Slaughterhouse* about the USDA's failures to enforce the law. More recently, a number of press exposés at slaughter plants have revealed federal workers either not paying attention to the fate of the animals or simply ignoring abuses in plain view. In almost every investigation conducted by animal-welfare groups, serious problems have been uncovered. Together, these random samplings and federal audits paint a picture of chronic deficiencies in the inspections program—a systematic failure to enforce the law.

In 2001, the *Washington Post* reported on problems in slaughterhouse enforcement, with an award-winning article recounting the observations of line worker Ramon Moreno, who was assigned the task of cutting the hocks off carcasses as they sped past him at a rate of 309 an hour at a beef plant in Pasco, Washington. He told the *Post*'s Joby Warrick that even though a person in the line ahead of him was assigned the task of cutting off parts of the body, the animals were sometimes alive when they got to him. "They blink. They make noises. The head moves, the eyes are wide and looking around." They die, he said, "piece by piece."

At Hallmark, according to Adam, the vet appeared in the live-animal area at the plant at the same time every day. He made no unannounced visits, and he had been inspecting this plant for more than a decade, so his routine was well known to the plant's management. When he arrived, slaughterhouse workers were usually on their best behavior, according to Adam. And they had little to worry about from the other USDA inspectors inside the plant, because they never ventured outside to make spot checks on the animals, though they did sometimes oversee what happened in the knock box.

In 2005, an outside USDA veterinarian visited Hallmark to perform a humane handling verification review and he documented a failure to provide animals with water, excessive use of hot shots to herd the animals, and improper stunning. These are all "egregious violations" by USDA's own standard, but the vet issued only a "noncompliance report"—the equivalent of a slap on the wrist. Beyond those infractions, there were no blemishes on the company's record as far as the government was concerned. In that same year of 2005, the USDA had named Hallmark/Westland its "Supplier of the Year" for the National School Lunch Program. And, in addition to the USDA inspections, the plant's management hired auditors to inspect their animal-handling operations. In fact, according to Hallmark's president, in 2007—the year we conducted our investigation—there had been seventeen third-party audits, and every one of them resulted in the highest marks for the plant.

If you relied upon the accounts of the USDA and the private auditors, and took the word of the plant's management, this was a model facility. But our man Adam had a quite different experience, and his camera showed what all those audits and prizes were really worth. He recorded dozens of cows unable to walk and being tormented by workers forcing them to stand. These were so-called downer cows—they were down on the ground and couldn't get up on their own because of broken limbs, lameness, and other debilitating conditions. They had arrived in the trucks already down, gone down in the pens, or broken down on the way to the knock box. In each case, slaughter workers employed an arsenal of tools to compel them to stand. If they were down in the pens, just before the USDA man took his post, workers knew just when to harass them so they'd be on their feet to make it by the inspector.

Adam said the worst case he saw was a cow who went down in the truck and couldn't stand. Workers attached one end of a chain to her leg and the other end to a Bobcat tractor. It took thirty to forty-five minutes for them to drag her out of the truck and toward the slaughter area, and she was bellowing most of the time. "It got to the point where the animal quit struggling and bellowing. There was just silence. She just gave up," Adam told us.

The pen manager simply refused to permit any cow to be treated as a downer. He instructed pen workers to force every animal to stand, claiming that every time one was condemned as nonambulatory, the plant lost money. Workers used hot shots on the cows' genitals and eyes. They rammed downed cows and pushed them with the blades of a forklift. Adam even recorded them placing high-pressure water hoses in cows' mouths to simulate a water-boarding effect. The point was to cause the animals so much distress that they used their little remaining strength to rise and flee from their abusers—which again was enough to get past inspection.

## The Hypocrisy of "Happy Cows"

KATHY MILANI, WHO AT the time was our chief of investigations, said it was urgent that I see the footage Adam had sent. She and her deputy sat me down for a screening, where Michael Markarian, our chief operating officer, joined us. Whether it is a tape of animal fights, puppy mills, seal clubbing, or any other form of abuse, I always come to these rollouts of investigative footage with a mix of anticipation and dread. Of course, I don't want to see animals suffering in any way, and it is infuriating to watch. But after a while you learn to steel yourself, and I figured if men and women like Adam had the guts to see it in person, then my colleagues and I could watch every frame and hatch a plan to do something about it.

Before sitting down to watch Adam's tapes, I came in with my biases about the meat industry. After twenty years of work in the field, visiting farms and slaughter plants and studying the topic exhaustively, I felt that the meat industry had lost whatever humanity it once had, adding more and more miseries to the lives and deaths of animals raised for food. The modern livestock industry is never more resourceful than when it is cutting corners or evading even minimal regulations. Among many other wrongs it commits, the industry has betrayed the old standards of animal husbandry, which at their best showed a decent respect for the natures and needs of farm animals and obvious ethical boundaries. That sense of responsibility has been discarded by industrial agriculture; at every stage the animals are typically treated as unfeeling commodities and units of production. It can make for some sorry sights, and that's what I was prepared for when I sat down to see what Adam had recorded.

Yet for all that, the images still left me rattled. The abuse was just appalling. As if it were not hard enough for these afflicted creatures going to their death, their last moments were filled with beatings, shouting, and fright. The workers showed not the least sense of restraint or concern. Whatever compassion they might have felt for the animals had clearly worn away in the violent routine of the job.

One of my first thoughts was the disconnect between what I was seeing and the California dairy industry's "Happy Cow" advertising campaign—a high-profile, multimillion-dollar promotional effort financed by a two-cent surcharge on every gallon of milk. The ads play up the idea that happy cows produce good milk, evoking pastoral images of cows in green fields bathed in sunshine.

We now had in hand more than enough evidence that the industry's marketing campaign was a fraud. These cows at Hallmark had never set their hooves on grass; their path, from the day they were born to this day of their death, was one of confinement, privation, and misery. They passed their days standing on concrete floors or knee-deep in a dark brew of mud and manure before they came to the knock box at places like Hallmark.

My only concern was that the scenes were so painful to watch that the news media might not put them on the air—a common problem with issues of cruelty. In the view of a lot of producers and editors, there's enough upsetting news every day without forcing scenes of animal cruelty on their viewers and readers, and so they take a pass. The temptation to ignore the story is only stronger when it concerns farm animals and meat production and many viewers may feel a touch of guilt—since the average American consumes 220 pounds of meat a year.

This is true even though we seem to be in a period of social awakening about our food and its origins. In best-selling books, Michael Pollan has pointed out that corn is in so many foods, that processed foods are a fixture in our diets, and that industrialized agriculture abuses animals, who themselves are typically fed massive amounts of corn and soybeans. Pollan's general dietary advice is: "Eat Food. Not too much. Mostly plants."

In *Fast Food Nation,* also a best seller, Eric Schlosser showed how fast-food companies are "super-sizing" their products, leading to obesity, diabetes, and a range of other health problems. Men, on average, weigh seventeen pounds more than they did in the 1970s, and women weigh nineteen pounds more. Americans are simply eating more, and much of what they eat is doing them harm. Pollan and Schlosser, among others, have given us a window into what is happening in slaughterhouses and on farms, revealing the harsh, merciless world of modern food production.

At the time of the Prop. 2 effort, people were beginning to recognize that it doesn't have to be this way. All of us have choices. And to make those choices, we need to see our options clearly and take an unflinching look at the moral costs of industrial livestock farming. What Adam and his camera had captured would offer that clear view, and at just the right time. It always takes the right set of circumstances for a campaign to break into popular consciousness, and here those circumstances were aligned. A series of food safety scares in the months before the Hallmark case had made the issue a live one in Congress and among the public. This shocking story out of California happened to involve kids and school lunches—Hallmark's primary market—so parents would demand action. It also involved a mainstream group like HSUS, and a courageous undercover investigation that added some drama to the story. And, finally, the footage that Adam recorded had a digestible amount of violence: Mistreated animals. Not too much blood. On its way to kids.

Once we watched all the Hallmark footage, it was obvious we had evidence of criminal cruelty to animals against the workers and the company. But it was a state matter, not a federal one, since federal laws to protect farm animals are virtually nonexistent and there is no federal anticruelty law. There were clearly violations of the Humane Methods of Slaughter Act, but that's not a criminal statute—the enforcement tool for the federal government is to temporarily shut down the operations of the slaughter plant, and it was now too late for that. Moreover, no federal statutes existed to address the criminal behavior of the people involved.

That was just part of the reason we didn't make the USDA our first stop once we had assembled the case. We also had no confidence in senior USDA officials to act in a decisive way and demand reforms. Almost as a rule, high-ranking political appointees at the USDA come straight out of corporate positions in the meat industry. And they don't see much difference between their jobs in government and their jobs in the industry. Dale Moore, chief of staff at the agriculture department during the Bush administration, had previously been a lobbyist for the National Cattlemen's Beef Association. The meat industry had always seemed like a management training track for senior USDA work, and it's been that way for a very long time.

All these problems were compounded by the USDA's own failure to prevent the abuses at Hallmark. Everything had happened under the USDA's regulatory authority. And I thought it highly unlikely that the department would condemn its own employees' performance and highlight the poor oversight procedures at the plant.

Our only serious option was to hand over the footage in mid-December to the office of Michael Ramos, the very capable San Bernardino County district attorney. Ramos's deputy, Debbie Ploghaus, had a reliable record of prosecuting cruelty cases in the county and clearly cared about enforcing the law. Along with the footage, we also gave Ramos's office a legal analysis arguing that the conduct of the plant personnel violated California laws forbidding animal cruelty and the California Downed Animal Protection Act. Unlike many other states, California does not exempt routine or customary agricultural practices from its anticruelty statutes.

Even so, no state had ever successfully prosecuted slaughterhouse workers, so bringing Hallmark to justice was no sure thing. The case was set in a community with a strong agricultural heritage—another burden we'd have to overcome in a jury trial. We needed to produce an overwhelming case on the facts—one reason why Adam stayed undercover many weeks gathering more and more information.

Our lead farm animal attorney, Peter Brandt, had worked with deputy D.A. Debbie Ploghaus before and arranged for a meeting to hand over the evidence. Soon thereafter the district attorney's office informed us that it had opened an investigation. Days went by, and then weeks, and the district attorney's office gave us no definite word on the progress of the case. We could not know whether they were actively investigating, or if they had put the issue aside in order to handle other matters. By mid-January, Ramos's office had had the footage in its hands for nearly a month. Our team was getting impatient, and I began to worry that Ramos's office was going to bury it. Maybe the case was just too controversial for an elected D.A. in an agricultural county. Maybe pressures were being brought to bear from other sources.

WE KNEW THAT RELEASING the footage to the media would put pressure on all parties—the management at Hallmark, the USDA, and the district attorney. So on January 29, 2008, I called a reporter at the *Washington Post* and offered an exclusive to the newspaper. We'd release the footage the next day at the press conference when the story appeared in the *Post*. This was a case about animal abuse, and also a disgraceful example of the failure by our government to regulate the meat industry. I wanted the staff at the USDA and in Congress to see the story, and what better messenger than the *Washington Post*?

The *Post* story was powerful and validated the story, and now other reporters wanted to hear more and to see the footage for themselves. At the press conference that same day, I announced that we had found gross abuses of downer cows at a Southern California slaughter plant, that these downer cows were entering the food supply, and that this plant was the second-largest supplier to the National School Lunch Program—providing fifty million pounds of ground beef for schoolchildren all across the nation. Those facts—combined with the images, including one of a forklift rolling a cow—made a visceral impression, and reporters left that room determined to learn more.

We distributed the footage that morning and posted it on our website. It didn't take long for CNN, Fox, and other networks to start running it, and the story became the lead item in the news cycle that day. We also sent the footage and our documentation over to Ed Schafer, the new agriculture secretary, who'd been sworn in just the day before.

There were immediate aftershocks. No one could deny that something was seriously wrong at Hallmark, and even the USDA and industry trade groups recognized that right away. The USDA pulled its inspectors from the plant almost immediately. Given that federal law requires that meat fit for human consumption be USDA approved, that action alone was enough to shut down the Hallmark slaughter lines.

Major news outlets stayed on the story for days. Upon learning that Hallmark supplied ground beef to school cafeterias, parents across America were making themselves heard. They worried about E. coli and mad cow disease. Then they started to panic, and to press for action. And school districts began taking meat out of their cafeterias.

The release of the footage also lit a fire under the San Bernardino County D.A.'s office. On February 14, we heard from District Attorney Ramos, and the very next day I joined him for a press conference at his office, where he announced he was filing charges against two of the workers seen in the video committing the worst acts of cruelty against the cows.

Ramos also announced that he had charged Daniel Ugarte Navarro, a plant worker, with five felony counts under California's anticruelty statute, and three misdemeanor counts including allegations of using a mechanical device to move downer cows. Convictions on the felony charges could bring a sentence of up to fifteen years in prison and $100,000 in fines, plus additional penalties on the misdemeanor charges. A second Hallmark worker, Jose Luis Sanchez, was charged with three misdemeanors involving the mistreatment of downers.

To his great credit, Ramos also declared his reasons for bringing charges, in words that the entire industry needed to hear: "I need the public to understand that my office takes all cases involving animal cruelty very seriously. It doesn't matter whether the mistreated animal is a beloved family pet or a cow at a slaughterhouse. Unnecessary cruelty will not be tolerated, and will be prosecuted to the fullest extent allowed by law."

Agriculture secretary Ed Schafer, who was still getting to know his staff, treated the case seriously and said he was appalled by the mistreatment of the animals at Hallmark. But he also joined the meat industry in its counterattack against HSUS. "For four months they [HSUS] sat on that information," he told CNN. Schafer and the industry all but blamed us for the cows suffering for the last few weeks, and for the food safety threats, because we didn't deliver the footage immediately to the USDA.

Meat industry executives were very much on the defensive, without much of an explanation, so they attacked the messenger. We anticipated their counterattack. I'd already sent the USDA a timeline of our investigation, and by the reaction we were getting from the press and public, we had handled it just right. I reminded the public that the abuses happened on the USDA's watch, not ours, and that we were tired of seeing the cozy relationship between the meat industry and an agency charged with upholding federal law and serving the public interest.

With cruelty charges filed by Ramos and with much of the nation watching, USDA had little choice but to act. The USDA announced a voluntary recall by Hallmark of all meat from the plant distributed in the last two years. That came to 143 million pounds of ground beef, making this the largest meat recall in American history.

Much of the Hallmark meat had already been consumed, but millions of pounds were in freezers in schools and homes across the country, and school administrators were dumping it. Andrew Weinstein, a board member of HSUS, told me he'd recently toured an aircraft carrier and learned that the ship weighed seventy-five thousand tons. That was about the same weight as the recalled meat from Hallmark.

Spokespersons from the agriculture department called it a Class II recall, which was not the highest recall level, and said it was acting out of an "abundance of caution." They insisted there was no danger, but that rules on the handling of downer cows had been violated. Yet if there was really no danger, why impose a recall of this magnitude?

Confidence in the USDA was hardly at a high point. Concern was growing in Congress that our food safety mechanisms were obsolete, that the functions were balkanized among too many federal agencies, and that the agriculture department itself was too close to industry and compromised by conflicts of interest.

Also on everyone's mind were the public health risks of sending downer cows into the food supply. Because they were wallowing on the ground, the cows were covered in manure when they were slaughtered and were thus more likely to cross-contaminate equipment and spread *E. coli*. And plenty of evidence from Europe showed that downer cows are much more likely than ambulatory cows to have bovine spongiform encephalopathy (BSE). The human variant of BSE is Creutzfeldt-Jakob disease, a fatal brain-wasting condition.

As all of this unfolded, Steve Mendell, the president of Hallmark, seemed to be in hiding. There were eight congressional hearings centering on the Hallmark case, and it took a subpoena from the House Energy and Commerce Committee to get him to one. Mendell testified, under oath, that he had known nothing about the cruelty that had come to light, and that he was disturbed by the evidence on tape. He assured the committee that no downer cows were slaughtered, and that the ground beef the company sold was safe for human consumption.

Representative Bart Stupak, a Democrat of Michigan, a tall, imposing, yet youthful-looking former sheriff who chaired the Oversight and Investigation Subcommittee, had studied the tape and had a different view. He directed Mendell and a jam-packed committee room to turn their attention to the monitors on the wall, and then asked that our footage be shown. The scenes looked familiar to all by then, but Stupak focused the room on one particular moment when a downer cow was dragged into the knock box. "Do you now deny, sir—and remember that you are under oath—that downer cows were slaughtered for human consumption?" Stupak demanded. Mendell relented, conceding what everybody had just seen with their own eyes.

Soon, the crisis went global. America's trading partners became agitated about the safety of American beef. Our footage had now been broadcast on network television stations around the world, including in South Korea, the second-largest importer of American beef. There, the South Korean people saw how at least some American beef was getting to the table, and it did not inspire confidence. There were actually riots, and at one point some seventy thousand people took to the streets of Seoul. The rioting lasted for days, and both the USDA and the office of the U.S. Trade Representative launched a hurried diplomatic effort to avert a complete shutdown of the South Korean market for American beef.

Some months later, in response to the investigation, California lawmakers upgraded the state's Downed Animal Protection Act, prohibiting slaughterhouses from selling meat from downer cows and imposing penalties of up to $20,000 in fines and imprisonment of up to a year. I also urged Secretary Schafer and his department to ban all meat from downed animal products in the food supply. He hedged, claiming it wasn't necessary, but eventually announced nearly four months after the investigation broke that he favored a ban on slaughtering downer cattle in the food supply.

The problem was, the Bush administration's USDA took no final action. There was a promise, but no emergency measure to stop downers from entering the food supply. And when the president's term came to an end, nothing at all had been done to ban the slaughter of downer cows. It wasn't until the Obama administration took over, along with a new agriculture secretary in Tom Vilsack, that a federal ban on processing adult downer cows was imposed. Such a simple, minimal standard—that sick and lame animals shouldn't be abused any further—and yet in Washington it took two decades of debate and a series of national crises to finally make it law.

## Endless Denials: The Challenge of Reform

THAT IS THE SORDID tale of the Hallmark meat plant, and it explains why this site was the first stop on my tour of the Inland Empire with Maggie Jones of the *New York Times*. We showed up at the shuttered plant in July of 2008, just seven months after HSUS had released the final results of our investigation. The $110 million company had been out of business since the USDA first pulled its inspectors. Its president, Steve Mendell, had left Chino behind and retreated to his $4 million home in Orange County. When Maggie and I parked outside the main entrance to the plant, the only person we saw was a security guard, who would not leave his guard shack even after I politely motioned to him to come and speak with us.

With the watchman seemingly confined to his post, we walked around the side of the plant and peered over a five-foot wall surrounding the facility. We climbed atop the wall, and from there surveyed the place where hundreds of thousands of cows had been killed through the years. Now it lay abandoned, almost peaceful. The building was completely intact, and everything there still seemed to be ready for operations. I suggested to Maggie that we call Adam so he could tell us what to look for. After a couple of rings, he picked up the phone and I put him on speaker.

We described the area in front of us, and Adam told us we were looking at the holding pens, the first stop for the cows once they were off-loaded. He then directed our attention to the chutes that led up to the kill box—the lip of the large plant where the dismembering and packaging started. I gazed at a covered area where, as I recalled, workers had tormented one downer cow by placing a stream of water in her mouth to make her feel like she was drowning. What kind of man would do that to such a helpless, unoffending creature, and what sort of industry instills such a spirit?

Maggie had a few more questions for Adam. She wondered, for example, how he got through the day given his strong feelings about animals and their suffering. Adam was characteristically modest and replied that personal sacrifices were required to bring animal cruelty to light. We signed off with Adam and jumped down from the wall.

As we walked back to the car, I told Maggie that Steve Mendell's indignant denials about his company's animal-handling practices before the House committee were par for the course for industry executives. Many had convinced themselves that all was okay, because the whole industry now depended on the denial of doubts and the concealment of facts. Their industry faced very limited regulatory oversight, and their very friendly relations with the USDA would make sure things stayed that way. One brave soul with a small camera was their worst enemy, and their greatest fear was an informed public.

President Obama's announcement of a permanent ban in March 2009 was a triumph, but it was long in the making. HSUS, along with the animal-protection group Farm Sanctuary, had long warned Congress and the USDA that they were endangering public health by tolerating downer cows in the food supply. We had championed legislation to ban the processing of downer cows for human consumption, but faced resistance at every turn from the National Cattlemen's Beef Association and its allies on Capitol Hill. In 2001, the House and Senate each passed amendments to ban the slaughter of downers, but a conference committee led by leaders of the agriculture committees dumped the language from the 2002 Farm Bill.

We picked ourselves up and tried again. The U.S. Senate again approved a downer ban in 2003. This time, agribusiness made its stand in the House, sparking a debate on the floor over the provision during consideration of the agriculture department's spending bill. Representative Gary Ackerman, a Democrat from New York and the author of the amendment, showed his colleagues an enlarged photograph of a sick downer cow and warned that this was all it would take to disrupt worldwide beef markets and to shake consumer confidence in the American beef supply—one sick cow sent to slaughter. Congressman Ackerman argued that it made no sense to slaughter two hundred thousand or so downer cows, when thirty-four million cattle were going to slaughter every year, and that it was inhumane and wrong to allow these stricken animals into the food supply.

The chairman of the House agriculture committee, Republican Bob Goodlatte of Virginia, argued strenuously against the amendment, as did Texas Republican Henry Bonilla. Their colleague Charles Stenholm, also of Texas and the ranking Democrat on the agriculture committee (who lost a reelection bid and then moved to where he belonged anyway, in the meat industry as a lobbyist), protested against any talk of a downer ban, assuring members of the House, "That sick animal [in Rep. Ackerman's photo] will never find its way into the food system. Period."

With at least some House members trusting the assurances of

their colleagues most allied with agriculture, the House defeated the Ackerman amendment by the narrowest of margins—202 votes to 199. If just two members had switched their votes from "Nay" to "Yea," the amendment would have carried and downed animals would not have legally gotten into the food system.

It was just before Christmas of that year that USDA announced that a downer cow had tested positive for mad cow disease, or BSE, in Washington State. It was the first known case of mad cow disease in the United States. The sick animal had been processed for human consumption, and meat from that plant was distributed around the country. It was front-page news for two weeks.

In the United Kingdom, there had been more than one hundred deaths due to mad cow disease since the first outbreak occurred, with the disease having a frightening and ominously long latent phase. Mass culling of cattle and other livestock had ensued, with the bodies burned in huge pyres. It was a public health crisis and economic calamity, and the fear was that we in the United States might suffer the same fate. Nervous consumers started eating less meat, and more than fifty nations closed their borders to American beef.

In response to that crisis and to the fears of our trading partners, Ann Veneman, then agriculture secretary, announced a series of emergency policies, including a ban on downer cows in the food supply. If an animal was down on the ground and couldn't get up, then that animal had to be euthanized on the spot. In its proposed rule on the subject, the USDA cited a Swiss study establishing that nonambulatory cattle are forty-nine to fifty-eight times more likely to have BSE than cattle identified through passive surveillance.

It required all of this to compel the enactment of a policy that had been debated in Congress for a decade, and even the industry went along with the new policy in the days after the crisis. Folks in the industry had wanted to settle down the public and our trading partners. But they wasted little time in trying to undermine even these minimal reforms. Not long after the story faded from the daily news cycle, the livestock industry began working the USDA to undo Secretary Veneman's decree.

Undetected even by HSUS, much less by the American people, bureaucrats within the USDA were on the case, and in short order managed to revise the policy. It was July 2007 when the agency published a final rule on the downer cow issue, slipping in a loophole allowing some of these lame or sick cows to be processed for human consumption. Specifically, if a cow was approved for slaughter but then collapsed after the inspection, that animal could be slaughtered, as long as a USDA inspector recertified the animal as fit for processing.

It was this mysterious revision in policy that cleared the way for all that followed at the Hallmark meat plant: If the plant could just get the downers up, even for a few moments, then they might pass inspection after all and be slaughtered. If the weakened animals went down after passing inspection, that was no longer disqualifying. All that was needed was another okay by the inspector— assuming the plant even bothered to call out the vet. If he came out again and took another look at the animal, he could send her to slaughter.

In both cases, the meat industry was driven by a short-term mind-set and habit of corner cutting that didn't even serve its own long-term business interests. Here were animals so lame and sick as to be unable to walk to their own deaths, and still the industry leaders could see only the potential of lost profits. They disregarded even the safety of millions of people, and in the end brought on themselves troubles and losses far worse than the ones they were trying to avoid.

It's hard to summon much sympathy for an industry that lets cows live and die without even a touch of human kindness. To allow employees to drag these creatures to slaughter with ropes and chains, or to push them along the cement floor with forklifts—all for a few extra dollars in a multibillion-dollar industry—is beneath us all. And though the Hallmark folks probably still feel aggrieved, being run out of business is the least punishment anyone should expect for that kind of ruthlessness.

Even if minimal compassion fails to motivate the ranching and dairy industry trade groups and the slaughter plant operators, then in the case of downers, at least, enlightened self-interest should have been enough. The industry could have avoided the Washington State mad cow incident and the Hallmark fiasco had its leaders just accepted a sensible no-downer policy. A 2008 report revealed that slaughtering the downer cow in Washington State ultimately cost the industry $11 billion. The Hallmark case certainly cost a couple of billion more. Since that time more than a dozen BSE-positive animals have been found in Canada and the United States—nearly three-quarters of which were downer cows—confirming common-sense assumptions that the livestock industry once dismissed as ridiculous.

## *"The New Agriculture":*
## *The Ways of Industrial Farming*

IN RESPONSE TO OUR 2008 campaign to pass Proposition 2, the meat industry would trot out some familiar arguments. They said everything was just fine on veal, hog, and egg farms. We asserted that systematic cruelty occurred on these factory farms, and that it just wasn't safe for the public, much less decent and fair to the animals, to cram tens of thousands of creatures into harsh confinement.

The industry treated our arguments in favor of Prop. 2 with the same disdain and disregard as they had our admonition that downed animals shouldn't go into the food supply—saying we were naive, misunderstood basic economics, and should mind our own business and just work on dog and cat issues.

I had brought Maggie to Hallmark to give her just one glimpse of an arrogant, cruel, and often corrupt industry. With all those awards and commendations from its peers and its regulators, this one meat plant had been held up as a model for the industry. It turned out to be exactly that.

From the Hallmark plant, Maggie and I headed farther east to Yucaipa, in Riverside County, for another snapshot of modern animal farming—what the industry likes to call "the new agriculture." We drove into a subdivision with a patchwork of neat and well-manicured homes, alongside others that were a bit rougher around the edges. We turned right into a dirt driveway and came to a stop where Dave Long greeted us.

My colleague Paul Shapiro had read about Dave in a small Riverside County paper, which reported on testimony he'd given at a local hearing complaining about the number of chickens at the egg farm next to his home. Paul called him, learned more about him, and told him about our campaign to crack down on the egg factory farms. Dave agreed to meet with Maggie and me to share his experiences.

He had a small home that was part ranch house, part trailer house, with motors and metal parts strewn all around his yard. He was obviously a guy who worked with his hands. By the looks of his place, I guessed that he traded in junk metal, spare parts, and machinery. Dave's work was outside, so he could not escape the factory farm next door and its dreadful effects.

He was a trim yet sturdy fellow, probably in his early fifties, with flecks of gray hair on the sides, and that day was dressed in jeans and a T-shirt. He had lived on this property for more than thirty years, and when he moved there, the egg farm had about 40,000 birds. Now it had 763,000, and he told Maggie and me that it was ruining everything he liked about the area. When the wind blows in his direction, his property is overtaken with blackflies. "I am a Republican, I am a gun enthusiast, and I eat meat," he said.

"I am supporting Proposition 2 because it is just not right to cram those birds in there."

Maggie and I chatted with Dave for a while, and I asked if we could take a closer look at the farm. "Absolutely," he replied, and we headed over to his John Deere ATV.

Dave jumped in the driver's seat, Maggie took the passenger seat, and I stood up in the rear holding on to the cross frame. Dave gunned it, and Maggie and I held on for dear life as we listened to him recall how the chicken farm grew house by house, until finally he found himself living across a fence from three-quarters of a million closely confined birds.

We sped down the driveway, took a few quick turns, and passed the side of the egg farm opposite Dave's place. We went up a dirt road, crossed a small field, feeling every bump, and then came to an abrupt stop. From there we had a clear view of the egg farm.

Dave pointed to two large barns and said that they were modern hen confinement buildings, with the birds kept entirely inside. It was a battery cage facility, and it had football-field-length rows of cages stacked four or five high. Six to eight birds were stuffed into each cage, and each bird, under industry standards, got about sixty-seven square inches of living space—roughly two-thirds the size of a standard sheet of paper. Dave guessed 150,000 or so birds were in that building alone, and he pointed to the huge fans that regulated the temperature.

The egg farm also had about twenty smaller, old-style chicken houses, long structures open on all four sides, and with five columns of cages running the length of each long barn. The columns were stacked two high, with two wire cages on each row, and two birds in each cage. The birds had hardly any room to move. Dave estimated each structure had about twenty thousand birds, so with about twenty in all, there were more than four hundred thousand birds in these older-style cages.

Inside, the floors tilted slightly to allow the eggs to roll onto a

constantly running conveyor belt that took them away for processing. In the open-air structures, the floors also tilted, but there was
no conveyor belt—the eggs were manually collected.

In any modern laying hen building, a conveyor belt removed
the waste that fell through the bottom of each cage. But in the older
confinement system, the cages were on stilts and the manure just
piled up beneath them. The birds in the top cages were whiter than
the birds below, not because those below were of a different breed
with a darker plumage, but rather because manure dropped on them
day after day from the birds confined above, falling through their
wire cages, soiling them, and collecting on the ground. Beneath
some of the cages, the manure climbed three feet high, to within
just inches of the birds.

The manure piles drew the flies. And they were everywhere,
in thick, dark clouds. These were the flies that ruined Dave Long's
days. When the wind blew in his direction, the stench could make
you gag. Any gusts in the other direction brought a brief respite,
except for the neighbors on the other side of the bird factory.

Dave has an uncomplicated view of the problem here, shaped
mainly by a basic sense of fairness. "When I look into the chicken
houses and see the manure piled high, I feel like it shouldn't be done
that way," Dave said. "The manure should be removed a lot more
frequently, and the chickens should have a little more stomping area,
as they don't have enough room to maneuver inside those cages."

Dave had us get back on the ATV and then took us over to the
other barns. It was more of the same. Massive numbers of birds,
with no people around. He said there were a handful of workers,
with a ratio of about one worker for eighty thousand hens. He said
they cleaned the manure only when it was bumping up against the
cages, but that it wouldn't take long for the piles to build back up.
Basically, they get around to cleaning away the manure when it
becomes a problem for them, when it gets in the way of operations,
and not out of even the least concern for the birds or the neighbors.

"The reason they cram so many chickens into those cages is for profit. It has nothing to do with anything else."

Dave drove us back to his place, and we discussed Prop. 2. I asked if he had other neighbors who were upset about the factory farms. He said many had problems with it, but that they just kept their complaints to themselves. He was one of the few to speak out, and he didn't think he could help us attract any outspoken support. He had some fight in him, but he clearly didn't expect to win, and he was tired and frustrated. How could I blame him for feeling almost powerless, with a megacity of birds right next door, and corporate neighbors who couldn't care less whether he liked it or not?

I thought about telling Dave that though this egg factory was a disaster, things could have been worse. He could have found himself next to a hog factory. The average hog excretes three times as much waste as a person, and the waste goes untreated, typically funneled into massive, toxic lagoons. Rural residents are dealing with this all over the country—from North Carolina to Missouri to Ohio. And if Dave thinks his corporate neighbors are impossible to deal with, he should meet some of the characters running these animal factories.

During the second half of the twentieth century, the pig industry went through changes similar to those in the egg industry—with traditional farming practices supplanted by an industrial model and massive numbers of animals packed into sheds and barns. The breeding sows have it the worst, confined in individual gestation crates—two- by seven-foot cages—that prevent them from turning around, stretching, or even comfortably lying down.

These curious, social animals—who in the outdoors would root around with their noses, forage, build nests, and wallow in the mud—cannot engage in almost any of their normal, instinctive behaviors. For nearly their entire four-month pregnancies, they can do nothing but stand and lie on concrete slatted floors, breathing in toxic ammonia emanating from massive waste pits beneath. There is no positive stimulation and no social interaction. In such extreme confinement, sows' muscles atrophy, they lose bone strength, and they are often afflicted with lesions that are indicators of their terrible health, their unsanitary living conditions, and the psychological torment they experience from such severe restriction.

Before giving birth, the sow is moved to a farrowing crate—an iron enclosure much like the gestation crate, except that it curves out at the base so that piglets can squeeze under the bars and not be crushed by their immobilized and often lame mothers. In that modified cage, the sow nurses her babies for a fraction of the time that nature intended, until they are prematurely weaned and begin to get fattened for slaughter. The sow, meanwhile, is reimpregnated and hauled back to the original crate for another four months of confinement. Whatever the name for the crate or the process, for the sow it's always the same—a life of privation and unrelieved misery.

Although pigs can live to fifteen years or more, breeding sows are summarily culled when their usefulness has passed, or else sent to slaughter like the spent dairy cows. The sows who can still keep the piglets coming escape culling for a while. And the reward for that is to endure seven or eight cycles of pregnancy, being dragged back and forth from one small crate to another, without any relief, any time to rest or recover, any chance to move freely or be outdoors, or anything resembling human care or even pity.

In November of 2006, a coalition of organizations, including the HSUS, passed an initiative in Arizona to ban the confinement of pigs in gestation crates. For that fight, we were lucky to have a formidable friend in Maricopa County sheriff Joe Arpaio, who helped make the case in our TV ads. The self-described toughest sheriff in America, he was an unlikely ambassador for our message. He's a stern man in his treatment of felons and has no tolerance for cruelty to animals. With Joe's tough-guy reputation and his great love for animals, he was just the man to blow past all the stereotypes about overly sentimental animal lovers and get people to take an honest look at a serious issue.

With Arpaio making our case in the ads, we won big in Arizona, capturing more than 62 percent of the vote and sweeping thirteen of fifteen counties, including a county hosting a farm with seventeen thousand pigs crowded into gestation crates.

After that victory, even Smithfield Foods, the world's largest factory farmer of pigs, with 1.2 million of America's 5 million sows in gestation crates, announced it would begin phasing out the crates on company-owned farms over the next decade or so. Soon after, Canada's largest pig producer, Maple Leaf Foods, announced it would do the same. And within just a few months, the American Veal Association, a trade group emblematic of animal abuse in agriculture, pledged that its producers would eliminate veal crates within a decade.

## "Meat Science": Turkeys Who Can't Walk, Pigs Who Can't Move, and Other Victims of Agribusiness

ALL OF THIS SHOWS the possibilities of change in livestock agriculture, at least when the people themselves are given a say in the matter. A few serious blows at the ballot box, and suddenly we have the industry's attention. Up to this point, industry executives and their lobbyists had it all wired; with the right people in the right spots on the relevant legislative committees, they could brush us off and dismiss any talk of reform as ridiculous and out of the question. It's harder to treat the electorate that way, and clearly they're just now adjusting to the inconveniences of democratic debate and majority rule.

It's fair to say that when the industry started losing the good opinion of men like Joe Arpaio and Dave Long, it was losing Middle

America. The ground was shifting, and it turned out many people had gotten their fill of "the new agriculture" the industry was so proud of. Like Dave, millions of citizens now have their own stories to tell of corporate abuse, arrogance, and complete disregard for the standards and character of rural communities. Agribusiness used to take the political support of these people for granted, but those days are passing as factory farming spreads. For good reason, these same rural citizens now count themselves as victims of industrial farming, and they are among the most compelling witnesses against the whole corrupt business.

As we parted, I told Dave that if we won this fight in California over Proposition 2, it would be in part because of guys like him, who stood up in what had seemed a hopeless cause and saw a better way. I reminded him that he wasn't alone in the fight, though it might have felt that way for a while, and we stood a good chance of winning this thing. Even better, victory here would send a message to all America, and more reforms would surely follow. I said we'd be back in touch, and Dave promised he'd keep spreading the word about Prop. 2.

We said our good-byes and hit the road. Maggie and I had another long drive ahead of us. After seeing Hallmark and this monstrosity of an egg factory farm, I wanted Maggie to understand that this is the "new normal." And if one place like Hallmark could cause national outrage and panic, how would Americans feel to learn that this same crass and merciless spirit was at work at factory farms and slaughterhouses everywhere—that Hallmark was the face of modern agriculture.

THE PROBLEM ISN'T JUST intensive confinement, although that is one of the worst animal-welfare problems on factory farms. Really, the confinement system is only a symptom of the larger problem of agribusiness controlling every aspect of the animals' lives and treating them like objects and commodities. Dave Long was right that for all the industry's talk about "scientifically designed" farms, and greater efficiency, and all the rest of it, the only motive is maximum profit at the expense of every other consideration. This is an industry that has lost its bearings and is willing to inflict any degree of neglect and suffering on animals if it can shave more off the cost and add more to the profit.

All farm animals with commercially valuable parts have been caught up in a system that more than ever disregards their well-being. I told Maggie that the turkey provides the sharpest contrast in terms of the basic architecture of the animal—a classic before-and-after case that shows how the industry has transformed animals to suit its designs.

Wild turkeys can run twenty-five miles an hour and, in short bursts, fly faster than fifty miles an hour. They live in flocks and forage more than twelve hours a day before retiring at night to roost in trees. They are so alert and fast that hunters dress in camouflage and face paint and use a device that mimics mating calls to lure the birds within shooting range. Despite these technological and strategic advantages, hunters often return home empty-handed because of the birds' awareness and elusiveness.

Contrast that with the genetically manipulated turkeys designed and used by agribusiness. With their massive breasts of fat and muscle, they are caricatures of wild turkeys. They are so bulky that toms have difficulty mating without harming the hens, so breeding facilities "milk" the males for their semen and artificially inseminate the females.

Their legs, so well adapted for agility and speed in the wild, can barely support their unnaturally enormous bodies. Walking and standing can be painful. They suffer joint and hip problems, and their tendons and ligaments can even rupture. Even if they were not confined in buildings at a "stocking density" of two and a half to four square feet per bird (compared with the five hundred acres they might call home in the wild), flight is impossible for these grossly obese and heavily muscled birds. They live sedentary lives in overcrowded sheds, crammed wing to wing atop ammonia-laden excrement.

And it's not only their physical form that's so exaggerated compared with wild turkeys. Their growth rate is insanely accelerated too. Factory-farmed turkeys now weigh three times more than their wild brethren and reach that size by just four months of age. A more compact "grow-out" period means producers can raise and process more turkey flocks within a year—which means more profits.

The welfare of farmed turkeys is so compromised that mortality rates of 7 to 10 percent are expected and accepted. Summing up agribusiness's priorities, an industry scientist wrote that the "use of highly selected fast-growing strains is recommended because savings in feed costs and time far outweigh the loss of a few birds." The "few" birds he referred to were actually eighteen to twenty-six million turkeys dying on factory farms in 2007 alone.

It's not much different for chickens raised for meat. They, too, are confined in massive, filthy sheds and are selectively bred for enlarged breasts and fast growth. These birds are slaughtered at less than seven weeks of age—down from twelve weeks seventy years ago. Just think of it. They now reach "market weight" in half the time they used to. According to researchers at the University of Arkansas Division of Agriculture, "If you grew as fast as a chicken, you'd weigh 349 pounds at age 2."

This rapid growth places further stress on their already distorted bodies, producing not only tendon strains, joint ailments, and crippling leg deformities, but also respiratory, circulatory, and pulmonary disorders in birds who are the age equivalent of juveniles. Diseases often associated with old age are widespread in animals just a few weeks into their lives. Poultry scientist Dr. Ian Duncan notes that "without a doubt, the biggest welfare problems for meat birds are those associated with fast growth."

The numbers are beyond belief. Aided in part by today's ever-faster rates of growth, U.S. producers now raise about nine billion chickens a year, killing one million individual birds every hour to serve the average American's diet of thirty birds per year.

The billions of birds killed for meat each year in the United States must be caught and crated before they're stacked on massive slaughter-bound trucks. "Catchers" manually gather the birds by physically grabbing them, carrying several at a time by their legs and even wings, and throwing them into crates. A single chicken catcher may lift and crate as many as fifteen hundred birds in an hour. With workers moving at such a frenzied pace, the animals often sustain severe injuries, including broken legs and wings, internal hemorrhaging, ruptured tendons, and dislocated hips. A number of studies report that as many as 20 to 30 percent of chickens suffer injury during the collection process.

Crated in the truck, the birds are denied food, water, and even protection from the elements. Wholly unfamiliar with the outdoors—wind, sunlight, rain, and the chaotic noises of transit—the birds experience shock and fear. Some even die en route from infectious disease, heart and circulatory disorders, and trauma experienced during catching and crating. Dead-on-arrival estimates for chickens range from 0.19 to 0.46 percent, which means an incredible seventeen to forty-one million birds die during transport every year.

Fifty years ago, animals were raised on small, diverse, family-run farms, and agricultural students learned "animal husbandry." Gradually, independent farmers lost out to corporate factories, and animal husbandry gave way to "animal science" or even "meat science"—a change in terminology that signaled the new industrial ethos in the way we view and treat animals. Animals became "production units," and agribusiness's goal shifted to "growing" billions of animals as cheaply and quickly as possible in the smallest amount of space—all at the expense of the welfare of these individual creatures who can suffer and experience joy just as surely as the companion animals with whom we share our homes.

## Stirring the Silent Majority

HOW DID THESE CHANGES—IN the rural areas of the nation, in many agricultural schools, and within farming industries as a whole—come about? Some of the causes can be seen in the basic economic trends within animal agriculture, which has consolidated dramatically in recent decades. In 1970, more than 870,000 farms had pigs. By 2009, that number had fallen 90 percent to fewer than 72,000, while the number of pigs raised for food eclipsed 110 million. Likewise, in 1910, 870,000 farmers raised 3.7 million turkeys. By 2007, more than half of the nearly 265 million turkeys slaughtered in the United States were raised under contract in factory farms for only three companies.

It's been the same story in the chicken, egg, and dairy industries. The beef cattle business is the rare exception; there are still three-quarters of a million cattle ranches that typically follow the formula of grazing cattle. The biggest change in the beef industry is that before slaughter, cows are now sent to feedlots and loaded up on massive amounts of corn, rather than their natural diet of grass.

Farming has changed to the point that the animals, the practices, and the whole ethic of it are virtually unrecognizable. Yet corporate farmers continue to claim they represent rural values, even as they impose the harshest industrial values on the countryside.

In California, the veal and hog industries are small. So the debate over Prop. 2 was bound to center on the confinement of the 20 million hens in battery cages—a good share of the 280 million egg-laying hens in the nation. Like other animals genetically manipulated for commercial advantage, laying hens now produce a mind-numbing yield. A century ago, a hen laid about 100 eggs per year; today's factory-farmed hen produces 250 eggs per year, and some lay 300 or more. Their systems are depleted by calcium loss and taxed by near-daily egg laying; their bones are so weakened by the virtual immobilization imposed by restrictive battery cages that nearly 90 percent of laying hens suffer from osteoporosis. This makes the birds highly susceptible to painful and often deadly bone fractures. One study found that fractures were a main cause of mortality in caged hens.

While those of us leading the movement against factory farming had cause for optimism, I told Maggie Jones that with Prop. 2 way up in the latest polls, we also knew that all of that could change. It depended partly on what resources big agribusiness would invest in the effort to defeat the initiative. Our opponents were raising money throughout the campaign, but a month after Maggie and I met up for our trip through the Inland Empire, the major players in the egg and poultry industries gave $5 million in a single day to the campaign to defeat the ballot measure. A cross-check of contributing companies would show that many of them had made a habit of cutting corners and paying fines.

Pilgrim's Pride, whose own whistle-blower employee had videotaped chickens being stomped on and thrown against the wall at one of its plants, donated $25,000 that day to Prop. 2's opponents. The United Egg Producers, which had settled with seventeen state attorneys general a false advertising complaint about its fraudulent animal-welfare claims and had to pay a penalty, ponied up $185,000. Moark LLC, which had previously paid $100,000 to settle criminal animal-cruelty charges, shelled out $504,000 against Prop. 2. Cal-Maine Foods, the nation's top egg producer, which had been cited for spilling chicken parts and manure into waterways and killing tens of thousands of fish, invested $600,000 in the effort to block reform.

On that day especially, when agribusiness interests laid down so much money, there was no doubting that they had a lot to lose and knew it. But our coalition had a pretty impressive campaign of its own, with a broad coalition of groups and thousands of volunteers proudly making the case for humane farming to neighbors and to church and civic groups. Whatever the financial means of the opposition, it came down to a contest between raw self-interest and moral aspiration, between fear and idealism, and voters could tell the difference. The industry could feed suspicion with attack ads, warn of higher costs at the supermarket, and try to cast reform as somehow hopelessly impractical and unrealistic. But the one thing our opponents could not do was to speak to the hearts of the people. There is no such thing as an eloquent, crowd-stirring case for factory farming.

By the time I delivered Maggie to her hotel, I'm sure she'd had her fill of all the details and had heard enough of Wayne Pacelle holding forth to last for a while. We spoke a few more times by phone, and her piece appeared a week before the election. Maggie reported that we were likely to win, but that "in a way, win or lose, [Pacelle] already had a victory in hand. Proposition 2 had helped push his overall message about farm-animal welfare well beyond California. In late September, the comedian Ellen DeGeneres—who held a fund-raiser that brought in $1 million for Proposition 2—had Pacelle on her show to talk about the ballot initiative. Then, in mid-October, Oprah Winfrey devoted an entire show to Proposition 2, with Pacelle and members of the opposition as guests. Instead of baby seals and whales as the darling cause of two of TV's most popular daytime shows, it was America's pigs, calves and chickens."

That was a generous sentiment, and much appreciated. But we were playing to win—and would have been devastated by a defeat. Our opponents had employed public relations firm Golin Harris to lead a $9 million campaign against Prop. 2. As predicted, they argued that egg prices would surge, food would become unsafe, and farmers would be driven out of business and out of California.

We countered that egg prices would rise only slightly, safety would improve, and family farmers would regain market share. In fact, our lawyers and economists found that the United Egg Producers had been running a massive price-fixing scheme. While they were lamenting the costs consumers would have to pay if the animals got more space, they had been controlling flock size to artificially inflate egg prices. We showed that they had jacked up prices more than 40 percent. After we turned over our findings to the Department of Justice, it launched an investigation, and more than a dozen class action lawsuits were filed on behalf of consumers throughout the country.

We struck back in other ways too. Our videos and advertisements included footage that one of our investigators had taken of massive numbers of flies buzzing around the egg farm next to Dave Long's place. And we included a brief scene from Hallmark, to remind voters of agribusiness's callous attitude toward animals.

I spent Election Day handing out leaflets at a precinct in Los Angeles. The polls had us winning more than 60 percent of the vote, but we know never to relax at such moments and wanted to persuade as many individual voters as we could. Eventually, I made my way over to the Hyatt Regency Hotel, where we had our election night gathering. California's Democratic Party happened to choose the same hotel to celebrate Barack Obama's election victory, so the atmosphere was electric.

As the secretary of state posted results from precincts and counties across California, our early numbers came in at about 60 percent and then slowly climbed throughout the evening. And when the last ballots were counted, we had won with 63.5 percent of the vote. In fact, we got more votes than any ballot initiative in American history. We won not only in coastal California, but throughout much of the state's agricultural interior. We took majorities in forty-seven of fifty-eight counties.

The first county returns I looked for were those of the Inland Empire. In the end, Prop. 2 got 62.5 percent of the vote in Riverside and 62 percent in San Bernardino. They proved to be the bellwethers we thought they'd be.

Agribusiness leaders had mounted a strong campaign, but they had failed to convince the public it was okay to cram animals in cages for their entire lives. The images did not comport with popular notions of what agriculture should be. Across the Golden State, we saw that the bond between humanity and animals is more than a nice thought or phrase; it is something real and deeply felt. If animals are going to be raised for food, and be slaughtered for us, then we owe them something in return, and it starts with a little more respect and kindness.

In the end, I think a goodly number of Dave Long's rural neighbors voted for this landmark reform in the treatment of farm animals. Maybe they didn't quite share his passion for the cause. But they shared his good instincts. They knew cruelty when they saw it, and like the great majority that day, they wanted no part of it.

# CHAPTER FOUR

# A Culture of Cruelty:
# Animal Fighting in America

MICHAEL VICK LOVES ANIMALS. At least, that's what he told me when we met at Fort Leavenworth Penitentiary. And he said it with a straight face.

It was late May 2009, and the former Pro Bowl quarterback of the Atlanta Falcons was nearing the end of his eighteen-month sentence for dogfighting and related crimes. In a few days he would be out of prison, but Vick knew that his troubles would not be over when his sentence was up.

It was a warm, pleasant Kansas afternoon, at least if you looked straight up and forgot that you were in a prison courtyard. By official classification, Leavenworth is today a "medium-security" facility, but it has retained the feel of the maximum-security prison it was for generations. If anyone has ever escaped from the place, it wasn't by tunneling out: the high walls also extend forty feet down into the ground.

Because I'd never met inmate Vick before, I had to get a letter of permission from the warden to be here. My escort was Judy

Smith, a former White House deputy press secretary and now a crisis-communications specialist. In this work, Judy had been called into action to help U.S. senator Larry Craig of Idaho during his travails, Monica Lewinsky when her name was in the daily news, and other notables whose reputations needed serious repair. Now she had taken on one of her toughest reclamation projects yet.

We had checked in with the guard at the visitor's desk, and he offered, "Let me guess. You're the guy from the Humane Society."

I showed him my identification to confirm his suspicion and we made small talk for a few minutes before Michael Vick appeared in the doorway.

He first embraced his fiancée, Kijafa Frink, who had made the trek from their home in Hampton Roads, Virginia, this one last time before Michael's release. From here they would drive home to Virginia. Michael wanted to spend his first days of freedom on the open road. Vick then spoke for a few moments with Judy Smith, and eventually turned to me. We exchanged cautious looks and shook hands.

"Good to meet you, Wayne."

"I've been looking forward to meeting you, too," I replied, a pleasantry that didn't really convey my mixed feelings about this encounter with a man who for years had been torturing dogs.

He led me outside where we spotted a small metal table bolted to the ground. "Is this good?" he asked.

"Absolutely. This works."

Throughout the courtyard, other inmates were visiting with friends or family, and nobody seemed to pay us much attention. Since many visitors had been here before, perhaps there was no novelty in seeing the disgraced former NFL star, who two years ago had been the highest-paid, and arguably the most thrilling, player in the league.

For my part, just a couple of years earlier, this was just about the last place I could have expected to be, and the last guy I could have pictured myself chatting with. When Michael Vick became America's most notorious animal abuser, I became his most determined critic. And I had made it my business to see that he ended up right where he was at this moment.

## Bad Newz on Moonlight Road: Dogfighting Comes Out of the Shadows

POLICE HAD RAIDED VICK's home on April 25, 2007, not long after his first cousin, Davon Boddie, was pulled over in Virginia and authorities found narcotics in his car. Boddie gave his home address as 1915 Moonlight Road in rural Smithfield, Virginia. That was the address of the house Michael Vick purchased in 2001, just after being selected first in the NFL draft and signing with the Atlanta Falcons for the tidy sum of $130 million, making him the league's highest-paid player even before he suited up.

Police had been tipped off in 2005 about Vick's potential involvement in dogfighting, but had never bothered to investigate. Boddie's arrest now gave them grounds for a warrant to search the residence. Kathy Strouse, a veteran animal-control officer, was part of the law enforcement contingent that raided the home. After their authorized entry that uncovered the dogfighting operation, she told a friend, "We got him. We got Michael Vick."

Vick and a few of his childhood friends had been using the Moonlight Road property as a staging ground for their dogfights. Vick paid one of his friends, Tony Taylor, to locate the property, equip it, and run the operation. With two other boyhood friends, Purnell Peace and Quanis Phillips, they ran a dogfighting enterprise and named it Bad Newz Kennels—adapting the nickname that locals had given the sometimes troubled city of Newport News.

According to papers later released by the U.S. Attorney for the Eastern District of Virginia, the complex included "sheds and kennels associated with housing fighting dogs and hosting dog fights; approximately 54 American Pit Bull Terriers, some of which had scars and injuries appearing to be related to dog fighting; a 'rape stand,' a device in which a female dog who is too aggressive to submit to males for breeding is strapped down with her head held in place by a restraint; a 'break' or 'parting' stick used to pry open fighting dogs' mouths during fights; treadmills and 'slat mills' used to condition fighting dogs; and other items."

Just two days after the raid, Vick denied any knowledge of dog-fighting on the property, claiming that he had rarely visited this home and that the people he allowed to stay there had taken advantage of his generosity. "It's unfortunate I have to take the heat," he told reporters in New York City on April 27, a day before the NFL draft. "Lesson learned for me." It would be several months before Vick would recant that lie and the many others he told following the raid.

Once Kathy Strouse of animal control described for us what she had seen at Vick's property, we knew that law enforcement had uncovered a major dogfighting complex, and in Vick had found one of America's worst offenders. It was then just a question of whether prosecutors had sufficient evidence and resolve to pursue the case and make it stick.

In hindsight, it might seem that prosecution of Vick and the others was a certainty. But in the weeks following the warrant and raid, it remained an open question. Animal-fighting prosecutions are typically handled by local prosecutors, and in this jurisdiction Gerald Poindexter, the Surry County Commonwealth's attorney, seemed to want nothing to do with the case. Poindexter's investigation was lackluster, and as time passed he was nowhere close to announcing an indictment. He acknowledged to the press that dog-fighting had occurred on the Vick property, but claimed that too many people were passing through the home and that sorting out the facts would be nearly impossible. He told the press that he did not know Vick, but that he had heard the young football star was a really nice guy. Poindexter was an elected prosecutor presented with a potential case against a very famous athlete. Clearly, the last thing he wanted was to be the man who put the hometown hero behind bars.

Sensing that Poindexter didn't have his heart in it, we at HSUS began to pivot toward a backup plan: a federal prosecution. A few years before, we had fought hard in Congress for an upgraded federal law banning interstate transport of fighting animals. And we were just completing another lobbying effort in Congress to strengthen that law and to make animal fighting a federal felony. In testimony before a congressional committee in February 2007, I had pointed out that the federal government had pursued only a very few animal-fighting cases, largely because the offense was a misdemeanor and always gave way to other priorities. Police and sheriffs' departments would see it differently, and feel that their investment of time and energy was rewarded, if animal-fighting offenses carried felony-level penalties. Why go to all the trouble of investigating and raiding animal-fighting operations when the most severe penalty is a very modest fine, and little chance of any jail time?

In the weeks before the raid on Vick's home, both the House and Senate approved the legislation to make interstate transport of fighting animals a felony. Coincidentally, on the day of the raid, the bill was sitting on the desk of President George W. Bush, awaiting his signature. It would have reached the president some weeks sooner, and therefore applied to the Vick case, but for the obstructionist tactics of Oklahoma senator Tom Coburn and one other, anonymous senator who placed "holds" on the legislation.

We had been passing along intelligence on animal-fighting crimes to federal authorities for some time in hope of getting them involved in at least the most serious cases. We had started to win them over by showing how animal fighting is often tied in with other criminal conduct, such as narcotics trafficking, assault, and murder. In the Vick case, there were other compelling circum-stances to draw their interest. For one thing, you had the drumbeat of public pressure and intense media interest. Here, too, was a high-profile suspect whose prosecution would send a signal to dogfighters across the country. Then there was the local prosecutor sidestepping his duty, the prior reports of Vick's involvement in dogfighting, and ample evidence that this was a major dogfighting syndicate with enough activity across state lines to justify a federal prosecution.

The county sheriff's office had asked for assistance from the U.S. Department of Agriculture's law enforcement division on the day after the raid, so the federal government already had an im-portant tie into the case. As Poindexter hedged, the feds' interest seemed to grow. We knew they were serious when federal authori-ties seized control of the dogs rescued from Bad Newz Kennels, removing them from state to federal custody. Then, working with FBI agents and staff from the U.S. attorney's office, USDA agent Jim Knorr and other law enforcement personnel obtained their war-rant to search the property. District Attorney Poindexter objected and scuttled that intervention in late May, and the warrant was not executed. But this would be only a temporary delay. The federal agents obtained a fresh warrant just a couple of weeks later, entered the property, exhumed dead dogs, and launched a major forensic operation at the crime scene.

At the same time, Assistant U.S. Attorneys Michael Gillan and Brian Whisler began lining up cooperating witnesses. These indi-viduals gave firsthand accounts of what occurred at Bad Newz Ken-nels, implicating Vick, Taylor, Peace, and Phillips in an operation extending from New York to North Carolina to Texas. Gill and the other federal attorneys announced indictments of the four men on July 17, nearly three months after the initial raid. The quarterback who had made a career of scrambling out of tough situations had nowhere to run.

The brick house on Moonlight Road was painted white, with

a white door, and a white fence in front. But behind the house, on fifteen acres of land, were five smaller structures—all painted jet black. These were the Bad Newz Kennels buildings where dogfights were staged and the dogs and dogfighting paraphernalia held. One cooperating witness told federal authorities that the fights often occurred after 2:00 A.M., and the black buildings helped make the whole scene hard to detect. Tony Taylor had designed the complex to obscure the group's criminal activities from any inquiring eyes.

Investigators reported bloodstains on the walls and carpets of the fighting pits (the carpets are used to give dogs better traction). The Bad Newz gang had also "baited" animals, throwing in smaller dogs and other animals for the pit bulls to kill to make them more aggressive. As if that wasn't bad enough, Bad Newz had also featured another revolting feature of dogfighting: the game testing of dogs.

It's a common practice for dogfighters to spar the animals before they wagered on them. Dogs who don't perform well are typically killed. One cooperating witness said he was with Vick and Peace in February 2002 and watched them "roll" or "test" some of their dogs at a Virginia Beach location. Peace shot one of the weak dogs with a .22-caliber pistol. But death by shooting was merciful compared with the other means employed at Bad Newz Kennels. Just a couple of weeks before the raid, Peace, Phillips, and Vick "executed approximately 8 dogs that did not perform well . . . by various methods, including hanging, drowning, and slamming at least one dog's body to the ground." Investigators exhumed some of these victims, and as the media publicized new details of the cruelties, Vick's public support all but vanished.

When federal authorities put all the evidence together, it clearly showed that the defendants were buying and selling fighting dogs, killing poor-performing dogs in malicious ways, staging fights after midnight and charging admission, traveling to fights across Virginia and other states, and wagering thousands of dollars on the outcomes. What made it all the more disturbing was that the multi-millionaire Vick wasn't doing it for the money, but because this was his idea of fun.

After a grand jury had indicted them, Vick's accomplices began accepting plea agreements—never a good sign for a big-name defendant. First, Taylor pleaded, presumably fingering the others in exchange for leniency. Then Peace and Phillips confessed. By then it was clear that Vick would be forced to plead as well, and there was nothing that his attorney Billy Martin or the rest of his high-priced legal team could do but ask for the court's mercy.

In August 2007, Vick pleaded guilty, and soon afterward began serving time as he awaited his formal sentencing in November. Vick not only admitted that he had lied to federal prosecutors, but also to NFL commissioner Roger Goodell and to Atlanta Falcons' owner Arthur Blank. Blank had no choice but to drop Vick from the team, and then Goodell, who was earning a reputation as a law-and-order commissioner, indefinitely suspended him from the NFL. In the meantime, our membership was demanding that Nike and other corporations terminate their marketing and licensing agreements with Vick. In so many words, we told Nike and the rest to "just do it," which they finally did after some hesitation.

Later in the fall, the U.S. attorney indicted a fifth member of Bad Newz Kennels, Oscar Allen, known as "Virginia O." In November, U.S. District Court Judge Henry Hudson sentenced Vick to a two-year sentence—two months' time served, eighteen months at Leavenworth, and an additional two months of home confinement—with three years' probation to follow.

Michael Vick broke the law, tortured dogs, and tried to cover it up. So when his publicist Judy Smith and his lawyer, Billy Martin, approached me some six months before his release—a little more than a year after Vick had entered Leavenworth—and said he wanted to talk to me, I scoffed at the idea. There'd be no discussions. Vick *was* bad news. He had done wicked things, and I had no interest in dealing with him.

At the same time, I'd begun to see the Vick case as a test, not just of whether prosecutors would pursue a famous man in an animal-cruelty case, but also of whether they would treat animal fighting as a serious offense. If the case did nothing more than finish off Bad Newz Kennels and penalize its owners, it would have been a lost opportunity. Vick and his codefendants were among tens of thousands of men involved in illegal dogfighting. We had to raise awareness about the larger problem of dogfighting, and the Vick case gave an opportunity to do that as never before.

Even in the earliest stages of the case, I spoke about Vick, but also about dogfighting as a grave moral and legal offense. The cruelty was bad enough—that it was done for the gambling and the thrill of the bloodletting made it worse. We shared dogfighting footage with television stations all across the country—giving the public an unobstructed view of a vicious world they hardly knew existed.

Despite the best efforts of HSUS, dogfighting had been a low priority within the humane movement for years. But now the Vick case had given all America a tour of the pit, and a nation of dog lovers had been introduced to the strange new terms of "rape stand" and "game testing." The public had always seen organized dogfighting as rotten and cruel, but only now realized how widespread and truly loathsome it really was.

By our best estimates, as many as forty thousand people were involved in organized dogfighting, and up to a hundred thousand people involved in the less organized street fighting of dogs. Vick had started off as a street fighter in the projects of Newport News, easing into the ranks of professional dogfighting as he became wealthy. That's why he brought "Virginia O" into Bad Newz Kennels. "O" taught the crew about paying attention to bloodlines and conditioning the animals. Through the facts of this case, we could do some educating about both subcultures—street fighting and organized dogfighting—since Michael Vick was a case study in both worlds.

As the months passed after Judy Smith, Vick's crisis manager, first approached me, I had the nagging, uncomfortable feeling that there might be some good in putting Vick to work in the public fight against animal cruelty. We had taken the outrage that Vick's conduct had caused and directed it toward reform. Maybe Vick himself could serve the same cause after getting out of prison. But this could only happen if the man was truly repentant and could speak from the heart about what he did and why cruelty to animals was such a serious offense. I could hardly stand the thought of being in the same room with the guy, much less in the same cause. But I couldn't shake the feeling that this story didn't have to end with Vick's disgrace and punishment. After all, sometimes the best witness is a redeemed sinner, and in the end that would be his choice to make.

At the same time, I also realized that it would be a huge switch for me to work with Vick. It would be especially jarring for our supporters, who had called and written protest letters by the tens of thousands to Nike and to the NFL. We had identified the good guys and the bad guys in this real-life drama, and it would be tough for people to accept that we might work with a man they so thoroughly reviled.

I was still skeptical of Vick and wary of the challenges ahead, but eventually I agreed to a sit-down—with the faint hope that there was more to this story and to him than just Bad Newz Kennels. The case had turned his life upside down, but there's no question that it had put the anti-dogfighting cause front and center. And now, if he was committed to turning his life around, Vick could speak in a compelling way to kids and young black men about the ills of dogfighting and animal cruelty and steer clear of this whole enterprise. That was reason to give this a try, so I called Judy Smith and Billy Martin and said I'd fly out to Leavenworth to hear Vick out.

For Vick and his advisers, this alone was something of a break-through, even without a guarantee that it would amount to anything. Judy knew that if Vick helped the cause of animal protection, he might face an easier pathway after he left Leavenworth. He was about to reenter society, and the Vick camp did not want us to pick up where we left off prior to his incarceration. If we demanded an extension of his NFL suspension, and perhaps organized protests at football stadiums around the country or did letter-writing campaigns to team owners and their sponsors, it might make him even more radioactive and doom his comeback. The man was a professional football player who wanted to resume his career, and he had a steep public-relations climb to get there.

## Michael Vick and the Contradiction of Cruelty

WHEN WE SPOKE THAT first time, Vick underplayed his professional self-interest in joining the animal-protection cause, much as I expected. "I deserve exactly what I got for what I did," he told me in the Leavenworth courtyard, blinking into the sunlight. "Some of the guys in here said I shouldn't have had to serve this kind of time for something like fighting dogs and that it wasn't a serious crime. I told them they were wrong—that they had no idea what I did and that I am here because of stupid things I did."

He said it had been no picnic on the inside. The food was bad, the dollar-an-hour work monotonous, and the prison's management stern. Since NFL players have famously short careers—an average of fewer than four years because of injuries and competition—it must have felt like extra punishment for Vick to lose two years in his prime. He could never reclaim those years, and with this lengthy separation from football and the damage to his image, he had to wonder if he'd ever be able to resume a successful career. Few athletes at the highest level of competition in any sport ever pick up where they left off after such a long interruption. Football had been Vick's way out of poverty, and the case had not only cost him his career and reputation, but also left him bankrupt.

On the day we met, Michael was dressed in the standard-issue drab brown body suit, just like the other inmates. He looked fit and powerful even in that loose-fitting attire, and I could immediately see how he came to be known as the most exciting athlete in pro football—a hybrid, new to the NFL, of a strong-armed quarterback with the legs and moves of a running back.

Prison had not broken his spirit, Michael assured me, and it had been a learning experience. "I am a better person now," he said. "My future is still ahead of me. I've learned a lot, and I am ready to step back into society.

"I did terrible things to animals, and I am sorry about it," Michael continued. "Animal cruelty is wrong, and in the future, I want to be part of the solution, and not part of the problem. I want to work against dogfighting and animal cruelty. I love animals."

I NEEDED TO HEAR an apology and an admission from him that he had done terrible things—with no hint of an excuse for his conduct. But the kicker in his statement—about loving animals—well, that was too much for me.

Why would he tell that to me, of all people? As he said it, I could only think about the U.S. attorney's indictment and the U.S. Department of Agriculture's investigative summary, and the accounts of what he and his Bad Newz buddies did to the dogs.

"Michael, you did horrible things to animals, and no one who 'loves' animals could do such things," I replied. "Your past behavior and that sentiment just don't square. That's not something you can say on the outside if you hope to win back people's trust."

"I know what you mean, Wayne," he answered. "But I really

do love animals. And for the life of me, I can't understand still today why I did these things."

Although I knew that any public expression by Michael Vick that he loved animals would only make matters worse, I let him go on, trying to figure out how the guy's mind worked. Here you had a man who did merciless things to animals but still harbored the self-assessment that he'd always loved animals. It was a common refrain I had heard from animal fighters, so it wasn't a complete shock to hear it.

"What do you mean when you say, 'you love animals'?"

"I have always loved dogs, always wanted them around."

"But what about the dogs you fought? You didn't treat them like loved ones."

"They were like gladiators. So strong, so fast, and powerful. And they didn't make any noise," he said, in searching for the words to prove his point.

"You mean, they weren't like German shepherds or other dogs who would snarl, growl, and bark when they fought."

"Right. They were completely quiet during the fights."

At this point, I was resisting the impulse to lecture him. I was here to make a preliminary determination about Vick aiding our anti-dogfighting efforts. But I am an advocate at heart and thought this could be something of a teaching moment.

"So you were fascinated by them. You admired them, in terms of their strength and courage."

"Yeah."

"Mike, I think that's admiration, in some narrow sense. But that's not love. You don't hurt somebody or some animal you love. You may be drawn to animals and you may have an interest in them. But love's more than admiration and fascination. It's about care and affection."

During my dealings with animal fighters, they had often talked about how much they loved the animals. They spoke of the animals' physical attributes, and their strength and valor, just as Vick did. The professional fighters clearly did admire the animals—they were, to a great degree, a central component of the lives of these men. They spent hours a day with them—training and conditioning them, feeding them a special diet, and just enjoying watching them. They built intimate relationships with these animals, even as they subjected them to lives of fear, injury, and torture.

This way of thinking among animal fighters reminded me of the same mind-set I'd come across elsewhere, in the self-justifications I'd heard before. Duck hunters, though a much more "highbrow" and educated lot than animal fighters, came to mind as I thought about what Michael Vick had said. These men may rise before four A.M. to get to their duck blinds well before sunrise, then stand in the freezing cold and call ducks for hours—in order to bag their daily limit. The unoffending victims are flying with their mates when suddenly they are hit with metal ammunition that must seem to come out of nowhere. The pellets pierce the ducks' bodies, but often don't kill them immediately, leaving them in pain and fear as they struggle to stay in the air. If the shot doesn't kill them, then the birds die when they slam into the ground, or when a hunting dog or a hunter finishes them off.

So many duck hunters I've known talk incessantly about ducks, maintain wood carvings of ducks in their homes or offices as decorative items, and mount painting of ducks on the walls. The U.S. Fish and Wildlife Service runs an annual duck stamp art contest—and often it is hunters who create the most stunning artwork. Ducks enchant them, and they perceive something glorious in these creatures that eludes even the most ardent animal lovers. They would be the first to tell you that they feel a bond and a special appreciation for ducks, and I have wondered at times if their effort to protect breeding grounds is a form of penance for the harm they do, as well as an effort to perpetuate the killing.

Of course, duck hunting as a sport is legal, the killing is commonplace, and the hunter stands at a distance, while animal fighting is illegal and underground, and the animal's suffering is seen at close range. But something about Vick's rationale sounded very familiar. He was one of many people who abused animals while telling themselves they loved them.

I didn't want to hammer away at this contradiction too forcefully with Vick. I wasn't his psychologist, and I am not sure I could untangle the problem anyway. He clearly didn't understand what was going on in his head either—he was still trying to figure it out for himself. So we moved on to other questions.

"Well, Mike," I said, "I am going to have to get to know you better before making any judgment about whether you can be involved in our anti-dogfighting work. But let me say that, even if I can find a way to get comfortable with you working with us, I would not want you to simply appear in a Public Service Announcement or in a few media reports where you talk about the issue. That's too easy, and you wouldn't reach the right audience anyway. I think you need to demonstrate commitment and to invest some serious time in the program."

"Okay, what did you have in mind?" he asked.

"We are concerned about the growth in street fighting in urban communities around the country. We've started programs with former dogfighters and gang members and they talk to at-risk kids on the streets and try to steer them away from involvement in dogfighting. They are mostly African American kids, but a person of any race or ethnic background can be involved. PSAs on television or radio would never reach these kids, but our on-the-ground programs do."

"That sounds really good. I like working with kids, and think I can make a difference."

"But, Mike, I am worried you'll forget about us when you get hired by a team and when everything gets back to normal. You won't feel the pressure to stick with it."

"No, I won't forget. This is what I want to do. I'll get a day off a week during the season, and I can do stuff against dogfighting on that day. Then there's off-season when I'll have plenty of time."

"I am talking two or three years, Mike. I am talking about a commitment to this program."

"I am into it for the long haul. I want to fight cruelty to animals for the rest of my life. When I am done with football, I want to be a wildlife conservation officer and have a refuge for abandoned wild animals."

*Here he goes again with his love of animals,* I thought. Despite my interest in moving on, I couldn't help but spend a few minutes with him on this, too.

"Mike, what do you mean, you want a refuge for abandoned animals?"

"I would like to run a place where I can have animals in a sanctuary. I had pet birds. I love watching the Discovery Channel and nature shows."

He was either trying to con me, had gone completely insane, or had more nuance to his emotions than I had assumed. I was beginning to think that maybe Michael did feel a real bond with animals, even if it had been twisted and corrupted into something ugly and merciless.

"Mike, I need some full disclosure, especially if we are going to consider giving you any sort of connection with the HSUS, if you are doing a community service sort of program. The HSUS is the nation's leading animal-welfare charity, and there are obvious risks for us in working with a convicted felon and animal abuser. People hate what you did, and they're still very angry. So I need to know the worst things you did to animals, and I need to know the scope of your involvement in dogfighting."

Vick took a deep breath and replied, "I've been involved with dogfighting for a long time. I wasn't just financing dogfighting. I was involved in the actual fighting itself."

"What about the reports of culling. All true?"

"Yes, true. The worst thing I did was that I drowned dogs. But I never electrocuted any dog. That's not true."

"How did you drown a dog?" I asked.

"I had a bucket of water and I put his head under."

"Did he struggle?"

"Yeah, he was struggling." Vick looked down as he recalled the scene, but he spoke in the tone of a man telling the truth.

"I look back, and I can't believe I did these things," he continued. "You know, man, I was just about out of this. I was thinking I had to get out of this situation. I knew dogfighting was wrong, and that it was no longer the thing for me. I just didn't have the strength to get out of it. Then, just a week later, the police showed up."

There's really no telling what Vick might or might not have done had law enforcement not arrived to limit his options. Nor was I entirely convinced that this was real remorse I was hearing, and not just a very impressive act. He looked penitent now, but this is the same man who with his own hands had drowned a dog struggling for his own life, and taken pleasure in watching other dogs tear each other up. This man had been right on the scene, up close and personal. In any case, in the weeks ahead, I would have to decide whether HSUS should have anything whatever to do with him.

I told Mike I'd think about our conversation and get back to him with our answer. We stood up and headed back into the visitor's lounge to let Judy know we were done. She walked out to see him, leaving me and Mike's fiancée, Kijafa, to wait in the drab little visitor's room.

Kijafa was clearly relieved that this would be her final visit to Leavenworth, and I asked her if she really thought Mike would abandon dogfighting for good. She said he definitely was done with that, and she wouldn't stick around if she ever suspected otherwise. I didn't ask her how much she had known before the raid on their house, though throughout the trial that question had hung in the air.

Kijafa asked me what HSUS did. She mentioned that she had watched a segment of the *Oprah Winfrey Show* about puppy mills and she was surprised to see how badly the dogs were treated. I had been one of the guests that day on the show, and I told her that puppy-mill owners treated the dogs like a cash crop—that they had so many dogs it was impossible to care for them properly. I said it was different from dogfighting because the animals don't get torn up like they do in fights. But it was inhumane because the breeding dogs live in lifelong confinement, often in overcrowded and filthy conditions, and they get lonely and depressed.

"Dogs need love and companionship—they were bred to love us and to spend time with us," I told her.

She asked what else we worked on, and I told her about our larger campaign against animal fighting, which also included cock-fighting; about combating factory farming and the killing of animals for fur; about protecting marine mammals and other vulnerable animals. She didn't seem to know about any of it. It reminded me that I live immersed in a world where animal cruelty is a familiar topic, but that many people have scarcely heard of the problems of animal cruelty, much less our campaigns against them. That was the case with dogfighting before her fiancé put it in the news: few had any idea what was going on, and many who did know were unaware just how awful it really was.

Without prompting, Kijafa said Mike was always watching Animal Planet. She also mentioned he had pet birds at his house. "They would make a mess all over the house, but Mike didn't want them confined in the cage and wanted to let them fly around." This all seemed completely authentic; if this was a con, she was a natural.

After a few more minutes of conversation, Judy returned, and it was time to say our good-byes. I left Leavenworth behind, as Vick himself would in just a few days. Whether we would meet again, I still wasn't sure.

## Blood Sport Through the Ages

THE VICK CASE HAD been the biggest momentum builder in the history of the movement to eradicate animal fighting. Between the raid on Vick's property and my first sit-down meeting with him two years later, we had worked with our allies to upgrade the federal animal-fighting law yet again, in the fall of 2008—strengthening its felony provisions and making it a federal crime to train or possess fighting dogs. People convicted of animal-fighting offenses would face a penalty of up to $250,000 and five years in prison. That upgrade of the law—the third one since 2002—would not have happened without the Vick case.

We had also worked to upgrade more than twenty-five state laws against animal fighting—making dogfighting a felony in every state. Arrests of animal fighters had more than doubled, and we had expanded community-based programs to attack the problem. The landscape had changed dramatically for the better, leaving animal fighters at greater risk than ever of being caught and punished.

Animal fighting is hardly a new vice. The same thrill that drew Michael Vick into the underworld of dogfighting had a hold on many others before him. In America, as long ago as 1866, the anticruelty crusader Henry Bergh exposed a dogfighting ring in New York City known as "Sportsman's Hall." The state legislature had banned animal fighting a decade earlier, but then as in our time the law was seldom enforced. Bergh's new organization, the American Society for the Prevention of Cruelty to Animals, embarked on a campaign of daring raids to change that. By 1880, the ASPCA had a hand in 510 arrests for dog- or cockfighting, and, by 1895, its agents could proudly report a noticeable decline in organized fights. Although animal fighting would continue in the dark corners of society, more and more it was treated as a serious offense to be rooted out.

Go back to most any time and place in history, and you will find staged fights between animals. Archaeologists have found evidence of animal fights tracing back at least three thousand years, apparently begun by tribes in the Indus Valley, where semiwild jungle fowl were pitted against each other. From there cockfighting seems to have migrated eastward into Persia, India, and China. The Greeks added their own twist to the blood sport by attaching sharpened blades to the spur behind the legs of contending birds. We're told that before battles, the Greek general Themistocles even staged cockfights to get his men worked up and ready for violence.

The ancient Romans carried these cruel recreations to another level. The cages beneath the Colosseum were reserved for exotic creatures captured from every corner of the empire—elephants, bison, lions, bears, and even seals. These animals fought to the death—against each other and against gladiators—in front of cheering crowds. This was the fate of hundreds of thousands of animals in the years of empire. And along with most everything else Roman, organized animal fighting eventually caught on across Europe.

It isn't a coincidence that the humane movement got its start in England: it was there that the cruelties of animal fighting became commonplace. In Medieval Britain, bull and bear baiting with dogs were common village entertainments. In A.D. 1050, Edward the Confessor presented a fight between a bear and six mastiffs: as spectators huddled around a ring, the dogs were thrown in and set upon the bear tethered to a pole in the center. During her reign, Queen Elizabeth I granted her royal patronage to several bull-baiting rings and entertained foreign ambassadors with bear-baiting spectacles.

Cockfighting also flourished in Britain from the twelfth century as entertainment for both the landed gentry and for commoners. Though British authorities and the Roman Catholic Church frowned upon such pastimes, the blood sport was too popular to be outlawed. King Edward III attempted to ban cockfights in 1365, as did Oliver Cromwell three centuries later, but both times to no avail. Ultimately it would prove more deeply entrenched than bull baiting or dogfighting, and cockfighting would be the last among them to be banned.

When Henry VIII took on the Catholic Church, the church's moral influence against cockfighting also fell away, and among his successors James I, Charles II, William III, and George IV were all cockfighting enthusiasts. Such was the popularity of this "sport" that in the late eighteenth and early nineteenth centuries, cocks bred on Cock Lane, near St. Paul's Cathedral, were equipped with silver spurs made on Cockspur Street, near present-day Trafalgar Square, and sent to fight in cockpits across town. Three pits stood in the city of Westminster, where Parliament is located. Animal fighting had powerful fans and friends.

Dogfighting was a favorite pastime in what one author has called "Bulldog Nation." The original English bulldog, with its strong lower jaw and deep-set nose, was bred for the baits of the early modern era. In the nineteenth century, it was crossbred with more agile and speedy terriers to create the modern bull terrier or "pit bull" breed. This was the animal that flourished in the dogfighting pits of the era, but also was used in interspecies fights, such as badger drawing and rat killing.

Over time, fights between different dog breeds ensued, as breeds other than pit dogs were matched against dogs specifically bred for the sport. Dogs and monkeys were sometimes thrown together too. In the 1820s, a monkey named Jacco Manacco fought dogs and kept winning until his final match, in which he and the dog both met a violent end.

Around that time, the humane movement in England was gathering strength and calling for the abolition of animal fighting. Between 1800 and 1835, Britain's Parliament considered eleven bills on animal cruelty, most seeking to prohibit animal fighting. The bills enjoyed the support of the great William Wilberforce, and in 1826 secured fifty-two petitions of support from across the nation, but they still went down in Parliament amid a hail of ridicule. Yet the reformers persisted, and in 1835 and 1849 passed bills outlawing various forms of animal fighting.

English and other European immigrants brought cockfighting, dogfighting, bull baiting, gander pulling, and a few other blood sports to the American colonies, where they flourished, especially in the South. In the Northeast and the mid-Atlantic, Puritan and Quaker reformers enjoyed a degree of success in efforts to prevent animal fighting. The 1641 Massachusetts "Body of Liberties" included an implicit prohibition on bull baiting, and in 1687, the Puritan minister Increase Mather condemned cockfighting as a "great inhumanity, and a scandalous Violation of the Sixth Commandment."

As in England, some of the strongest opposition to animal cruelty came from various strains of Protestantism, which regarded animal fighting and related activities like gambling and drinking as sinful. In 1682, the Pennsylvania Assembly, influenced by the colony's large Quaker population, banned cockfighting, bull baiting, and other "rude and riotous sports." The First Continental Congress even temporarily banned blood sports, in part to encourage the virtues necessary for success in the revolution. That progress was short lived, however, and after the revolution, cockfighting quickly caught on again.

In the United States, of course, prohibitions against cruelty necessarily had to proceed state by state, and in a few cases anti-animal-fighting statutes appeared even before the founding in 1866 of the ASPCA. Cockfighting had even been an issue in the presidential election of 1828, when Andrew Jackson was careful to assure voters that he had not been near a cockfight in thirteen years. The first state prohibitions against cockfighting came in Pennsylvania (1830), Massachusetts (1836), and Vermont (1852). New York prohibited all forms of animal fighting in 1856, and other states followed. By the end of the century, a majority had outlawed animal fighting—making animal-fighting prohibitions the first victories of the humane movement in America.

Yet after those early successes, it would be decades before further progress came. And even though broadly criminalized, animal fighting continued. The pit bull debuted in America sometime in the 1860s or 1870s, and by the early twentieth century this type of dog was the victim of choice for animal fighters. As late as 1881, the Ohio and Mississippi railroads advertised special fares to dogfighting matches in Louisville.

By the 1930s, as high-profile organizations such as the United Kennel Club (which had previously sanctioned dogfighting) dropped their endorsement, animal fighting lost its institutional support, but it was still widely practiced. Enforcement was inconsistent and half-hearted, however, and so it fell to modern humane societies and national animal-protection groups to carry on the campaign against animal fighting. The HSUS arrived on the scene in the mid-1950s and has been on the case ever since.

## "The Feathered Warrior"

IT WAS IN MISSOURI—JUST over the Kansas border from Leavenworth—where we launched a major campaign against cockfighting back in 1997. At the time, few humane organizations were doing much in the way of public policy, and I had come to HSUS three years before intending to change this. I knew that most humane organizations devoted the bulk of their resources to animal rescue, sheltering, and animal care. That is crucial work, but ultimately it addresses only the by-products of cruelty and not the underlying problems. The obvious way to stop those profiting from cruelty was to take the money out of it—by setting bright-line standards and by imposing penalties in the form of fines and jail time. The goal was to prevent cruelty in the first place, and often that meant reforms in law, enacted by lawmakers or, in some instances, by direct voting of the people.

In my first few years at HSUS, we stepped up our activities in state legislatures and Congress and began developing a serious public-policy agenda. But when we tangled with major industries, like agribusiness or the hunting lobby, we often ran into legislative committees dominated by rural lawmakers beholden to these constituencies. These lawmakers had the power to stop our reforms cold, no matter how popular our measure might be with the public, and they used that power to the full.

Like other reform movements before ours, we ran into a lot of powerful legislators who didn't have much use for us and wouldn't give us the time of day if they had a choice in the matter. That's when a winning cause starts looking at the full range of options, and in our case that led to a national strategy of taking matters directly to voters. We launched a series of ballot initiatives, mainly on wildlife issues such as trophy hunting, bear baiting, and trapping with steel-jawed traps. We won most of our battles, even against tough, politically astute opponents like the NRA. The animal-use industries we were up against had a lot of money and political influence on their side. But when the issues appeared on election-day ballots, suddenly these industries and their lobbyists didn't look so powerful anymore. The public was with us, and with the animals. The folks on the fringe, it turned out, were the people involved in the abuse, and the men with the gavels in those legislative committees.

When I first tried to get the lay of the land on animal fighting, there was only one place to turn, and that was my colleague Eric Sakach. Dogfighting was banned in all fifty states back in 1975, when Eric started with HSUS, but it was not a felony anywhere. There were no felony cockfighting laws, either, and cockfighting was legal in more than a half-dozen states. Very few humane organizations focused on animal fighting, and law enforcement generally did not consider this misdemeanor offense worthy of their attention. Thanks to Eric and to Ann Church, then HSUS's director of state legislation, our organization and others began the campaign to upgrade penalties. Progress was slow and halting, but they did manage to persuade some states to enact felony-level penalties.

Passing better laws was the straightest path to getting rid of organized animal fighting. Although we could educate young people and try to warn them about dogfighting and cockfighting, there were tens of thousands of hard-core "dog men" and "cockers" intensely, almost religiously, committed to their sports. Only the threat of arrest and serious penalties would compel them to give it up.

Eric gave me a tutorial and suggested I acquaint myself with the animal-fighting subculture by reading its magazines, especially the advertisements. There were three major national journals: *Grit & Steel,* published in South Carolina, and *The Gamecock* and *The Feathered Warrior,* both based in Arkansas. These were monthly, full-color, aboveground subscription magazines, written by and published for cockfighters—each with thousands of paid subscribers. Some years later, you could subscribe to them on Amazon.com, and they were among Amazon's 150 top-selling periodicals.

The cockfighting culture had three principal areas of commercial activity beyond gambling: the sale of fighting birds, breeding birds, or their offspring; the sale of stimulants and other performance-enhancing drugs; and the sale of cockfighting implements—knives or curved ice picks called gaffs, which are affixed to the birds' legs to deliver lethal killing blows during the bouts.

There were no aboveground dogfighting magazines, but at least ten underground publications, such as the *Sporting Dog Journal* and *American Game Dog Times*. Like cockfighting, the sale of pit bulls was the big moneymaker. But there was also commerce in growth-promoting drugs, treadmills, spring poles, and other paraphernalia of dogfighting—the kind of stuff found at Michael Vick's house. You could learn the basics of becoming a "dogman" by purchasing the videos and books advertised in the magazines and on the Internet.

The "gamefowl breeders," as they were known, raised fighting birds in states throughout the country, but most advertisers operated in jurisdictions where cockfighting was legal—a tier of southern states including Arizona and New Mexico in the West, Oklahoma and Missouri in the Midwest, and Louisiana in the Deep South. There were also thousands of gamefowl breeders in states where cockfighting was illegal—the magazines ran ads for sellers from Connecticut to Oregon, with a good many in Texas and California.

The ads were customized but had the same basic elements. Typically, the advertisers boasted that their birds had competed at major cockfighting derbies and won in these round-robin fighting competitions. Only cockers with winning birds could sell their fowl at a higher price point, so there was not a class of "breeders" and a separate group of "fighters." To make money, you had to do both.

For example, David Mitchell of Rattlesnake Game Farm in Robards, Kentucky, was a regular advertiser in *The Gamecock,* offering "battle cocks" for $200, "brood cocks" for $500, and, for $750, "grey trios"—one rooster and two hens for breeding purposes. Mitchell provided his name, address, e-mail, and even his pager number and identified himself as the president of the Kentucky Gamefowl Breeders Association—the statewide pro-cockfighting trade group.

Cockfighters also advertised stimulants and steroids for the birds. Doping improved performance. A product called "Pure Aggression" was advertised to "do more than any other single last-minute pit aid to prepare your cock for his time in the pit." "Strychly Speed," which is a strychnine product, "speeds up a bird's reflexes, making him 'quick on the draw.'" A product called "Insulator" was said to "insulate your cock from shock, with all natural, extremely potent ingredients." Each magazine was a shopping guide for drugs for fighting fowl.

This is the world Eric Sakach came to know well. Standing six feet four inches, strongly built, and possessing a thick shock of salt-and-pepper hair, Eric could easily pass for many of the county sheriffs he worked with on animal-fighting cases. When he came to HSUS in his early twenties, he was an investigator and an undercover infiltrator of the animal-fighting underworld, and he had emerged through the years as the nation's preeminent expert on the subject. He was a court-certified authority, and he had been involved in dozens of raids on dogfights and cockfights. He testified before state legislatures on upgrades of animal-fighting laws and helped to train law enforcement on the ways of the industry.

But Eric was fighting an uphill battle. In the 1980s, he began working in our West Coast regional office and eventually became director, covering a large swath of western states and animal issues. He was shoehorning his animal-fighting work into a busy schedule, while confronting an industry with commerce in the hundreds of millions of dollars. He, Ann Church, and others were fighting to upgrade dogfighting laws to make them felonies in a number of states, but arrests were still infrequent and there just weren't enough resources to turn around the situation. It was even tougher on the cockfighting front. Florida banned cockfighting in 1986, and the Kentucky Court of Appeals in 1994 interpreted its anticruelty law to prohibit the fighting of roosters, but that still left five states with legal cockfighting and a raft of others with anemic laws that police and cockfighters essentially ignored.

We were in an odd political circumstance. Most people considered animal fighting a vice—it was largely a settled moral question. Yet here we were dealing with a vast and thriving underground—and, in some cases, cockfighting rings operating in plain view. The industry had lots of money behind it, a political operation to fend off further restrictions, and more participants than anyone would expect. And although there was enough law enforcement activity to keep the animal fighters alert, it was hardly enough to threaten their industry.

Traditionally, the people involved in dogfighting were rural whites, many of them "professional" fighters who focused on the bloodlines of the dogs and made a livelihood by gambling and selling dogs to other dogfighters. There were even growing international markets for the dogs in Eastern Europe, Russia, and Italy, and American dogfighters sold their breeding stock all over the world.

At the same time, the growing problem of street fighting had emerged in the 1980s, with young men like Michael Vick in urban neighborhoods fighting pit bulls for bragging rights and gambling money. The growth was fueled by hip-hop music and the fad of pit bulls as powerful, walking weapons. It didn't take long for young men to square up their animals to see which one was stronger, in alleys or abandoned buildings. The dogs were easy to get, either by breeding them, stealing them, or even going to animal shelters to adopt them under false pretenses.

Cockfighting, like dogfighting, had a strong rural component— and fewer risks for the participants since the laws were either weaker or nonexistent. It was legal in a few states, and in many others the penalties were so slight that it was effectively decriminalized. The most troublesome region was what my colleague John Goodwin called the "cockfighting corridor"—a contiguous group of states stretching from Alabama into Tennessee and Kentucky and up into Ohio. These states banned cockfighting, but just barely. In Alabama, the maximum penalty for fighting roosters was a $50 fine. In Ohio, it was $250, hardly a deterrent when you could sell a trio of fighting birds for $1,000. These people were also gamblers, and if they'd risk $5,000 on a fight, they'd risk $250 on getting caught. For them, it had all the sting of a parking ticket. Eric and John told me that some local law enforcement officials were clearly on the take—allowing cockfighting to occur in their jurisdictions with impunity, thanks to the protection money doled out by those running the house.

There was no better example of this culture of corruption than in east Tennessee, which boasted the nation's largest cockfighting pit in, of all places, Cocke County. The Del Rio pit had been operating for decades, and almost everybody in town knew that the pit was hosting weekly fights before a packed house. State authorities raided Del Rio in 1988, arresting four hundred people, and prompting Tennessee lawmakers to adopt a felony penalty for animal fighting. But it was short lived; the state representative for Cocke County led a successful campaign to reduce the penalty to a misdemeanor. After the raid, Del Rio had only a short interruption before it was back in business. It took another seventeen years for the pit to be shut down—and federal authorities had to do it. When the FBI raided the pit in 2005—finding violations of animal-fighting laws, prostitution, and gambling—they arrested not only the pit's owners, but also top officers from the Cocke County Sheriff's Office, who had been running a protection racket for the operation.

On some fight nights, there were crowds of six hundred to seven hundred. A cooperating witness for the federal government noted that there were "approximately 182 cock fights at the Del Rio cockfight pit in a single evening," and in each fight "between $2,000 and $20,000 was gambled by the spectators." The pit was drawing cockfighters from throughout the South, and for some it was a family outing: parents brought their children to watch and even to wager on the outcomes. Another "cooperating witness observed a girl approximately 10 years old with a stack of $100 bills gambling on several different cock fights."

As depressing and corrupt as all of this was, cockfighting in the United States was one vice bound up with other problems, and with consequences that in recent years have even put public health at risk. When the avian influenza outbreak in 2008 occurred in Southeast Asia, cockfighting contributed to its spread, yet even then many governments refused to crack down. In Thailand alone, there are reportedly thirty million fighting cocks. Its enthusiasts like to boast that cockfighting is the second-most-popular sport in the world, after soccer. New waves of immigrants from nations where cockfighting was legal were fortifying the industry in the United States. Once the recreation of rural whites, it was now Mexicans, Filipinos, Vietnamese, and others often filling the stands at cockfighting pits in the United States.

In the legal-cockfighting states, there were dozens of fighting arenas, and hundreds of gamefowl farms. Oklahoma alone had more than forty major arenas, some with stadium seating. Experts estimated the state had 2.8 million fighting birds, with some breeding operations keeping thousands of birds tethered to A-frame huts or barrels in long rows.

In 1976, Congress had passed a law against animal fighting, restricting the transport of fighting animals, but this statute was riddled with loopholes and light penalties. The cockfighting lobby saw to that, with Senator Wendell Ford of Kentucky doing their heavy lifting. The legislation originally approved by the House banned any interstate transport of fighting animals, but Senator Ford succeeded in amending it in a conference committee to allow the shipment of fighting birds to any state, U.S. territory, or country where it was legal to engage in cockfighting. That meant cockfighters could ship fighting birds to Arizona, Kentucky, Louisiana, Missouri, New Mexico, Oklahoma, Puerto Rico, the Virgin Islands, and many foreign nations.

In practice, they could really ship birds just about anywhere, and law enforcement wouldn't stop it. In the twenty years after this measure was signed into law by President Gerald Ford, authorities hadn't prosecuted a single federal cockfighting case. So Rattlesnake Game Farm and hundreds of other advertisers were brazenly trafficking in fighting birds, even in states with laws forbidding cockfighting. It was a free-for-all, and everybody in the industry knew it.

I wondered how the humane movement could allow these industries to operate so publicly, with magazines, advertisers, thousands of gamefowl farms, and hundreds of cockfighting arenas. If we didn't confront such gratuitous cruelty in a serious way, how could we hope to attack other, more politically challenging animal problems like factory farming, trophy hunting, or trapping of animals for fur? Our movement was rightly investing many resources in cracking down on these legal cruelties, yet here was a filthy, corrupt, and illegal industry permitted to flourish by a lack of enforcement. In the rare cases of police actually raiding animal fights, mainly in northern states, the offenders simply paid their fines and went right back to abusing animals and breaking the law.

This was the backdrop in 1997, when HSUS resolved to make it a top-tier issue for the animal-protection cause. I laid out our goals: outlaw cockfighting in the five states where it was legal; make dogfighting and cockfighting a felony-level offense in every state; make training or possessing a fighting animal, or attending a fight, a felony offense too; train law enforcement in investigating fighting operations; make all interstate commerce in fighting animals a federal felony; and shut down the major players and pits throughout the United States through aggressive law enforcement actions.

As long as cockfighting was legal anywhere in the country, that undercut enforcement efforts everywhere else. It allowed cockfighters to claim at least a shred of legitimacy and granted them legal standing to trade. Imagine trying to enforce anti-narcotics laws if drug use were legal in five states. And then suppose that federal law allowed growers and dealers in the other forty-five states to ship narcotics into those five, or to markets abroad. Any serious attempt at enforcement would be hopeless.

In the states where it remained legal—Arizona, Louisiana, Missouri, New Mexico, and Oklahoma—some politicians viewed cockfighters as a highly engaged voting bloc, and they didn't see a group on the other side of the debate as passionate about the cause. Many lawmakers from rural districts viewed cockfighting as a form of "alternative agriculture"; the cockfighters sold fighting birds, and that seemed little different from selling livestock for meat. In fact, the cockfighters liked to point out that their animals had a fighting chance, whereas millions of birds raised by Frank Perdue were doomed at birth to live in terrible conditions until a very certain death.

Three of the five states with legal cockfighting allow citizen initiatives, and we decided to launch ballot campaigns in all three— first in Arizona and Missouri and then on to a tougher battle in Oklahoma. To give you an idea of what we were up against in the Sooner State, we faced opposition from the Oklahoma Farm Bureau, and even the Oklahoma Veterinary Medical Association refused to help us. But in the short version, voters sided with us by wide margins in outlawing an obvious form of cruelty. And in their votes, in defiance of entrenched interests and their own state legislatures, they only further isolated the two states left without any prohibitions on cockfighting—Louisiana and New Mexico.

Just as we launched the Oklahoma campaign in 1999, I thought it was time for a concurrent campaign to persuade Congress to halt all interstate trade in fighting animals. I approached some unlikely allies to carry the legislation—Senator Wayne Allard, a Colorado Republican and former large-animal veterinarian, and Representative Collin Peterson, a Democrat from rural Minnesota and the cochairman of the Congressional Sportsmen's Caucus. Neither was altogether aligned with animal protection, and with strong ties to the hunting and farming lobbies they inoculated us from the charge that this was a first step toward outlawing hunting and farming—a common refrain of cockfighters in all of our legislative battles. To their great credit, Allard and Peterson agreed, and by March 1999, they introduced House and Senate bills. That set a complementary federal campaign in motion.

We faced some obstacles in Congress, too. The United Gamefowl Breeders Association—the national trade group for cockfighters— mounted a major lobbying effort. Aiding them were two former U.S. senators, Steve Symms of Idaho and J. Bennett Johnston of Louisiana, who together helped persuade an already reluctant House Agriculture Committee to take no action.

This time around, our best shot came in the upcoming Farm Bill—the legislation that Congress takes up every five years or so to set the nation's agriculture policies. We decided to offer our amendments to the Farm Bill in committee or, if necessary, on the floor. We asked Representative Peterson, the author of the House bill, to carry the amendments, but he refused under pressure from the committee's chairman, Republican Larry Combest, and the ranking Democrat Charles Stenholm, both from Texas. But knowing we had a strong chance of winning a vote on the floor, we went to one of our most trusted allies, Democratic representative Earl Blumenauer of Oregon, and asked him to lead the fight.

"The purpose of this amendment, Mr. Chairman, is to make sure that the Federal Government is not complicit in aiding and abetting this barbaric practice," Blumenauer argued on the House floor. "The Federal Government has no business undermining the laws in the 47 States by permitting the transfer of these birds across State lines. . . . Take the Federal Government out of the business of aiding and abetting this three-century legacy of shame."

Then Combest and Stenholm took to the floor, challenging the wording of the amendment and its alleged unintended consequences, while taking care not to go on the record explicitly defending cockfighting. But this time, it wasn't in the cards for them. They lost the vote and had to wait for another shot at it during negotiations with the Senate on the final bill.

Blumenauer's amendment succeeded, and sensing the strength of his position, he offered a second amendment to ban importing or exporting fighting animals, and also to make the penalties for any violations of the law a felony. That amendment passed, as well—in this case, with barely a word of objection even though its effects would be much more sweeping than the first amendment by barring all sales abroad and closing down the last of their markets.

In the Senate, we faced even less resistance. The leaders of the Senate agriculture committee, Democrat Tom Harkin and Republican Richard Lugar, were steadfast supporters, and they included the House-passed language in their committee version of the Senate Farm Bill. No amendment would be needed on the floor, and no senator had the will or the votes to try to strip it there.

Still, we knew not to celebrate too soon, and our instincts were right. On a major authorizing measure like the Farm Bill, congressional leaders appoint a small group of lawmakers—a conference committee—to reconcile differences between the House and Senate versions of the legislation. Because the House and Senate amendments on animal fighting were identical, the subject should not have even come up in conference. But Combest and Stenholm hadn't given up. They fought to open up discussion, and by taking advantage of some inexperienced Senate staff, they managed to nix the bill's felony penalties.

It was this little maneuver by Combest, Stenholm, and their allies that gave Michael Vick a much softer landing when he was arrested. If the leaders of the House Agriculture Committee had not weakened the penalties, Vick probably would have faced a much lengthier prison sentence.

On balance we had come out well ahead with a new federal law banning the transport of fighting animals, and things were looking up across the board. No state had outlawed cockfighting in decades; now we had banned cockfighting in three of the five states where it was legal. We had delivered major blows to an industry grown comfortable with its semilegal status and launched an aggressive campaign to abolish it altogether.

The time had come to turn to Louisiana and New Mexico. Because neither state allowed citizen initiatives, the only pathway was through their legislatures, which had blocked reform efforts for years. To win, we'd have to challenge the conventional political wisdom. A chance came when the U.S. Senate race in Louisiana rolled around in 2004. The incumbent, John Breaux, chose not to run for a fourth term, and two long-serving congressmen from opposite ends of the state were the frontrunners to succeed him: Representative David Vitter, a Republican from Metairie, one of the conservative suburbs east of New Orleans, and Representative Chris John, a conservative Blue Dog Democrat from Crowley, in the heart of the rural Southwest and the supposed cockfighting stronghold of the state.

John had been the cockfighters' point man in the House of Representatives, and he and I had sparred in the House Agriculture Committee when I testified in favor of legislation to close the loophole in the federal law in 2000. During that hearing, he accused me of threatening Louisianan traditions, from cockfighting to hunting. At one point, John said, "I strongly support the cockfighting industry in Louisiana" and that it was "an industry that is very important to America." He had previously told the *Baton Rouge Advocate* that "cockfighting is a cultural, family-type thing."

Congressman Vitter, on the other hand, had been an opponent of cockfighting while serving in the state legislature, though not an outspoken one. Among animal advocates, he was considered preferable by far to Chris John. But if animal advocates were going to make an issue of cockfighting in the Senate election, they'd have to demonstrate an ability to dry up votes for John, not add to his totals. It was a long-held assumption in Louisiana politics that cockfighting was popular with a subset of the electorate who cared deeply about this issue.

As in Arizona and Missouri, however, a statewide survey showed that view a false political assumption in Louisiana—there was no meaningful political support for cockfighting anywhere in the state. The poll revealed that 82 percent of Louisianans wanted cockfighting banned, and that a majority of voters would be less likely to vote for a statewide candidate if he or she favored cockfighting. Just 2 or 3 percent of the people would be more likely to support a candidate who defended the cockfighting industry. By and large, Louisianans did not regard cockfighting as some cherished, "family-type" tradition; they viewed it as an embarrassment and wouldn't mind at all seeing it disappear.

With these findings in hand, Humane USA, a political action committee I founded to work independently of HSUS, rolled out a grassroots campaign against Congressman John. The Democrat's bizarre quote on cockfighting as a venerable tradition and fun for the whole family was put to good use in TV ads and in leaflets distributed door-to-door. Just about every paper in the state joined us in condemning cockfighting and pronouncing it a disgrace.

After cockfighting became a campaign issue, Congressman John must have done some fast polling of his own, because he chose to say as little as possible on the subject. The cockfighters seemed to fall silent as well. Their bluff had been called. And it wasn't a bunch of radical outsiders trying to outlaw cockfighting—just the good people of Louisiana.

When the votes were cast in Louisiana's "jungle primary," with about a dozen candidates on the ballot, including frontrunners Vitter and John—Congressman Vitter won the election outright. He became the first Republican to win a U.S. Senate seat in the state since Reconstruction, and it was his support of animal welfare that had made the crucial difference. A postelection poll showed that 32 percent of white Democratic women voters—the group that Humane USA had focused on—had defected from John and favored Vitter. Almost overnight, some axioms of Louisiana politics and cockfighting had been turned upside down. Legalized cockfighting was bad for a state's reputation, and defending the practice was a bad idea for anyone trying to win votes.

Still, it took several years for us to complete the fifty-state strategy. New Mexico fell next, in the early spring of 2007, after Governor Bill Richardson finally came out in favor of a cockfighting ban during his Democratic presidential campaign. The cockfighters had done political fund-raisers for him in years past and claimed to have received a pledge from the governor to hold up legislation they didn't like. But the pressure on a national figure like Richardson forced him to finally take a stand, which he did just before entering the 2008 Democratic primaries. He had White House ambitions, and in presidential politics the ringing endorsement of *Feathered Warrior* doesn't get you too far.

In Louisiana, pro-cockfighting state legislators saw the writing on the wall and actually proposed phasing out cockfighting over three years—to preempt efforts to impose an immediate ban. Eventually, a one-year phaseout was negotiated, with the ban taking effect in August 2008. Our leading advocate in the legislature, state senator Arthur Lentini, a Republican from Metairie, anticipated this possibility and advanced a complementary strategy. At the same time he introduced his anti-cockfighting ban, he introduced a separate bill to ban gambling at cockfights. Since gambling occurred at every cockfight, that legislation killed half the fun for the crowd, and several of the state's major pits shut down just like that.

In August 2008, with the felony cockfighting ban in Louisiana set to take effect, I appeared at a press conference with Louisiana's attorney general and the heads of the state police and sheriffs' associations, to tell the cockfighters it was time to stop. It was a remarkable moment. We had helped to outlaw cockfighting in a state that just a few years ago had seemed its safest redoubt. Cockfighting was now illegal in all fifty states. Even though it took a decade—starting in Arizona and Missouri—democratic decision making had worked, reshaping the law to reflect the public's abhorrence for cruelty.

We had also finished the job at the federal level, when after needless delays the reform making it a felony to ship fighting animals across state lines finally reached President Bush in April 2007. And though the fortifying of that law came just days too late to apply to the Michael Vick case, Vick's offenses had an immediate impact on Congress in the months after the story broke.

Taking the lead this time was Senator John Kerry, who had

read enough details about Bad Newz Kennels to know that a lot of other kennels needed some policing too. His reform, prohibiting the training and possession of fighting dogs, was offered as an amendment to the latest Farm Bill. The main problem was that the House had already approved its version of the Farm Bill a few months earlier, back when Michael Vick was still best known for his running and passing.

So just as with the 2002 Farm Bill, it would all come down to the conference committee—only now, with Vick in the foreground, there would be no last-minute power plays from the usual suspects in Congress. Senator Kerry's animal-fighting provision not only survived the conference committee, but actually came out much stronger. House judiciary committee chairman John Conyers of Michigan broadened the Kerry amendment to cover all animal fighting, upgraded the felony provision, and made any training or possession of a fighting animal a federal offense. The new language allowed federal authorities to bring charges even when fighting animals were not being shipped between states. With the back-to-back upgrades in the law—in 2007 and 2008—a federal statute with teeth was now on the books.

## "I Won't Disappoint You"

WITH HIS SORDID DOINGS in the dark of night, Michael Vick had not only brought changes in his life that he could not have imagined, but also brought about changes in law that none of us in the animal-protection movement ever believed could happen so quickly. Whatever was in Michael's heart, whether or not he really was now a changed man, the things he had done had made him an agent of change in the world he had chosen.

And after his prison sentence was up, there remained the question of whether he could somehow be a good influence in the lives of young people who looked up to him. I still had my doubts about putting him to work for our campaign. I wanted to look him in the eye and ask some more questions before making him a messenger for our cause. So I agreed to meet with him once again, this time at his home in Hampton Roads, Virginia.

It was about a month after his release from Leavenworth, and the press was no longer camped out in front of his house. His publicist, Judy Smith, and I arrived to a friendly greeting from Vick and his two little girls. Kijafa was in the kitchen, and after we said hello, she told us that she was going to take the girls to a fair so we could talk business. As Mike and I sat down in the living room, the first thing I noticed was the electronic bracelet around his ankle. He was serving the sixty-day home-confinement phase of his term.

I wanted to hear more about how he got started with animal fighting, and whether his experience might translate well for kids who faced the same pressures he did to get involved. Vick seemed to be in a candid mood.

"I started dogfighting when I was eight years old. So many kids in the neighborhood were involved. We lived in public housing and we couldn't have dogs in our houses. But we got a hold of dogs and we kept them in abandoned buildings. We fought them during the day, and then at night, we used them to chase down cats."

"Eight years old? You were just a child."

"You grow up quick in Newport News."

"And then you were hooked at that point?"

"Yes, I was. It was exciting and I was really into it. I kept doing it as I got older, and nobody really ever told me to stop. I knew it was wrong, but I figured it was just a misdemeanor and I could get away with it. When I was at Virginia Tech, and we were fighting dogs in a field, I remember seeing a police car, and I was getting ready to run, and they just drove on by."

"And then you kept on with it when you were playing football with the Falcons?"

"Man, I went back to Virginia on our one day off a week just to fight dogs. I mean, it's crazy now that I think about it. I spent so much time on something so pointless."

He continued, "We met a guy and he really showed us the ropes, taught us about the industry. We just got deeper and deeper into it. It was a really big part of my life." That guy was "Virginia O"— Oscar Allen—indicted in October of 2007 and the last of the Bad Newz Kennels crew to be arrested.

By now I had made up my mind, but I wanted to lay down a marker. "Well, Mike," I told him, "I really want to give you a chance to turn this around. I believe in change—on a societal level and on a personal level. But if you backtrack, or don't show a continuing commitment, I'll be the first to call you out. Mike, I think you know you are out of chits. You've got to make this work."

"Wayne, I won't disappoint you."

We continued the conversation, and then I walked out to allow him and Judy to speak privately about other matters. I stepped into his backyard and made a phone call. But for the most part, I sat quietly and looked out into a field, waited, and thought about the mission of HSUS and whether Michael Vick could ever really fit in.

I still didn't know whether he had changed at his core. Perhaps I'd never know with certainty. But I did know that his words and behavior had changed from his days at Bad Newz Kennels. His public statements, and his pledges to NFL commissioner Roger Goodell about working with us, encouraged me. That was an insurance policy, by my way of thinking, and if he reneged, everyone would know it.

I knew a lot of HSUS members would be upset about any association with Vick. They hated what he did, and so did I. But I didn't see how shunning him would save the life of a single dog. I was there to think about what's best for the animals—even if that made for some uneasy alliances.

Just as Vick had paid $1 million for the care of the dogs seized at

Bad Newz Kennels, I thought he should now have to invest his time in community outreach programs to reach kids. Animal fighting was gaining new recruits in the African American and Latino communities, and we had to do a better job of reaching all those young men in time. We were trying to jump-start our outreach, and Vick's commitment could help.

We who advocate for animals are a movement of converts—former meat eaters turned vegetarians, trophy hunters turned wildlife watchers, onetime fur wearers now decked in cotton. We're each on our journey, trying to live with integrity, and just as in meeting life's other moral tests, no one is perfect or pure. Yes, Vick committed a terrible crime and did awful things to animals. But even for the guy who ran Bad Newz Kennels, there can be the good news of a second chance and a shot at redemption. Sometimes the change we seek comes down strange pathways. And sometimes it comes through the unlikeliest of people. Every great moral cause needs penitent converts and witnesses to a better way. If Michael Vick wanted to walk through that door, I would not be the one to close it on him.

A month later, after Vick's home confinement was served, we did our first event with him, at the Baptist community center in Atlanta. The audience was a group of about fifty-five kids considered at risk of falling into the dogfighting world. I spoke, and so did Tio Hardiman, an African American who leads our community-based programs. Then, Mike came up and spoke. The kids listened to his every word. A crew from *60 Minutes* was taping, and the stakes were high, for HSUS and for Vick.

"I want all of you to know that this haunts me, getting rid of dogs, it haunts me," he told them. "There were many times when I said to myself that I gotta stop this. Stop doing this. My intuition was telling me this ain't right but I couldn't stop doing it—peer pressure. Five days after I really had those thoughts, my cousin was busted and then the whole thing came down and before I knew it I was facing jail time.

"I spent 544 nights in prison and I cried on so many of those nights. It hurt and I knew I had to come out and make an impact. So during my time in jail, I put together a strategy with one main message, which is that the feelings we have for one another, for human beings, we should have for pit bulls.

"What's happening now with dogfighting is sick and it upsets me because I was a follower in all this, not a leader, and now I want to be a leader against dogfighting. I never gave them a chance and if I could go back I'd change it."

In Chicago, at a similar forum several weeks later, Mike carried the same message. "We need to be good to animals. Not just dogs, but all animals—horses, birds, cats, and all animals. Don't do what I did. I hurt animals for such pointless purposes. Now, I want to help more animals in the future than I harmed in the past."

As I sat and listened to Vick, on both occasions, I was thinking, *We have to embrace this kind of change. This is what we want. We should not look for reasons to reject it. We can't afford to.*

Our programs are designed to lead these kids away from dogfighting and to show them how a man is supposed to treat a dog— with respect, gentleness, and love for a creature who can love you back. After all, if an adult in Michael Vick's life had stepped up to teach him that, he'd have been spared a lot of grief, and many dogs would have been spared from the horrors he inflicted.

I don't know if Mike, now a star again in the NFL, will stay with the program forever, but he is with it now and I am happy about it. By the end of 2010, he had spoken to ten thousand kids in at-risk communities, and to a person, they'd listened to him with rapt attention. I can sense he is feeling some pride in being part of something good. I hope he'll stay with it for the distance. And as long as his heart leads him into our ranks, he'll always be welcome.

# CHAPTER FIVE

# For the Love of Pets

IN A LIFETIME OF being around animals, I'd seen some strange and interesting things, but here was something completely different: I was standing in the front galley of a Continental Airlines 727 that was soon to depart Baton Rouge Airport, surveying a plane packed with passengers—all of them dogs. It was like something from a Far Side cartoon, complete with chipper flight attendants serving dog biscuits and water. Given the desperate circumstances, everyone was quiet and well behaved, and the sound of the captain's voice had ears up and heads tilted. Not one of the 140 dogs on board was barking or whining—not even the ones stuck in the middle seats.

There was something comical in the scene, but also deeply touching, seeing all these frightened faces wondering what was happening to them. Yet however afraid they might have felt at the moment, every one of them was lucky to be leaving New Orleans in the days after Hurricane Katrina. These guys had already been through a lot. They'd been abandoned when the city was evacuated, left to fend for themselves as the waters rose, their food ran out, and they found themselves all alone in an empty house. When finally they heard a friendly voice, it wasn't their owners but one of the hundreds of rescuers who had come to help. Now they were bound for California, where others would take them in until one day soon, if everything worked out, their owners would find them and take them back home. What mattered most right now was to get them to safety and worry about the reunions later.

Their first stop after the rescue had been the Lamar-Dixon Expo Center, an equestrian facility which almost overnight had been transformed into America's largest emergency animal shelter. That facility in Gonzales, Louisiana, had a capacity of two thousand dogs and other animals and could surely have fit more had the management company renting the place to us given its permission. But two thousand was the rule, and they held us to it, no matter how great the need. So for every new arrival past that number we had to make space, and that meant moving animals somewhere else. With hundreds more arriving by the day, we were in a tough spot, and as usual it was the kindness of strangers that showed us the way. This is how I came to know Madeleine Pickens. She called out of nowhere to say she'd been reading about the abandoned animals of New Orleans, knew that they needed transport from the disaster area, and if I could get them to an airport, the planes would be waiting.

Not the kind to just cover the costs and leave it at that, Madeleine was there on the tarmac to help with the off-loading when the trucks rolled up. After the dogs had been placed in their seats, Madeleine walked up and down the aisle, giving each one a little dose of love. And just before that plane took off, headed first for San Diego and then Marin County, she assured me that there would be more planes as needed. I took her up on that offer too and returned to Lamar-Dixon relieved to know that in the unfolding crisis of Katrina we had a new and very formidable ally in Madeleine Pickens.

The amazing thing was how many people from all across our country stepped up to rescue Katrina's animal victims, and in time to reunite them with their worried, often despondent owners. With so many human lives in the balance, it took a while before the media turned their cameras on the plight of animals in the flood. But when they did, what America saw was deeply troubling and tragic in its own way—a dog swimming in the waters, seemingly with no place to go; a cat trapped on a rooftop, and exposed to the withering summer sun; a forlorn dog peering out of an attic window, abandoned and alone. As millions saw these images, they wanted action—and this outpouring of concern was like nothing we had ever experienced at HSUS.

Then an interesting thing happened—something I always hope for, but that you can never quite anticipate: the world saw that helping animals is a task that also helps people. The separate stories of the human rescue and the animal rescue melded together, and it became clear that if government authorities did not attend in a serious way to both problems, then both efforts would fail. We saw it play out time and again. Those who were forced to leave their pets behind went to heroic lengths to come back and find them. Others defied orders and simply refused to evacuate without their pets. They stayed behind in flooded homes or waded through the streets carrying their animals. Friends don't walk away from each other in a time of need, as they saw it, and whatever the dangers they'd face them together. It was a kind of revelation for the American public, and a defining hour for the humane movement. We saw the forceful pull of the human-animal bond, and how it reaches almost everyone. We saw the courage and love it inspires, even at the most dire moments.

Amid all of the tumult and stress of those weeks, I noticed something else—something that didn't happen. As a spokesman on the scene for HSUS, I had expected to hear a charge about all that was going on around me—the thousands of rescued animals, the hundreds of people who deployed, the millions of dollars sent to support the operation. Why help all these animals when people were suffering so terribly? No matter what the situation, or how desperate the need of animals, there are always those who say that caring for animals is a case of misplaced priorities.

But this time, the question never came—not once, in hundreds of interviews. And I think I know why: because people seeing these rescues on television instantly understood the need at hand and the goodness of our effort to save lives. Nobody demanded an explanation or sought to belittle our concern for these innocent creatures even amid so much human loss and sorrow. The animal-rescue efforts after Katrina were all a part of the same picture of how a kind and caring society responds to suffering—not helping one instead of the other, but helping human and animal alike. And if any cynics had been tempted to make light of our efforts, the best answer would have come from the many human victims of Katrina whose love for their own animals was so apparent, and in some cases so strong that they even put themselves at risk to save their pets.

## Katrina and the Human-Animal Bond

RESIDENTS OF NEW ORLEANS and surrounding areas were left to take those risks because there was so little in the way of a plan to help people and their pets, and this brought confusion from the very beginning. One of the few acts of foresight, as far as animals were concerned, was to convert Tiger Stadium at Louisiana State University into a holding center for the pets of early evacuees. People fleeing inland before the storm were able to drop off their pets there, and staff and students of the veterinary college stood ready to care for them. This one precaution saved hundreds of lives, but it was a happy exception in the crisis to come.

In New Orleans and beyond, many people had fled their homes but left the animals behind. Most of them topped off large bowls or even filled their bathtubs with water. They left mounds of pet food and assumed the animals could cope on their own for a couple of days until the storm had passed. And, of course, some people did not leave at all, either unwilling or unable to evacuate. They hunkered down with their animals and braced for the worst.

That moment drew closer on August 29, 2005, when Katrina flattened some of the outlying islands south of the mainland before plowing into the beachfront communities of southwestern Mississippi and eastern Louisiana. The hurricane's eye had landed fifty miles east of New Orleans, not a direct blow but close enough to leave the city battered and damaged. The second blow didn't come from above, but when the rising waters overwhelmed the levees and released the floodwaters.

It didn't take long to realize outside help would be needed. It was a challenge far beyond the means of local humane organizations to deal with, though they gave it all they had. The Louisiana SPCA, for example, had already evacuated all 263 animals from its shelter and sent them to Texas. Just in time, too, since its main facility in the Ninth Ward of New Orleans was soon under water. Other animals in the high-impact zone had a different fate. I was sickened to learn that twenty-three dogs and cats held by the Humane Society of Southern Mississippi had drowned in their cages after vainly swimming for hours as floodwaters rose inch by inch until the air was gone.

So it fell entirely upon the HSUS and dozens of other non-profit organizations to save and shelter animals and to reunite them with their families. At such moments, there is no safety net for the animals, but after Katrina, we and allied rescue groups offered the closest thing to it. Police, firefighters, and other first responders had received no guidance at all for animal rescue, and even if they had known what to do, they had no equipment, training, or other resources to do the job. All of that would become our responsibility, and soon we found ourselves conducting the largest animal-rescue mission in American history.

Volunteers came from everywhere. There were professional rescuers from all over the country—from Rhode Island and New Jersey to California and Hawaii. There were veterinary teams— VMATs—to provide emergency care for the weakened and famished animals brought to Lamar-Dixon, in tents that looked like MASH units. And then there were the many hundreds of nonprofessionals who just showed up, took on any job they were given, and never stopped working except for a few hours' sleep on a good night. As tens of thousands of people were rushing out of New Orleans and surrounding areas—for good reason, and often under government order—these volunteers were rushing in to help. And just getting to the disaster area took a lot of doing. They were a selfless and courageous force, and there was no hardship they wouldn't endure to save a life.

Along with the volunteers on the scene, hundreds of thousands of Americans donated products and money to enable the rescue and relief. When it was all added up, close to $100 million was donated to local and national humane organizations in the weeks after Katrina hit. It was not anything like the billions given for the response to the human crisis, or the additional billions spent by the federal government, but it was a wonderful show of generosity, and it spared a lot of heartbreak.

The worries and fears of pet owners grew by the day, and for many our rescuers were their last hope. In the days after the levees broke, HSUS received some seven thousand phone calls and e-mails just from people in New Orleans and surrounding communities who pleaded for help in saving their animals. Other groups responding to the disaster logged thousands of calls too. Residents weren't allowed back into the city because National Guard troops had walled it off, barring anyone but emergency responders.

No one knew how long authorities would maintain the barricade preventing reentry, but seeing the chaos and destruction around us, we began to realize that it might be weeks or even months. That meant it was a race against time in rescuing thousands of animals trapped in homes, or struggling to survive on rooftops or car tops in parts of the city with no dry ground. Starvation and dehydration, to say nothing of distress and fear, would set in if help didn't arrive soon.

Small bands of rescuers went neighborhood by neighborhood, house to house, peering through windows or listening for the sounds of animals. If they heard or saw signs of life, they'd crawl through a window or force open a door. Somehow, many animals found their way onto rooftops, and rescuers clambered up ladders to reach them. After a room-by-room search, they'd leave large-lettered markings on the front of a damaged house to signal to other rescuers that it had been checked. By the end of the day, the transport vans were full with barking and meowing evacuees—and even the occasional shriek of a parrot or the quieter sounds of rabbits, reptiles, or other small animals stirring in their crates.

For all its distinctiveness, New Orleans was not so different from other cities in terms of pet keeping. Like the rest of the nation, some 70 percent of households had one or more pets. So just playing the odds, it was likely that rescuers would find evidence of pets in two of every three homes—either hustled away with their owners during the evacuation or left behind and alone. Although surveys reveal that our active supporters and members are predominantly white and tilt heavily female and older, the situation in New Orleans reminded us that there is a much larger universe of people who care deeply about animals. Concern and love for animals was spread across every demographic group, and hardly confined to any one class or race.

With their frightened passengers on board, the rescue drivers then headed about an hour inland to Gonzales to deliver them to Lamar-Dixon, where still more volunteers helped to off-load the animals. Volunteers noted the pickup address or location if they could get it, presented the animals to veterinary teams, and then placed them in makeshift stalls. End to end, each barn at Lamar-Dixon was about the length of a football field. One was reserved for horses rescued from outlying parishes, and two were for dogs and cats—with four or five wire kennels placed in each stall. There was no way to easily clean the cages, and this setup was a logistical challenge at every level. The lack of air-conditioning, the drainage and cleaning complications, and the absence of sleeping quarters for staff and volunteers took their toll on everyone, including the animals. When lightning storms struck—and they came often—it was like a fire drill and the volunteers and staff would rush to the community bathrooms, since those were the only protected structures at the site, and huddle inside until the worst had passed.

The biggest logistical hurdle of all was dealing with the cap of two thousand dogs imposed by the owners of the facility—a restriction they didn't apply to other animals. If three hundred more rescued dogs arrived, then three hundred had to leave, as with the first of the plane flights arranged by Madeleine Pickens. We reached out to shelters and rescue groups and shipped animals by truck around the country—to Florida, Missouri, Ohio, Texas, and even California. These rescue partners agreed to a holding period, to give people time to get their bearings and reunite with their animals. It had turned into a diaspora for the animals and people of New Orleans, and it made reunions much harder to pull off. The HSUS offered to cover the costs of sending people to their animals, or sending animals back home, but the matter of cost was the least of the problems.

AT LAMAR-DIXON, PEOPLE WHO had left animals behind in the rush to get out of town were beginning to return, walking the aisles in search of a friend. Among them was John Wallman, who with his wife was searching for their blind cat. "We left before Katrina, on August 27, figuring it was like all the other evacuations, and we'd be gone for three or four days," John told my colleague Chad Sisneros moments after finding his cat. "We left her with three weeks' worth of food and lots of water and took our dog with us. Because she [the cat] is very old and blind, we were afraid of the stress of being caught in traffic. And then we couldn't come back.

"We finally came back on September twenty-eighth and we were delighted to find on our front door somebody had marked 'cat rescued.' This is the first place we came to find her. It feels like a miracle."

Jeremy Campbell, a man in his late twenties, also showed up at Lamar-Dixon and was reunited with his two cats. "I was out of town working a job before I knew that a hurricane was even in the Gulf. And so it wasn't even an option for me to come into the city. . . . And my roommate was there with my pets so they were secure. But the storm hit and it was worse than everybody thought it was, and after about three days, my roommate had to leave."

Jeremy said his roommate left all the food that they had stored, and water. "I started putting in calls to the Humane Society, frantic calls—please, can you help? And I've been a nervous wreck looking for my cats ever since.

"I actually snuck into the city today to get them myself. And when I got there, they weren't there and I knew I hadn't been looted because none of my stuff was missing.

"I came here today with a group of three friends who love these cats as much as I do. My parents were like, 'Go! We know it's not really safe for you to be in New Orleans right now, but you've got to get those cats out of there.' So I'm so glad that I came here and I found them! You guys were one step ahead."

These happy endings gave us the encouragement and morale boost we needed to carry on. But for many people, tracking down their pets was a much longer journey, if they found them at all.

That's how it was for Richard Colar, a forty-six-year-old construction worker who lived right next to a levee in the Ninth Ward. Richard's home was overtaken by twelve feet of water, and he and his dog, a Siberian husky named Princess, stayed in the attic for three days, just above the water line, until it became unbearable. At that point, Richard actually built a small raft, gathered up Princess, and took his chances floating through the streets.

When authorities found Richard and Princess drifting along, they ordered him to evacuate but without the dog. At that point his only option was to entrust Princess to a neighbor, who himself was soon forced to evacuate. An animal-rescue group happened to be nearby, and that's how she ended up at Lamar-Dixon, and eventually at a shelter far from New Orleans.

It took a lot of computer searches, but after weeks of separation, Richard—by then resettled in North Carolina—finally reached someone who could lead him to Princess. He called HSUS and spoke to Cory Smith, then our program manager for animal sheltering. "I think Cory was sent to me," Richard said later. "I'd lost everything. I wasn't working. I was trying to feel my way in a new place. Princess was all I had."

Cory and her team eventually traced Princess to the shelter in Ohio. The shelter managers were reluctant to give her up, because she cowered from a man at one point in her stay and triggered suspicions of past abuse. Cory believed that Richard was sincere and had taken good care of the dog. She pressed the issue until the shelter staff relented. A team of drivers volunteered to handle successive legs of the journey, with Cory herself finishing the trip to North Carolina.

"They were kissing each other over and over; it was so sweet," Cory remembers. "Princess was just so happy to be with him. Every time he sat down for a second she sat on him or curled up inside his arm." Richard and Princess have since found their way back to New Orleans, where Richard is rebuilding a home that would never have seemed complete without her. "That dog is my child. I know I'm blessed."

People like Richard refused to give up on their animals, and the animals didn't give up on them, either. A fellow animal rescuer in Louisiana tells of a poodle mix whose owner was shot in a dispute in the harrowing days after the storm. She had to rescue the shivering, growling companion with a catch pole because the distraught creature wouldn't leave his owner's side. It had been five days since the man died, but still the little dog faithfully stood guard.

When I came across this story, it brought to mind a story from the 1920s about a Japanese professor and his faithful Akita. Professor Hidesaburo Ueno would leave his dog, Hachikō, home each day, but always returned to find the dog waiting for him at the train station. One day, Professor Ueno never came back—he had suffered a fatal stroke. Friends adopted out Hachikō, but he routinely escaped, and they always knew where to look for him: on the platform of the train station. This went on every day until Hachikō himself died nearly a decade later. The bond runs that deep, and sometimes it is beyond the power of death or separation to break.

That was the case in perhaps the best-remembered incident involving people and their animals caught up in Katrina. Every tragedy has its iconic moments, and during Katrina it was the story of a little boy and his dog, Snowball, in a scene captured by the Associated Press. The boy and his family had left their home for the Superdome to find safety, but that arena, of course, turned out to be anything but a secure place. When they were then put on a bus to another location, police would not allow Snowball to go with his family. They confiscated the little white dog, ignoring the boy's pleas. He cried hysterically when they grabbed the dog, and then he vomited, helplessly calling out "Snowball! Snowball!" as they took his pet away.

Here was a boy displaced from the hurricane and clinging with all his might to a beloved companion for comfort. One of the few things he still had was taken away by people who said they were there to help. Everyone saw that this seizure compounded the injury. For the American public, Snowball's story sealed the case and proved beyond a reasonable doubt that our disaster policies had failed both people and animals.

So much of the havoc and heartbreak we at the HSUS witnessed during Katrina came as a consequence of the failure of government, at every level, to provide for the safety of animals in an emergency situation.

A few animal-protection advocates sought to cast blame on those who left animals behind, and there were certainly a few owners who acted irresponsibly. But generally speaking, that was a misreading of the situation. The government simply had no policies to guide the rescue of animals and offered little or no help to people trying to save their pets. So many people did their best in a crisis situation, and they had to make quick, tough choices—people like Richard Colar who stayed behind until ordered to leave, or like the family of Snowball who were traumatized by heavy-handed and entirely unnecessary government orders.

Basically, the government's plan presumed that when things got really bad, and citizens had choices to make, they would be willing to leave pets behind without delay or complication—seriously underestimating the power of the human-animal bond. It was a reflexive, bureaucratic mind-set that assumed you could seize or abandon any animal and pay no attention to the whimpers or the tears.

As it turned out, when they were put to the test, most people had more character than that, more loyalty. They weren't about to turn their backs on dogs, cats, and other animals they considered family. The official policies didn't just sell the animals short—they sold the people short, and in the process undermined the entire rescue operation. Pet owners across the Gulf region were prepared to leave behind everything they had and cherished—but by God they were not going to forsake their pets.

After seeing these problems firsthand, I vowed that never again would we at HSUS meet a disaster of this magnitude less than fully prepared. And we made it a priority to reform the law so that people in distress would never again be forced to abandon their pets.

We worked with our allies in Congress on legislation to make certain the situation in Louisiana would not be repeated. Instead of hearing rescuers say, "Leave your pets behind," evacuees should have had the option of saving both themselves and their animals. Their hearts told them that abandoning helpless animals to a horrible fate was the wrong thing to do, and the law should not say otherwise. Transportation should have been available for everyone. At least some evacuation shelters should have had capabilities to house people with their pets, or else co-located shelters for people, so they did not have to be separated.

Through the years, I've seen Congress take swift action when there's a public clamor, and it happened again on this issue. Before our rescue mission in Louisiana had even ended, we drafted, with our top supporters in Congress, the Pet Evacuation and Transportation Standards (PETS) Act—a federal requirement that every local or state disaster response agency receiving funding from FEMA have in place a disaster plan for animals. In this one case, lobbying Congress on this issue didn't seem all that difficult. Members had watched the drama play out on television and had seen that the broader disaster response was undermined by a failure to account for the needs of pets. We and other animal-protection groups had been trying for two decades to enact such a reform. After Hurricane Katrina, the idea didn't require much explaining.

On the House floor, Congressman Tom Lantos, a Democrat of California, made the case for the PETS Act this way: "I was watching television one night, Mr. Speaker, and I saw a seven-year-old little boy with his dog. His family lost everything, and all they had left was their dog. And since legislation such as ours was not yet on the statute books, the dog was taken away from this little boy. To watch his face was a singularly revealing and tragic experience.

"Many pieces of legislation we pass in this body are the result of months and years of study and research and preparation. Not this bill. This bill was born the moment the seven-year-old little fellow had to give up his dog because there was no provision to provide shelter for his pet."

The late congressman Lantos saw the bill through the House, and even lawmakers who had a history of indifference or hostility to animal protection added their enthusiastic support. It passed in fall 2006 by large margins in both chambers and was promptly signed into law by President Bush—who himself had told a reporter that, if faced with a Katrina-like disaster, the first thing he'd grab before fleeing would be his dog, Barney.

Around the same time, some sixteen states also enacted laws to require disaster planning for animals. Out of an awful situation came new awareness, and then new policies, and a new determination to address the problem. Never again would we be so ill equipped. Never again, when it came to disaster preparedness, would animals be completely overlooked and left behind.

## The Animals in Our Lives

AMONG ALL THE ANIMALS left behind in New Orleans was a pair of puppies found wandering the streets together. They had been abandoned, and after the rescue they went unclaimed until a New Orleans couple, Paul and Christine Fowler, gave the little dachshunds a new home and new names, Bella and Deiter. Two years later, the Fowlers moved to Port-au-Prince, where Christine, a health worker employed by Tulane University, had been sent to help in the cause of AIDS prevention. Naturally, Bella and Deiter came too, Paul Fowler having made a vow: "I promised them I would never abandon them."

On the afternoon of January 12, 2010, Paul was in the back-yard of their apartment building in Haiti, holding their new baby daughter, Victoria, when suddenly they were thrown into the air by one of the strongest earthquakes ever in the Western Hemisphere. "I was sure we were going to die," Paul recalled. "It sounded like hell was opening up. I told my baby good-bye." Somehow the building was left standing and Paul ran upstairs to get the dogs, who were cowering under the bed. It took three days for the family to walk to the U.S. embassy, where they were offered a flight on a C–17 transport plane—but without the dogs. After some fast thinking, Paul dropped them off with a friend who helps run a local orphanage, God's Littlest Angel. But it tore him up, and as soon as the plane landed in Florida, Christine called the HSUS. As Paul later told a reporter, "I can still see their faces in total confusion as I left. Broke our hearts."

Within a couple of weeks, our HSUS rescuers had found Bella and Dieter at the orphanage and flew them to Miami, where a joyful reunion awaited them in the arms of the Fowlers. They were in pretty good shape for two dogs who had survived a Category 4 hurricane and then a 7.0-magnitude earthquake. "I left them," said Paul, "really thinking that I would never see them again—thinking to myself that I had reneged on the promise that I made to the two Katrina puppies, and that was that I would never abandon them." We were happy to help Paul make good on his word, and with the reunion the family felt whole again. "It's like we haven't missed a day."

It was an awful lot of trouble to go through, but nobody involved doubts that it was all worth it. Disasters put people in extreme circumstances they never imagined having to face, and they test a lot of qualities, including loyalty—as in the case of these dogs twice left behind and twice saved. And even under the force of two of the greatest natural disasters of our time, the human-animal bond was stretched but did not break.

When it was all over, Katrina proved to be a turning point in our recognition of the human-animal bond, and of the responsibilities that come with it. It was a wake-up moment for America, as the Michael Vick case would be less than two years later. Vick's offenses revealed a human capacity for cruelty and betrayal worse than we knew; Katrina revealed a bond of loyalty deeper than we knew. In the one case, there was universal condemnation; in the other case, universal affirmation; and in both cases America was moved to action by the same great reservoir of compassion and decency.

In our country and in others, this empathy and respect for animals has been building up over time, even as many forms of animal exploitation have become more severe and widespread. And surely these very cruelties help explain the growing concern for animal protection in our day. The more abusive the conduct, the more apparent the need for reform. The blindness of the cruel only sharpens the vision of those who can see.

In our time, we see animals and learn about them in ways that other generations didn't. And, at last, we have begun to see them on their own terms. Consider, to take one example, the vast difference in the television programs that give us glimpses of the animal world. Among the most popular shows on Animal Planet in recent years has been *Animal Cops,* which shows investigators and police confronting cruelty. Today the highest-rated show on the network is *Whale Wars,* which follows the exploits of the daring crew of the Sea Shepherd Conservation Society in their pursuit of Japanese whale-hunting ships. Millions of people across the world watch the show, and not many of them are rooting for the whalers. As TV programs go, this is a long way from *American Sportsman,* which aired on ABC in the 1970s and presented trophy hunters as the stars and lions, elephants, and other animals as mere props for the hunters' phony heroism. This marks a deep shift in our way of thinking, and especially in the thinking of a new generation. With the exception of the hunting shows on some cable networks, today's animal shows are generally appreciative and respectful, and the heroes are those who protect animals instead of those who torment and exploit them.

Animal Planet itself, for that matter, marked quite a dramatic shift in attitudes when the Discovery Channel introduced it in 1996. It was hardly the first animal-friendly programming for a large audience. Walt Disney had long before that given America his weekly television programs and films—and even before that produced *Bambi,* which charms audiences to this day and still stands as an enduring indictment of hunting. And then, going back to the early 1960s, there had been nature shows like the popular *Mutual of Omaha's Wild Kingdom*, which gave millions of Americans their first acquaintance with wildlife from around the world. But Animal Planet offered a much wider lens than any of these, covering wild and domesticated animals alike. It's all animals, all the time, and the first network devoted exclusively to that purpose.

National Geographic television, about the same time, stepped up its own animal programming. And it has had its own share of runaway successes, not the least of which features the "dog whisperer" Cesar Millan, who counsels dogs and their guardians for the benefit of eight million viewers every week. It's a safe bet that many are looking for tips on how to improve life for their pets, or to better negotiate the terms of their bond with the dog in their lives. There are roughly 171 million dogs and cats in America's homes—almost three times as many as in the mid-1970s. Then there are the millions of rabbits, hamsters, gerbils, fish, and birds. A census that took note of animals would probably turn up more pets than people in the United States.

Something in the human heart seems to crave their company, and a lot of pet owners will tell you that their lives and family wouldn't seem quite the same without the nearness of an animal. We let our own animals sleep in our beds, and usually it is their idea. We cook for them and refine the menu according to their reactions. We even buy them pet health insurance, to avoid the high veterinary costs that most people will pay anyway to save or improve the life of a pet.

The pet products and services industry is a more than $45 billion business. PetSmart, a big-box store for pet supplies and services, was not started until 1986 and now has nearly twelve hundred stores. Customers there have the option of rounding up their payments, and these transactions have made PetSmart Charities the largest animal-welfare foundation in the country, in terms of annual giving. Its chief competitor in the marketplace, Petco, also has more than a thousand stores, and it, too, has its own foundation, which also provides millions of dollars to support dog and cat welfare programs.

When Petfinder.com went nationwide in 1998, it provided a whole new world of opportunities for shelter pets. For the first time, those looking to acquire a pet could see animals for adoption at their local shelters from the comfort of their own living rooms—people who otherwise might never check the shelter because of fear it would be depressing or concerns over the health or behavior of animals. PetFinder offered a showcase for homeless pets, and fifteen million dogs and cats have been adopted through the site these past twelve years.

A whole new business of pet sitting has arisen. And pet owners who take off for a few days have the option of high-end "pet resorts" to look after their dogs and cats. Some employers allow people to bring their pets to work and offer day care for pets on workdays.

There are pet cemeteries, as well as grief counseling programs for pet owners who lose their animals. My friend Dwight Lowell lost his beloved dog Chrissie, a rescued border collie mix, and then became clinically depressed. He stopped exercising, withdrew socially, and experienced intense sadness and a feeling of loss he couldn't shake for months. Millions of Americans know the feeling, and they keep a place in their hearts for animals they once shared their lives with.

Americans now want the best in medical care for their animals, including care for their emotional trouble. Many of the nation's twenty-eight veterinary schools—which graduate about twenty-five hundred students a year—have developed specialty training programs that prepare students for the new demands of people deeply and emotionally invested in their animals. Dogs, cats, rabbits, hamsters, and other pets are anything but throwaways for millions of people.

From 1984 to 1996, total veterinary income grew at a rate of 4.9 percent per year (from $4.4 billion to around $7 billion). From 1996 to 2006, veterinary income grew at almost 8 percent per year (from $7 billion to $11.2 billion)—so, in other words, veterinary spending has nearly tripled just in the last quarter century. That's a sign of rising health-care costs for pets, but also an indicator of the rapidly accelerating demand for these services. An individual can now spend thousands of dollars to prolong an animal's life, where euthanasia formerly looked like the only logical outcome if an animal had been afflicted by a severe medical condition. Banfield pet hospital runs the largest network of veterinary clinics—some seven hundred, with many co-located at PetSmart outlets. Veterinary hospitals now have oncology centers and offer a range of other services that we used to associate only with human medicine. Dental care is a growing part of veterinary practices. Repair of the cruciate ligament (in the dog's "knee") is now an emerging business. Dogs themselves—not just their human guardians—are now treated with antidepressants, and an increasing number of behavior specialists make a living treating aggressive and anxious dogs and maladjusted cats.

There are just a few downsides to America's interest in pets, problems all the more troubling because they affect the animals we often know and love the best. They fall into the category of too much of a good thing, showing that feelings of love are not enough unless they're backed up with consistent care and responsible management. Every day across America, thousands of dogs and cats disappear in shelters never to be seen again, despite the best efforts of often heroic shelter workers and volunteers. And many of the dogs and cats in our lives are themselves a product of a system of commercial production that is itself often irresponsible and abusive, while also contributing to the enormous challenges of shelters. These pet producers present themselves as exemplary caretakers and friends of the animals, but in truth are just opportunists trading on the genuine affection and love that the rest of us feel. If we saw for ourselves the needless euthanizing of healthy animals, or the commercial breeding of animals in often squalid and miserable conditions, it would break our hearts as much as anything we witnessed during Katrina.

## In the Name of Mercy: Animal Shelters and the Problem of Euthanasia

LOCAL ANIMAL SHELTERS OPENED their doors in many American communities in the latter part of the nineteenth century. Born principally out of concern for the horse, humane organizations had to adapt to changes in our culture and economy, devoting over time an increasing share of their attention to dogs as both pet keeping and animal homelessness grew and the mistreatment of horses waned with the arrival of the automobile. In addition to the private humane societies, governments established municipal animal control agencies, largely to conduct rabies control, stray collection, and euthanasia—with community health and an interest in public order driving the work of these agencies more than any particular concern for the strays themselves.

Over time, private citizens and governments built shelters in towns and cities throughout the nation, and today there are roughly thirty-five hundred brick-and-mortar facilities. In recent years, an estimated eight million dogs and cats entered these shelters—with cats entering and being euthanized in greater numbers than dogs. Upwards of $1 billion a year is spent operating these facilities, the money coming from charitable giving or from local tax dollars.

Aiding their work are thousands of breed rescue organizations, feral cat allies, foster networks, and adoption groups operating without shelters, each one saving as many lives as it can. They are the underground army that provides care and hope for so many animals in need of attention and love. These groups often partner with shelters, to take animals to spare them from euthanasia and increase their prospects for adoption.

Many decades ago, the local shelters, especially the public animal care agencies, took up the grim task of euthanizing dogs and cats in order to create space for more animals coming in the front door. Most of them had open admission policies, not turning away any animals in need. That created pressure to euthanize in order to prevent overcrowding. In the 1970s, American shelters killed thirteen to fifteen million animals every year. Thanks to vigorous efforts to promote spaying and neutering, to promote adoption, and to keep animals in homes by resolving behavior problems, the number of animals euthanized has been in steady decline. Today, shelters euthanize just shy of four million dogs and cats—of whom some three-quarters are healthy and adoptable.

Over the last twenty years or so, there's been an emerging no-kill movement, which was first viewed as heretical, its goals unachievable. The rise of that movement was a wake-up call to the shelters of America, calling into question why organizations with a mission to protect dogs and cats would in fact conduct the largest share of killing of companion animals, even if the task was done with regret and by the most humane means available.

The man who provided the intellectual spark for the no-kill movement is Ed Duvin. A student of the civil rights movement, he sought to bring some of its spirit and convictions to the cause of animal protection. Early on, he recognized that although the people who worked in the field of animal sheltering were intensely dedicated and working hard to reduce pet overpopulation, there was something discordant about their work, given how many cats and dogs were being condemned to die.

Duvin was supported in his work by John Hoyt, who ran the HSUS at the time, and was given latitude to express himself freely. Hoyt, to his credit, thought that any serious social movement should support independent thinkers. And in Duvin he saw a clear thinker willing to face the problem squarely. Duvin began writing a regular essay, called "Animalines," on movement dynamics and mailing it—these were the pre-Internet days—to interested parties.

At the time, the vast majority of philanthropic and government funding devoted to the broad cause of animal protection—probably upwards of 90 percent—went to local animal shelters, and though the euthanasia rates were declining, that was small comfort to the millions of animals who entered shelters and never left. The traditional view held that while the killing was unfortunate, it was a necessary evil. The fault rested with communities that professed to care about the problem but did little or nothing to fund the solution. Shelter personnel had no choice but to clean up the mess created by the rest of society—and they'd be the first to celebrate the end of euthanasia once people sterilized their animals to reduce total numbers, provided lifelong homes for them, and got a larger share of animals from shelters.

The situation dragged along, and even a new generation of activists didn't have much to say about it. Peter Singer in his landmark work *Animal Liberation,* published in 1975, addressed pet homelessness and euthanasia only in passing, directing most of his attention to institutionalized cruelties such as factory farming and animal experimentation. Singer's work prompted the formation of animal organizations in communities and on campuses throughout the country, and some new national groups such as PETA and

In Defense of Animals. These newly inspired activists generally subscribed to Singer's utilitarian view that their energy was best spent in dealing with the hundreds of millions, or even billions, of animals used in science and food production rather than just a few million homeless cats and dogs. Even if they were killed, they were not suffering all that intensely, since shelters generally used humane means of euthanasia. The issue of sheltering seemed both old school and, in relative terms, inconsequential.

Shelters and other more traditional humane organizations did attract the attention of other activists, not because of the moral urgency of euthanasia, but because of their focus on dogs and cats. Many activists got agitated about shelters serving meat at events conducted for the purpose of protecting animals—in fact, HSUS's national convention was picketed for offering a meat option to attendees in 1987. The shelters were held up as a symbol of society's general moral laziness and selective standards when it came to animal protection. Singer didn't say as much, but the implication was that the animal-welfare movement's focus on sheltering was part of the problem—not because of the euthanasia per se, but because, as critics saw it, too little money was spent on the campaigns that would affect the greatest number of animals at risk.

For his part, Ed Duvin completely understood the evils of factory farming, experimentation, and abuses of wildlife, and he often wrote about them. But he thought that for the humane movement as a whole, shelters had the greatest physical presence in communities and the most contact with the general public. Our movement was, implicitly, sending the message that euthanizing healthy and treatable animals was okay. We built facilities and called them "shelters," only to kill millions of animals when they got there. It was self-defeating at the very least, he argued, and corrosive to our cause.

"Whether strays or surrenders," wrote Duvin, "these animals inescapably experience the kind of psychological trauma and terror that we find unacceptable for caged zoo and laboratory animals. Euthanasia might be a relatively painless end to this journey of terror, but each death represents an abject failure for all of us—not an act of mercy."

"Shelters," he said in a 1989 essay entitled "In the Name of Mercy," "represent the last line of defense for homeless animals, and if they fail to wage a full-scale war on behalf of these beings, they cannot rightfully call themselves a shelter—which, by any definition except that of our movement, is a *safe* haven."

He added later, in a separate essay, "Shame is what I feel, shame for being in a movement that justifies the institutionalized killing of healthy beings 'for their own good.' Indeed, such an assertion is obscene on its face. The mass killing 'manages' an animal control problem, but only a morally bankrupt movement would participate in this madness. Healthy animals deserve more from us than 'gentle' deaths, and those who continue to embrace a 'killing them kindly' ethos perpetuate the tragedy they are charged with ending. If as much energy and resources had been expended on anti-breeding programs as on controlling and killing the excess, the slaughter would have long ago ended."

George Orwell observed that the first obligation of the intelligent man is sometimes to restate the obvious, and this is what Duvin did for the shelter movement. He showed the power of a simple truth spoken plainly. Much as Singer's condemnation of factory farming galvanized an entire movement, Duvin raised his voice to demand that animal shelters be true to their name.

It fell to shelter operators to put high principle into everyday practice, and some were committed to showing that it could be done. Rich Avanzino, of the San Francisco SPCA, took a major step in the right direction by providing an "adoption guarantee" for every animal he took in. On the other coast, meanwhile, director David Ganz of the North Shore Animal League on Long Island built his operation around the same idea. With a brilliant marketing campaign, he also showed that if you raise the banner of true shel-tering, without killing healthy animals, the public will get behind the effort. In fact, Ganz took a small suburban shelter and turned it into what, for a time, was the largest animal-welfare group in America. Far from being apathetic, people were ready to embrace this more hopeful objective, and to shake off the mind-set that ac-cepted euthanasia as the best we could hope for.

Today, many of the major national animal-protection orga-nizations have embraced the goals of the no-kill movement, in-cluding HSUS, the ASPCA, the National Federation of Humane Societies, and Best Friends Animal Society. They are just the best known among hundreds of such organizations, and many shelters do not explicitly brand themselves as "no kill" but are still striving to reduce and end euthanasia of healthy animals. Rich Avanzino now runs the largest pet foundation in the country, Maddie's Fund, named for the beloved dog of founders Cheryl and David Duffield. The bond these two generous people had with their dog Maddie has inspired a wonderful charitable endeavor that is saving tens of thousands of lives, in pursuit of the goal of a "no-kill nation." It pro-vides $20 million a year—second in grant making only to PetSmart Charities—for community-wide programs built on the conviction that homes can be found for every healthy dog and cat. An increas-ing number of major city shelters are following the no-kill objec-tive, including the Mayor's Alliance for NYC's Animals. What was once the far-off goal of a few scattered shelters is today the opera-tional model for America's largest city.

This progress in New York, moreover, is part of a fairly dra-matic change for the better across much of the Northeast. For one thing, many shelters—and not just in that region—are a world away from old and dilapidated structures that we still associate with the dog pound, somewhere down by the railroad tracks or the town dump. There's been a building boom in shelters, and when you walk through these state-of-the-art facilities, there's a new sense of pride and hope. Just as important, many shelters are now run by professionals who've been at it for a while and over time have developed a mix of strategies and programs that have their communities within striking distance of the goal of a home for every healthy animal. They know, better than anyone, that making no-kill policy a reality is not just a matter of flipping a switch. It takes low-cost spaying and neutering; adoption efforts at various locations, instead of just the shelter at the edge of town; keeping pets with their families for behavior training, instead of relinquishing them; and developing a community-wide network of adoption and foster groups all working in sync. These are the basics of the formula that is making such a difference in the Northeast, where euthanasia rates have declined rapidly; some shelters are actually importing shelter dogs from elsewhere to meet local demand—not only finding homes for those dogs, but at the same time creating space for strays in those other regions.

Duvin knew that the path to progress did not involve disparaging the leaders of more traditional shelters. They were in the fight, but in desperate need of more allies, more funding, new strategies, and a fresh approach to the problems that overwhelmed them. And too often their critics had only contempt and derision to offer instead of moral support or material help. The effect has been to slow the very progress they demand and to add a needless taint to the rallying cry of saving every life.

Despite the boxes we put ourselves in, we all share that goal. But to meet it we have to understand just how far many shelters and their communities still have to go. When I toured the shelters in Louisiana and Mississippi a couple years after Katrina, I learned that some of them were euthanizing 80 percent of the animals taken in. They were overwhelmed, and it was not hard to see how they got into such a terrible situation. One had drop boxes in front of the facility for after-hours relinquishment, and it was completely full on the morning I went there. This poorly funded facility had a hard slog ahead of it. And who can fault the folks who work there for being deeply discouraged, when they're dealing with people who leave a box full of puppies or kittens at the front door and think that that's the end of the problem?

Especially tough challenges confront urban centers beset by large numbers of abandoned and relinquished pit bulls. Many come with behavioral problems because people acquire them for all the wrong reasons—such as a fighting instrument or a macho display—rather than as a companion in need of care. The nation also has tens of millions of feral cats, and often the only local policy is to trap and kill them. The far better alternative is a policy of trap-neuter-return (TNR), which means sterilizing feral colonies and leaving them right where they are. The goal is to manage the populations and, over time, to shrink and eliminate them altogether. It sounds like a lot of work, but I can assure you that there is an army of cat lovers who are up to the job. You'll find them everywhere, skillfully setting traps in alleys and abandoned buildings across America, and this is compassionate work they feel called to do. They even have national leaders in groups such as Neighborhood Cats and Alley Cat Allies.

This is the willing spirit that every group needs in the cause of saving strays, especially the shelter groups that today feel overwhelmed and misunderstood by their critics. One thing they can be sure of is that, if you know where to look, you will always find people who care and who want to help. They'll find as well that by changing to the best practices demonstrated elsewhere—by creating a clean and inviting shelter atmosphere, by handling adoptions in the best spirit of customer service, and by going out into the community and making it convenient to adopt—they can begin to turn around even the worst situation.

The major national organizations, including HSUS, the World Society for the Protection of Animals, Maddie's Fund, and others, have set a great goal: to end the euthanasia of healthy dogs and cats in America by the year 2020. A lot of people on the ground believe we can get there, and I am one of them. One statistic in particular supports this conviction. Right now, slightly less than 25 percent of all dogs in American households come from shelters or rescue groups. That means that roughly three out of every four dogs come from other sources—from pet stores, puppy mills, small-scale breeders, or friends adopting out a litter. There's still a stigma associated with shelters, the vague, sometimes snobbish, and always uninformed view that something is wrong with shelter animals. In America, of all places—the country of the second chance—you wouldn't expect to find that attitude, but somehow it survives. And the result is that millions of loyal, loving, and perfectly healthy animals—dogs and cats down on their luck after their owners moved, got divorced, or lost their jobs or homes—wind up at shelters through no fault of their own.

As disappointing as that 25 percent statistic might be, it also shows us the way out of the problem. It's a simple matter of arithmetic for shelters—all that's needed is a modest increase in adoption to end euthanasia of healthy pets altogether. If another 20 percent of pet owners acquired their next dog from a shelter—or a total of 45 percent of all people with dogs—we would solve the problem, and every healthy dog would in time find a home. With a decent marketing campaign and some money behind it, along with a lot of hard work, there is no reason we cannot get there by 2020, or even sooner.

## Factory Farms for Dogs: The Tragedy of Puppy Mills

NOT ONLY ARE THERE many misconceptions about shelter animals and their fitness for adoption—there are all sorts of illusions and articles of faith about dogs from pet stores and puppy mills. And sometimes even the best-intentioned people, with a great and sincere love for animals, have no idea where the animals they buy have really come from.

When I was a teenager, my uncle Stan, my mother's brother and a man with a wonderful heart for animals, bought a West Highland terrier for our family from a local pet store. He thought West Highlands were adorable, which they are, and he purchased other Westies for several of my aunts and uncles. We named our little dog Randi and pointed with pride to the papers from the American Kennel Club (AKC) vouching for her purebred status and Heartland lineage. Somehow it only made it seem more exciting that she had come to us all the way from Kansas. Only later in life did I realize that Randi was almost certainly from a puppy mill and that her AKC papers provided no assurances of proper care whatsoever. If Dorothy's Toto had been in Kansas in more recent times, he would have almost certainly started life in a small, overcrowded cage, exposed to the elements, like lots of other toy and terrier breeds at puppy mills.

When Uncle Stan first dropped her off at our home, Randi would dash into a bathroom and hide behind the toilet, with her ears down and her eyes wide. She was shy and fearful, undoubtedly the consequence of little or no socialization as a puppy. Early on, bolting out of some protective corner, she would often engage in manic behavior, running around the house until exhausted. Eventually, we worked through these initial problems, and she was a fabulous companion. She greeted me every day when I came home from school, and I was always so excited to see her.

Even so, there were still physical problems that proved more difficult to overcome. Randi had skin problems and other allergies, and she was plagued with them throughout her life. She constantly chewed on her skin and had severe hot spots that we tried to medicate. She looked both funny and ridiculous wearing an Elizabethan collar, but that was the only way to inhibit her self-destructive behavior. We managed that problem, too, and she had a very good life with us, until she passed away at about fourteen years old.

Today as I look back upon it, I wonder about the choices we

made as a family. Here we were—a family that loved animals—yet we had no idea we had supported a puppy-mill operation by patronizing a pet store. We had obtained a dog bred and born fifteen hundred miles away, but there was a city animal shelter less than a quarter mile from our home—you could actually see it from our front door. There, we could have also gotten a great dog. Randi was a dear companion and we loved her with all of our hearts. But another friend, whom we never met, was waiting for us right around the corner.

I have learned, during my time at HSUS, that for those who want purebreds, shelters have them too—about a quarter of the animals they take in are "pedigree" dogs. And now, on Petfinder.com, you can search for just about any dog you want. Local shelters and rescue groups post available animals, and you can find the perfect companion.

Despite a general disdain for puppy mills, the public has unwittingly allowed this industry to grow and expand, especially in the last two decades. With the rise in pet keeping, puppy-mill operators have capitalized on that interest—supplying pet stores with adorable puppies and, now quite commonly, marketing directly to the public through deceptive websites. When you fall for these marketing efforts, you consign the parent dogs to a lifetime of breeding in confinement, and you enrich the mill owner, who will just churn out another maltreated dog for the pet trade. It is the cruelty my uncle Stan supported with the purchase of Randi without even knowing it.

The HSUS assists police and local humane authorities throughout the country in their efforts to enforce the law. One puppy mill we raided in 2008 in Tennessee—Pine Bluff Kennels—was an Internet seller with a beautiful website with pastoral images. "We have a small farm . . . about 90 acres," read the site. "We love the setting and so do our animals as they have plenty of room to run and play without being a bother to our neighbors." In reality, the owner never let anybody come to the "small farm" and instead shipped dogs by air to customers or sold puppies in parking lots or at flea markets.

After working with an informant and raiding the location with sheriff's deputies, we found something quite different than the website described. We found 450 dogs, almost all the smaller breeds, in raised, crowded, squalid hutches and makeshift kennels in an overgrown field and hidden back in the woods. Another 250 dogs, mothers and puppies, were in the filthiest trailer you could imagine. None of them was being properly fed, and 90 percent had no access to water.

Puppy mills attach nice-sounding names to their facilities to give the appearance of quality care, like a hellhole we raided in Pennsylvania called "Almost Heaven." These operations are factory farms for dogs, and the dogs produced are a cash crop—the business model being to produce the most dogs at the lowest cost. But factory farms for food animals are standardized operations, with the same confinement systems used from one location to another. At puppy mills, the arrangements are improvised, with many variations on the confinement theme.

At one particularly sickening mill in Quebec that we shut down, the couple operating the mill occupied a perfectly respectable living space, with two or three pet dogs living on the ground floor and the second floor in great shape. But in the basement, they had 110 dogs living in an ammonia-filled room that required our workers to conduct their operation with gas masks. These dogs were living in that environment day after day. We even found two puppies inside a closet, in a large Tupperware tote with holes in the top, which meant those dogs were living in total darkness most if not all of the time.

The dogs who have it worst in the puppy-mill industry are the breeding females. The mother dogs are conscripted to serve as breeding machines, producing litter after litter. The puppies are sold at

eight weeks, but a mother may stay for eight years or more, sometimes even being sold at auction once a mill decides she's no longer
valuable. Puppy millers have applied an agricultural model to companion animal production, and the results are similar scenes of squa-
lor, privation, and cruelty. We estimate that there are more than ten
thousand puppy mills in the nation, with Missouri, Oklahoma, Iowa,
Kansas, and Arkansas being the top producers and worst offenders.
In Virginia, the U.S. Department of Agriculture listed seventeen
licensed commercial breeders, but we found nearly a thousand—
exposing the enormous gaps in the current federal inspections program. Most puppy mills today are not inspected at all, either by state
or federal regulators. In all, two to four million puppies are churned
out by mills each year.

In 2009, on the morning of one raid in Arkansas, our animal-
rescue team followed sheriff's deputies down the long dirt road
leading to a man they intended to arrest. They knew they were
close when the unmistakable stench of animal filth filled the country air. What they saw was a familiar scene at puppy mills: hundreds
of dogs confined to rusty wire cages, wallowing in their own waste,
in various states of mental and physical disrepair. Many of them had
matted fur, urine burns on their paws, and any number of other ailments. The house itself had all the telltale signs of an improperly kept
breeding facility, complete with stacks of American Kennel Club
registration papers—and an owner who suffered from compulsive
hoarding, a common psychological disorder among irresponsible
animal keepers. Hoarding is an odd rupture of the human-animal
bond, in which people who purport to care about animals actually
neglect them and inflict terrible harm.

Inside that house in Arkansas were another hundred dogs confined to more wire cages. Their cages were stacked on top of a
urine-soaked carpet and surrounded by waist-high piles of sales
records and books. Rescuers, wearing breathing masks to walk
through this swamp of filth, found a litter of day-old pups, all barely

moving except the runt, who lay seemingly lifeless on fouled newspapers. It was hard to believe that anyone would want to buy these dogs to begin with, but once these little ones were cleaned up, an unsuspecting buyer would have no idea of the hell the poor creatures had gone through in the weeks and months before.

Housed in the center aisle in two rows of kennels was the most pitiable sight: a massive, 130-pound, aging Akita who seemed to be blind and deaf. The ten-year-old dog had lived out his entire existence in this small "alley" with a concrete floor. At many places, he would have been killed or auctioned off. But this puppy-mill operator had the collector's mind-set, and that spared him from death but not misery.

Many dogs left to endure such an existence would be aggressive, so he was approached with extra caution. But this big guy was as gentle as a lamb. When finally coaxed from the rear of his pen, he walked as far as the door, and then stopped abruptly, too scared to leave the prison that had been the only thing he had ever known. Our staff then took him to be examined and fed a proper meal, and for the first time in his life, he was given a name: Gentle Ben.

Although veterinarians eventually had to remove Ben's sightless eyes, due to extreme pain, he's doing very well in his new life. He was taken in by Akita Rescue of Western New York and has now found a permanent, loving home where he's spoiled every day. In his time, Ben has experienced the worst instincts of humanity, and the best. Ben's a happy old guy, and he sure deserves it.

## The AKC and "the Maintenance of Purity"

TED PAUL HARDLY LOOKS the part of an agitator. Everything about him suggests moderation and control. Approaching the age of eighty but looking fifteen years younger, he is six feet two inches, trim and fit. His full, well-groomed hair is gray on the sides and black on top. In wire-rim glasses, and a perfectly tailored suit, he looks like a million bucks. He still runs a business near Salem, Oregon, Beautiful American Publishing, which produces travel books on cities and states—with pictures that show the natural wonders and landmarks of these destinations. He's a man who appreciates beauty and has always found it in dogs.

Ted's temperament matches his appearance—self-possessed but warm and without a hint of guile. He is a member of the Humane Society of the United States, but he's also steeped in the "dog fancy," as insiders term the breeding and showing of dogs. He's had purebred dogs, mainly collies, since 1960, and had a kennel at one point with a dozen breeding animals. And he is also past president of the Collie Club of America, past president of the Purebred Dog Breeders and Fanciers Association, and a longtime AKC judge. He's raised a champion collie and even authored a book called *The Christmas Collie*. He's judged dog shows all over the nation, and for all breeds, but especially collies rough and smooth, Welsh corgis and Cardigan corgis, and Shetland sheep dogs. He tells me that you'd be lucky to judge best in show in your specialty once in a lifetime, and he's done it three times for just one breed.

Such associations and experience would seem to have equipped him to comment on a bill before the Oregon legislature to require humane dog-breeding standards—maybe a bit too well for the American Kennel Club and other opponents of the 2009 bill. The HSUS state director knew of his participation in the fancy, and also of his intense dislike for puppy mills, and asked Ted to review the bill and to show up if he thought it was something he could support. He asked for the bill and, after reading it, announced that he would provide his enthusiastic support for it. He said he'd be ready to testify at the opening hearing before a House committee.

"I arrived at the hearing and saw so many old friends and went up beaming to talk to them," Ted told me. "And each of them looked the other way." But it only got worse from there. After he testified, citing his nearly fifty years in the world of purebred dogs, he was deemed a "traitor," with many fellow breeders calling for his "suspension" from the show-judge world for "conduct prejudicial to the best interest of the American Kennel Club and to the best interest of the sport of purebred dogs, pursuant to the Constitution, Bylaws and Rules of the American Kennel Club." The Dog Press and other websites wanted him drummed out of the show-dog world.

With that reaction, you would have thought the bill was designed to ban purebred dogs, or to make life so onerous for breeders that they could not operate. In fact, the bill probably didn't affect a single one of the dozens of breeders who attacked Ted. The measure stipulated that dogs used for commercial breeding purposes not be kept in cages too small for them, that the cages not be stacked, and that they have solid flooring rather than wire. The dogs had to be let outside for exercise for one hour a day, and they had to be out of the cage when their waste was cleaned, which also had to happen once a day. The bill forbade any breeder from having more than fifty sexually intact breeding animals. It exempted any breeder with fewer than ten sexually intact animals from any inspections, meaning that only a tiny percentage of breeders would have been affected by any provision in the bill.

The bill was enacted, and Ted's offense was that he was, as a dog breeder, actively supporting HSUS-backed legislation to crack down on puppy mills. Since the hearing he's basically been blackballed as a judge, and members of the dog fancy promised they would not show their dogs if Ted was in the chair.

"I was deeply saddened by the sharp reaction from breeders I know in Oregon," Ted said. "Yet I recall seeing this type of reaction forming among dog breeders many years ago—as various pieces of legislation began to surface around the country to curb excess breeding and puppy traffic," which he argued "contributed so much to the overpopulation of dogs in the United States."

The AKC is not best known for its political lobbying, but for its role in the sport of dog fancying. Founded in 1884, a few years after the Kennel Club in the United Kingdom, the AKC pledged "to do everything to advance the study, breeding, exhibiting, running and maintenance of purity of thoroughbred dogs." It is the largest and best known of the breed registry organizations, although there is an alphabet soup of them now, such as the UKC, or the United Kennel Club. At official shows, dogs are judged on how they conform to the standards set forth in the official registry of the AKC. The group recognizes 167 breeds, and partially recognizes about a dozen others, breaking breeds into seven major classifications, including sporting, working, terrier, and toy groups.

For all of the investments AKC members make in breeding dogs, the payoff is recognition within the fraternity through competition at show events across the country. By securing "best in breed" or "best in show," breeders can then sell the offspring of championship dogs and make money. For most of them, it's a hobby, with the proceeds from the sale of the offspring typically not offsetting their investments in the animals. At some level, it's more sport than business, and these people certainly admire their dogs and often give them top-of-the-line care. The Super Bowl in the world of sporting dogs is the acclaimed Westminster Kennel Club Dog Show in New York, and for several nights each year, millions of Americans tune in to watch pampered dogs and their anxious owners or handlers prance around the floor of a sold-out Madison Square Garden.

Even though AKC papers are no proof that a dog has been humanely treated, much less pampered, the club in recent years has committed to several animal-welfare programs. It gave money for the rescue of dogs during the Katrina crisis, and it has a kennel inspections program for high-volume breeders. As far back as 1995, the AKC Canine Health Foundation began contributing to research in breed health problems and now donates about $1 million a year.

For all of that, however, the AKC doesn't welcome outside opinion on matters relating to the care and breeding of dogs. They think they've got it all under control. As they see it, the least concession to animal-welfare standards from outsiders will only invite more and more regulations, inevitably bringing an end to their entire industry. If you think the NRA has mastered the "slippery slope" argument, you ought to hear the executives and lobbyists at the AKC.

The club and breeding groups have opposed just about every legislative effort to address puppy-mill problems in a serious way or to establish minimum humane breeding standards. In recent years, lawmakers in Connecticut, Louisiana, Oklahoma, and other states have considered and passed bills relating to puppy mills, and in each case the AKC opposed reform. They not only opposed a ballot initiative in Missouri that pretty closely mirrored the Oregon reform—the one supported by Ted Paul—but actually worked with agriculture groups on a countermeasure in the state legislature. That proposed amendment to the state constitution would have affirmed "the right of Missouri citizens to raise animals in a humane manner that promotes the health and survival of the animals without the state imposing an undue economic burden on their owners." The term *humane manner* in this language is just a throwaway line, and in practice it would have meant that animal-use industries could operate as they pleased, with any real attempt at reform disqualified as "an undue economic burden." It would have reduced all issues of animal welfare to a calculation of costs. It would not only have prevented the puppy-mill measure or any subsequent reform of factory farming, but also nullified the state's anti-cockfighting law approved by voters in 1998. And fortunately, the AKC's opposition was not persuasive to voters considering Proposition B, the Missouri ballot measure to impose humane standards on large-scale dog-breeding operations. Voters approved it by a comfortable margin in November 2010, enacting reforms for the care of dogs in the state that has one-third of the nation's puppy mills and churns out perhaps one million dogs a year for the pet trade.

The registration fees that the AKC gets for new litters of puppies explain in part its reluctance to crack down on mills. Every time the group registers a litter of puppies, it collects a fee, and this arrangement accounts for a large share of AKC revenue. A puppy producer can use the AKC papers as supposed evidence of a superior breeding operation, but the papers mean little more than that the breeder knows the lineage of the animal—the parents, grandparents, and other forebears. The papers do nothing to guarantee the animal is healthy, or that the breeding occurred in a high-welfare setting. There are no meaningful health standards for the breeds in the AKC registry. The standards relate to the outward physical appearance of the animals, known as the conformation, not to their overall well-being or fitness. "The best use of pedigree papers is for house-breaking your dog," says veterinarian and animal behaviorist Michael Fox, for many years a staff expert at HSUS. "They don't mean a damn thing. You can have an immune-deficient puppy that is about to go blind and has epilepsy, hip dysplasia, hemophilia and one testicle, and the AKC will register it."

Not a thing in AKC papers on a dog provides the least assurance that the animal is not the product of a puppy mill. Although the group says it's tough on substandard mills, they don't back it up with much. In fact, several of the major mills raided by law enforcement and the HSUS had dogs registered by the AKC. And even if the AKC denied registration to a mill, the owners of that operation will simply turn to another respectable-sounding registry for their "papers," without changing a single thing to benefit the animals. If the AKC's inspection program were as sound and serious as its leaders insist, then why do ten thousand or more puppy mills exist in America? And why, when local police and HSUS emergency personnel raid these mills, do they invariably find miserable, squalid, and cruel conditions?

One basic problem here is that the AKC is a captive to its revenue model. Any real standards and enforcement would disrupt the whole profitable enterprise. The AKC wants the money, the mill operators need the stamp of approval, and supporting them both are millions of Americans duped into thinking that the registry is meaningful and that the mills are humane.

It used to be that the AKC was the only breeder registry that anybody paid attention to, and the only brand that mattered. But now there is a long list of other registries, and all of these groups compete against one another for registration fees. The result is a kind of race to the bottom, in which competition doesn't raise standards but lowers them. It's become strictly business for the breed registries, with no one wanting to offend or lose fee-paying breeders, and the proof can be found in those thousands of mills that the AKC has sought to protect from even the most modest of reforms.

With the AKC and other such groups doing so little to protect the welfare of dogs in the commercial pet trade, one might think that at least federal regulators are doing their part. Under the federal Animal Welfare Act, the USDA is supposed to conduct inspections of large, licensed commercial breeders. It turns out that more than half of the ten thousand mills are not required to get a license and thus are not inspected by USDA at all. A loophole in the regulations exempts breeders if they sell dogs directly to consumers right from the farm or through a website, rather than selling them through a pet store, and thousands of operators have been quick to exploit this loophole. So if you go shopping for a dog from a commercial breeder on the Internet, chances are you're dealing with a seller operating free of any inspection process at all, unless state law has a requirement of its own.

Even for licensed breeders, however, it doesn't take much to clear inspection. There is a requirement for clean water and proper sanitation, but no prohibitions on lifelong confinement in small wire cages, cage stacking, overcrowding of the dogs, denial of exercise or socialization, or unrelenting breeding of the females. If you kept your own dog in a small wire cage twenty-four hours a day, month after month, in every extreme of heat or freezing cold, and gave her no company and human affection, but did once a day drop some food and water in the cage and clean up her mess now and then, you would meet the federal standards that apply to puppy mills.

It's hard to believe, but a large percentage of licensed breeders cannot meet even these minimal rules. The inspector general of the USDA, in a 2010 report on examining the department's inspection program, seemed astonished by the widespread disregard for basic standards of humane care, and by the casual treatment of repeat offenders. As the Associated Press described the USDA audit: "The investigators visited 68 dog breeders and dog brokers in eight states that had been cited for at least one violation in the previous three years. On those visits, they found that first-time violators were rarely penalized, even for more serious violations, and repeat offenders were often let off the hook as well. The agency also gave some breeders a second chance to correct their actions even when they found animals dying or suffering, delaying confiscation of the animals."

The report describes one case, at an Oklahoma puppy mill, in which a USDA inspector found twenty-nine violations of the law, but imposed no penalties. A subsequent inspection at this same facility found five dead dogs, while surviving dogs had resorted to cannibalism. USDA still took no enforcement action and left the dogs behind. It took the deaths of twenty-two more dogs before the USDA finally revoked the breeder's license. The government's investigation of USDA's program found that problematic dealers were repeatedly violating the law because the department took little or no enforcement action. As the inspector general reported, "At the re-inspection of 4,250 violators, inspectors found that 2,416 repeatedly violated AWA [the Animal Welfare Act], including some that ignored minimum care standards. Therefore, relying heavily on education for serious or repeat violators—without an appropriate level of enforcement—weakened the agency's ability to protect the animals."

This indictment by the department's own inspector general is only the latest of four similar audits on animal-welfare enforcement problems dating back nearly twenty years. And it completely discredits the pet trade's last-resort defense that at least they are "USDA approved." When the puppy-mill operators are not holding out their AKC or UKC papers, they're proudly displaying their USDA certification. Among pet store owners, likewise, "We only use USDA-licensed breeders" is the standard assurance when a customer asks if the dogs on display came from a puppy mill. We now know what a USDA license is worth, because the USDA inspector general has told us for the fourth time—exactly nothing.

## Nothing Fancy: The Costs of Reckless Breeding

MANY MEMBERS OF THE AKC and of other breed clubs obviously are fascinated by their dogs, and they certainly have genuine affection and pride for them. It's their version of the human-animal bond, and it's real enough. But they seem to lose sight of the bigger picture—valuing their own convenience and satisfaction over the larger problems within the field. They pamper their own animals, but stand in the way of a more universal approach to protecting dogs. They're like those puppy-mill operators in Quebec, attentive to their own pets on the first floor of the house while somehow remaining indifferent to all of the miserable, mistreated creatures crowded in the basement.

It is not just their tolerance for puppy mills that is a problem, but an absolutely obsessive focus on conforming to the standardized breed type, which results in inherited disorders and congenital health problems that beset almost every breed. These inherited diseases, disorders, and body malformations are one of the most significant companion-animal-welfare issues there is—not as bloody as dogfighting, not as plainly obvious as abuses on puppy mills, and not as final as euthanasia. But in a different way, they are every bit as detrimental to the well-being of dogs because of the duration of the suffering, the shortening of their life span, and the ubiquity of problems among the many purebreds with inherited disorders.

Each of the breeds has its alluring qualities that attract the interest of fanciers. Many terriers are efficient hunters, able to dig out prey. Border collies herd sheep. The Labrador retriever swims out and returns birds shot by hunters. In looking at the physical diversity of dogs, it's hard to believe that a tiny, hairless Chihuahua and a Great Dane descend from the same common ancestor—the wolf. A study in the *American Naturalist* compared the genetic diversity among dogs across the entire order Carnivora. The researchers found more difference between the skulls of a Pekingese and a collie than between those of a walrus and a coati, a member of the raccoon family native to South America.

One thing purebreds all have in common, to a greater or lesser extent, is a long list of physical problems afflicting them. These troubles are the result of a remarkable degree of inbreeding, with brother and sister or father and daughter bred to achieve "the perfect look" and to match the standard of the registry. It would be unthinkable to countenance this sort of incest in the human community, partly because of the detrimental genetic consequences. "We have allowed some breeds to become too heavy, some too short-faced, some too heavy-coated, some others short-legged, others too short-lived," says historian David Hancock, "all in the pursuit of cosmetic points, not sound anatomical points."

German shepherds too frequently have hip dysplasia that cripples the animals early in their lives. Greyhounds seem the model of strength and fitness, but as they get older they develop cancers with unnerving frequency. My little West Highland terrier Randi was beset by skin problems and allergies, as are so many others of the breed. Cavalier King Charles spaniels have an inherited heart disease and a skull malformation called syringomyelia, a condition in which fluid-filled cavities occur within the spinal cord near the brain. The condition causes intense pain, and sometimes "a dog's brain swells beyond the space provided by her skull." Some 30 to 70 percent of Cavaliers develop this condition.

English bulldogs "may well be the most extreme example of genetic manipulation in the entire canine world." They are susceptible to any or all of the following problems: cataracts, demodicosis (a skin disease), elbow dysplasia, entropion (an abnormal rolling in of the eyelid), hip dysplasia, hypothyroidism, neuronal ceroid lipofuscinosis (a congenital disease where fatty pigments are deposited in the brain and cause brain dysfunction), and von Willebrand's disease (a type of bleeding disorder caused by defective blood platelets). In short, these animals are suffering because breeders are pushing in their noses, extending the length of their backs, narrowing their hips, compressing their skulls, and doing so many other things to them for cosmetic show purposes.

In the United Kingdom, Crufts is to the Kennel Club what Westminster is to the AKC. The BBC ceased broadcasting the show in 2009 because of the Kennel Club's failure to establish standards to prevent inherited disorders. One recent winner at Crufts, a pug named Danny, had to lie on an ice pack as his owner was conducting a giddy victory interview. Danny had overheated from the limited amount of walking he had to do in the show competition. His face was so flat that he could not pant enough to cool himself, and to help him breathe better Danny had undergone surgery to clear his palate of excess skin. That's a heritable disorder that would likely pass on to his offspring as a prized Crufts champion.

"When I watch Crufts, what I see is a parade of mutants," Dr. Mark Evans, the top veterinarian with the Royal Society for the Prevention of Cruelty to Animals (RSPCA), told the BBC. "It is some freakish, garish beauty pageant that has nothing frankly to do with health and welfare. The show world is about an obsession about beauty and there is a ridiculous concept that this is how we should judge dogs."

Until very recently, the AKC and the Kennel Club have had no health and welfare standards in their judging contests, just conformance standards for the breeds they recognize. For instance, Rhodesian ridgebacks get their name from a distinctive ridge on their back that is thought to be a mild form of spina bifida. Yet if the dog does not have that characteristic, the animal is disqualified from shows, and some breeders are not sentimental about what to do with them. Breeder Ann Woodrow told the BBC that the culling of ridgebacks without the ridge was getting tougher. "We do have trouble nowadays with the young vets who tend to see everything in black and white and won't put them down," said Woodrow. "Usually we have to go to an old vet that we've known for years and just quietly put him to sleep." So all of these dogs are killed for the simple reason that their ridges weren't quite up to show specifications. And by her reckoning, a quick and quiet death administered by the older vets, who understand how these things are done, is preferable to putting the dog "in the hands of the fighting people, who are appalling."

In the wake of the BBC exposé *Pedigree Dogs Exposed,* the Kennel Club and the Dogs Trust, an animal-welfare group, jointly commissioned an independent inquiry led by Cambridge professor emeritus Sir Patrick Bateson. That report affirmed the findings of the BBC broadcast and a similar report commissioned by the RSPCA. In response, the Kennel Club has since banned registration of puppies from closely related parents and revised some of its breed standards. The AKC has taken no similar steps, for all of that research they've funded through the Canine Health Foundation.

Reform in the world of breed fancy in the United States is long overdue. The gene pools of certain breeds must be infused with new genetic material, and freakish conformity standards must be overhauled. King Charles Cavaliers, for instance, descend from just six individuals—and the situation for these dogs continues to get worse with every successive case of inbreeding. Indeed, most breeds today descend from just a few founder animals, and as they breed relatives together, they further shrink the gene pools. Shows must recognize not just body type, but health, agility, and certain functionality. The focus on appearance has been a disaster for dogs.

If the breed clubs were concerned more about the welfare of dogs, and less about the sport of dogs, they could readily turn this situation around. A failure to do so puts the AKC in the dock alongside the more familiar perpetrators of cruelty and abuse.

Some breed clubs are showing the way. Ted Paul told me that when he was president of the Collie Club of America, he worked hard with researchers at Ohio State University to solve the eye disorders of collies. The Clumber Spaniel Club of America and the Portuguese Water Dog Club of America have done the same—jealously guarding the breed not just from puppy mills, but also, through responsible breeding practices, from inherited disorders.

The AKC must join that movement, just as the Kennel Club in Britain has done under pressure. Better breeding practices could solve most of these issues. Animals are not art, to be obsessively reshaped according to some dilettante's vision of what they should look like. They are living beings, and to afflict them in a highly predictable way with chronic pain and discomfort and to cut their lives short because of some frivolous fashion is disgraceful. When it comes to dogs, the true best in show are the ones who are healthy, happy, and protected from cruelty of every kind.

# CHAPTER SIX

# The Cull of the Wild

I'VE SPENT TIME IN most of America's fifty or so national parks, and each stands out in its own way—the sensational colors of ancient wood at Petrified Forest; the brilliant red rock formations at Arches; the shaggy, white mountain goats bounding up the almost vertical faces at Olympic. But for me, one park holds a special place in memory. It's Isle Royale National Park—a rocky, out-of-the-way archipelago, washed on all sides by the cold swells of Lake Superior.

I arrived at Isle Royale in the summer of 1985 as a ranger for the Student Conservation Association (SCA). Over the next four months, I hiked almost all of its 165 miles of trails that scarcely mar this classical boreal forest. Breezes off the big lake chill the air even in midsummer, and it rarely gets warmer than 80 degrees. The same breezes stir the leaves of white pines and quaking aspens, sending a gentle flutter through the stillness of the park. There's a solitary beauty to the place, and I found it everywhere—on long hikes, canoe trips, and very quick dips in the cleanest, coldest water in North America. For all its beauty and tranquility, Isle Royale also tested my reverence-for-life ethic, with swarms of blackflies and mosquitoes that welcomed me to the park in their own way.

The park was established by an act of Congress during the presidency of Franklin Roosevelt, but even by the 1980s it saw only ten thousand or so visitors a year—nothing compared to the eight million or so who entered the Great Smoky Mountains National Park or the three million who went to Yosemite. The park is so remote that at first even wildlife had a tough time getting there. Wolves loped across an ice bridge to the park around 1950, doubtless in pursuit of an abundant moose population that descends from the hardy originals who made an incredible swim from Canada. Bears never made it, and deer didn't last. Red foxes and beavers did establish themselves, and I saw them often. Bald eagles nested there too, and the habitat is also perfect for northern loons. As a child, I had read about the wolves and moose of Isle Royale and always hoped to see the park firsthand. Researchers had discovered that wolves had an impact on the entire park ecosystem—not just the moose, but also the beavers and foxes, and in turn, the fish, birds, and even plants. Something about this cascade effect enthralled the naturalist in me—each little movement of nature affecting everything else around it.

There is a quiet in the boreal forest—it is not filled with the variety of life of a tropical forest or a coral reef. It is grudging and spare in sustaining animal life. That thrift appealed to me. It was more modest than ornate. There was no outsized feature—no huge mountain, ancient tree species, or powerful waterfall—to leave you in awe. Yet in its proportionality, the place commanded your respect.

The place had tugged at me for a long time, and when I got the SCA assignment, I was thrilled. I was just twenty years old, and soon I was not only on the island but showing tourists around like an old hand. I never saw any wolves—they were few in number, wary of people, and hard to spot in the thick forest. But I saw plenty of other wildlife. I'd pass time studying a mother moose and her calf, or watching loons make their awkward crash landings on the lakes. They were much more elegant in their movements on the water than in the air. They'd gracefully disappear beneath the surface, and I'd try to guess where they'd emerge, often on the opposite side of a large pond.

A mother fox gave birth to her kits under the cabin where six of us lived. I wondered if these foxes were driven by the same impulses as the first wolves who befriended humans—starting that fateful process of domestication. The foxes showed genuine interest in us, and we returned the favor.

One of my cabin mates, a maintenance man from rural Michigan, had been tough on me, especially when he was drinking. He often told me that my preoccupation with animals was a waste of time, and he delighted in recounting his hunting excursions, adding extra detail for my benefit. But I saw another side to him when the baby foxes appeared under our cabin. He watched them endlessly and talked about them excitedly. It was a lesson for me that the human-animal bond can touch anyone, even those who seem least likely to care.

That summer at Isle Royale came at a turning point in my life, when I was starting to think seriously about animals and nature and our duties to both. For the creatures who lived there, Isle Royale was a sanctuary from the unsparing pressures of humanity. All of the violence of nature was at work there every day, and no one clued in to the struggle between predator and prey could hold an idealized or sentimental view of nature's operations. But there is a world of difference between animals killing animals for survival, and humans killing animals for no reason at all, except money or sport. At least here, there were no trophy hunters traipsing into the picture to bring gratuitous death, no steel traps to bring needless and prolonged suffering, and none of the other destructive influences that only human beings, at their most careless, can inflict upon wildlife.

Human beings were welcome guests in the park, drawn there by a sense of exploration, fascination, and a spirit of stewardship. The handiwork of the ages, and of the Creator, had left us something that could not be improved upon. I felt proud that the American people had set aside this place and, in fact, an entire system of national parks—ensuring that some places would be protected for all time. Wild animals needed more places like this, and so did we. At summer's end, when I pushed off from Isle Royale for the last time, I felt that I had been given a glimpse of human stewardship at its best. And whatever work lay ahead for me, I wanted to see more of it, and I was committed to do my best to make that happen.

## Unfaithful Stewards: Betraying Yellowstone's Bison

YELLOWSTONE WAS A FAR more mixed and conflicted experience for me, the kind that shows you the world as it is, rather than how you want it to be. My first trip there, in the summer after college graduation, was memorable in the best way—a tourist's panning of the park's treasures. The place is like a fireworks display of the natural world—the herds of bison, the water of Yellowstone Falls disappearing into a froth in the canyon below, the panoramic views of the Lamar Valley and regular roar of Old Faithful. Hiking down to a lake in the park's center, I even spotted a grizzly bear a few hundred yards away. If Yellowstone had just a subset of these wild animals and geological and floral features, it would be impressive enough. But to have all of them in one place seems miraculous, and it helps explain how Yellowstone inspired the very idea of a national park system.

On my second trip to Yellowstone, it was winter and bitterly cold, and everything about the trek was more harsh and raw. I traveled there to help the bison, and to publicize a problem that surfaced there in 1989 but sadly remains unresolved more than two decades later. Montana agriculture and wildlife officials had decided to allow the sport hunting of bison straying from the park—even though Yellowstone had been meant as the final refuge for the few bison who had survived the market hunting onslaught of the nineteenth century.

These officials viewed the growth and roaming of the northern herd not as a sign of species restoration, but as a dangerous intrusion of trespassers. Yellowstone's northern boundary was a demarcation line, and no bison were to cross it. Not only did state officials think that bison would compete with cattle for grass and knock down the ranchers' fences, but, most important, that bison might also spread disease to the state's cattle industry—jeopardizing Montana's certification as a brucellosis-free state. When I became national director of the Fund for Animals in 1989, I went straight to Yellowstone to help shame state authorities into halting the hunt. Here was a creature best known as the classic example of human excess and callousness in the treatment of wildlife. And yet there were still men who looked at bison and wanted to shoot them.

To reach the fertile valleys beyond the park, the bison followed the groomed pathways, plowed roads, and trails packed down by snowmobiles. Their destination was a quilt of lands managed by the U.S. Forest Service, the Bureau of Land Management, and private ranchers. At a lower elevation than the park, these border areas had less snow, which bison could dig through to feed on the grass below. There were no cattle here during winter—and very few during the summer—but state officials were convinced that bison would pass on brucellosis, a bacteria that causes cattle to abort. About half the bison had the disease antibodies, but not the disease itself, and the idea of bison transmitting brucellosis to cattle was far-fetched—and a practical impossibility for the male bison, since brucellosis is spread through placental material.

The state planned a lottery hunt, with the winners of the $1,000 permits not only getting a freezer full of meat, but also a once-in-a-lifetime opportunity to shoot the last free-roaming bison in North America. Like Civil War reenactors, only with real guns, they could replay a defining moment in American history, slaughtering the bison for sport. State and federal authorities would choreograph the exercise, with Montana wildlife officials serving, in effect, as hunting guides.

I showed up at the park without announcing my purpose to the hunters or state authorities, though they found out soon enough. I came with video camera in hand and said I was there to document the first hunt of free-roaming bison in the twentieth century. A number of journalists had arrived, from the *Los Angeles Times, New York Times, Chicago Tribune,* and other newspapers, so my presence, while suspect, was not conspicuous.

On the hunt's first day, I hopped in the back of a state truck with the hunters and a few reporters and made small talk as the vehicle bounced down gravel roads toward scattered groups of bison. We passed through extraordinary country—broad mountains covered in thick blankets of snow, the Yellowstone River with its fast current cutting through the valleys, and the famous open sky of Montana. But the majestic scenery was all backdrop that day. All concerned were focused on the hunt, but for different reasons: the state personnel wondering how the nation would react to this experiment; the hunters calculating where to fire their first shots on such massive quarry; and me dreading the carnage about to unfold.

Montana wardens had been sent out as scouts to locate the bison and then to radio their whereabouts to our driver. Once the vehicle was close enough to the bison, he stopped the truck, and the hunters hopped off. One by one, they spotted the bison, readied themselves, and took aim with their large-bore rifles. The sound of gunfire filled the air, in what seemed a "hunt" only in name: it looked more like an execution. The bison didn't have a chance, and one after another they fell in a heap.

A few survivors ran away after hearing a shot or seeing a herd mate fall, but most just stood there befuddled. These animals were used to seeing people, and before that moment had never been harmed by them, much less slaughtered in a sneak attack. In a second large volley, hunters shot an entire group of about a dozen bison, and the animals were all splayed on the ground, with blood streaming from their bodies and then soaking into the snow and the exposed ground. I walked with the hunters through the field, surveying the warm carcasses steaming in the cold winter air. One hunter had shot a pregnant bison. I walked toward him as he started cutting her up, and watched as he put his arm inside and pulled out the fetus of an already well-formed bison. As I looked on wordlessly, he grinned and said, "It's a lot of fun." You'd think that such a sight might have awakened some capacity for remorse in a man, but not this fellow. He didn't seem troubled at all and apparently considered the unborn calf a kind of bonus.

I got back into the truck, and the state officials drove a fourteen-year-old boy to a clearing unobstructed by trees. They instructed the young hunter to shoot at one of three bison standing about two hundred yards away and offering us a perfect profile. It was a long shot, and it struck me as madness that this kid was being told to attempt a feat that would test the marksmanship of even the most highly skilled adult.

The boy took aim with his scoped rifle, squeezed the trigger, and sent a bullet in the bison's direction. The gun's kick knocked him back a step. Several moments passed before we realized the bullet had struck the middle bison. What a shot. She went down but was not dead. It seemed that she had been shot in the spine. Bison are famously tough creatures, and she tried to get up, only to fall back down again. She made another valiant but unsuccessful attempt. She kept at it, again and again. Stan Grossfeld, a reporter and photographer with the *Boston Globe,* was standing beside me, and he started counting her vain efforts to stand up—he logged more than forty failed attempts to rise.

The men seemed unfazed by the spectacle—they just stood around watching the crippled animal struggle. The kid was elated, and the group stood around for a good ten minutes reliving the moment and congratulating the young marksman. At length, they got into the truck and made their way over to the dying animal. About twenty minutes after the whole scene had begun, the boy completed his achievement with a final, point-blank shot to the bison's head.

When I wasn't documenting the slaughter, or publicizing it to the press, I was shepherding bison back into the park. I'd clap my hands and throw rocks behind them to herd them past the park boundary, where they'd be temporarily safe. Unbelievably, I saw National Park Service personnel doing the opposite, trying to scare them in the opposite direction. Here were the people entrusted by the American public with stewardship of the bison, and yet they were shooing the creatures out of the park and into the line of fire. These guys thought the bison herd had grown too large, and they were quite happy to herd them along so that the state could finish them off. I had always seen the Park Service as a force for good, but not this time. They brought discredit to their mission and to their uniforms—the ones bearing the National Park Service insignia, with the noble bison right at the center.

Hunters shot 569 bison in the winter of 1988–89, and the few surviving bison in the state retreated back to the park as spring approached. Montana had accomplished its short-term purpose of clearing out the animals one way or another, but the news reports had not been favorable. The whole spectacle had repulsed the public, who saw hunting bison as the sporting equivalent of shooting a zoo animal.

In the face of withering criticism, the state abandoned the public hunt the next year, worried not about the welfare of the bison but about tarnishing the image of hunting. The killing went on, with Montana officials taking over the task. In some years, few bison left the park, but every few years, there would be a larger exodus, and the public controversy would flare up again. Still, Montana authorities would not relent and, working with the National Park Service, actually began capturing bison to ship them to slaughterhouses. The state even contemplated an extermination of the entire Yellowstone bison population and repopulating the park with bison unexposed to brucellosis. There were congressional hearings after the major kill of 1989, and also in later years, especially as state and federal authorities began to manage the bison like a rancher handling a cattle herd.

In March 2007, nearly two decades after my first trip to Yellowstone, Representative Nick Rahall, a West Virginia Democrat and chairman of the House Natural Resources Committee, convened a hearing on the fate of the bison. His opening statement captured the thoughts of many Americans: "Is it any wonder then that the American public periodically looks on in horror at footage of employees of the United States Department of the Interior participating in the slaughter of Yellowstone Bison?" he asked. "The general public is under the impression that these animals are being sheltered and protected by the federal government, not rounded up and shot. And the obvious question is why. Why is the Department of the Interior murdering its beloved mascot?"

Ironically, one rationale for reintroducing wolves into Yellowstone was to bring back a major predator of bison, elk, and other prey animals to keep their populations in check. But the political reaction in Montana to wolf reintroduction was hostile and swift. Ranchers, hunters, and their allies did not want a single wolf to step into the state.

But this was a federal matter, with the Endangered Species Act legally requiring the restoration of species listed as threatened or endangered. And in the case of Yellowstone, the crown jewel of the national park system, the restoration of wolves was a powerful symbol of correcting a wrong of the past and making the park whole again. President Bill Clinton's interior secretary, Bruce Babbitt, an ardent environmentalist, argued for the reintroduction plan in the name of good stewardship and ecological integrity. So with great local controversy but also with much national fanfare and celebration, government officials trapped wolves in Canada and transported them to Yellowstone and parts of Idaho in 1995 and then released them in an experiment to restore what had been lost.

As a witness to the slaughter of bison, I had conflicted feelings about wolf reintroduction. I thought wolves should be restored because they were part of the natural ecology, and I regarded their elimination from the West by trapping, hunting, and poisoning as a despicable chapter in our history. In theory, I was all for reintroduction. But I also knew that soon their population would increase, and their range expand beyond Yellowstone, just like the bison's. I was sure that the same demarcation line would be set for them and the same clash of attitudes would surface. When they became abundant enough to lose federal protection, the aggressively prohunting and proranching states of Idaho, Montana, and Wyoming would take control. We would have honored the right principles, but, in the end, the political power always shifted back to hunters and ranchers, and when it was all over they'd have their way. Wolves would be hunted and killed. Their packs and families would be destroyed. State officials would kill pups in their dens. In the end, I just didn't think it would be a good outcome for wolves. But I stayed on the sidelines as the leaders of environmental groups and the Clinton administration charged ahead with the plan.

Wolves were meant for these lands. Mothers reproduced, and pups survived, and the population grew faster than even the experts had forecast. They had an immediate impact on the ecosystem, reducing elk numbers and allowing the forest's undergrowth to be restored. It wasn't long before competition inside the park caused pioneer animals to explore the areas beyond. As the number of packs increased, Interior Secretary Dirk Kempthorne, a former governor of Idaho, proposed removing wolves from the list of protected species in the northern Rockies. HSUS and a coalition of environmental groups held back their efforts in the federal courts, arguing that the states lacked plans to maintain sustainable wolf populations.

But that hardly settled the matter. In 2009, President Barack Obama's interior Secretary, Ken Salazar, formerly a U.S. senator from Colorado, proposed delisting wolves again. And after more than a half-dozen wins by our legal team in court, a judge balked and then allowed a preliminary delisting in two of the three states. State officials did not waste a moment in opening up hunting seasons for the first time in decades. In 2009, Idaho sold ten thousand hunting licenses—putting the ratio of hunters to wolves at more than fifteen to one. Idaho governor Butch Otter had previously declared that he'd be "prepared to bid for that first ticket to shoot a wolf myself." Montana followed suit, and hunters there killed nearly all the wolves from one of the most studied packs in Yellowstone, destroying its social structure and leaving orphans, a few other survivors, and not much else. Fortunately, the same federal judge issued a final decision some months later that rebuffed the administration, restored the endangered status of the wolves, and blocked the subsequent hunting season. Now U.S. senators from Idaho and Montana are seeking to enact legislation to delist wolves, trying to remake the law after repeatedly coming up short in the courts.

In this century, Yellowstone will again play a vital role in protecting wildlife. But the unhappy experience of wolves and bison reminds us that even the biggest of parks like Yellowstone are not big enough. The mere presence of wolves and bison and other wild animals in a single park is not sufficient, as long as they are hypermanaged and treated like walking museum pieces or as so many specimens needed only to fill out our postcard picture of the ideal park. They are not just populations to be managed, contained, controlled, or culled. They are individual creatures whose lives and travails matter for their own sake, especially when they are harassed or threatened by cruel and officious people. After all that these creatures have been through, they deserve much better than that.

To give them the chance to survive, or better still to flourish, we need a new model of wild lands protection—large, protected areas joined by corridors, allowing animals to freely roam as nature intended, to be where and what they were meant to be. The great goal would be a wilderness network of parks and corridors from northern Mexico, through the Rockies, and all the way up the spine of the mountains to the Yukon. Some of these pieces are already in place, with millions of acres set aside as parks and national forests. To complete this unified stretch of wilderness will require more public lands, more protection of wildlife on ranches and other private lands, and the same kind of vision and political will that gave us Yellowstone more than a century ago.

## Killing and Longing: The Lessons of TR

IF YOU WANT TO see wildlife at the Smithsonian's National Museum of Natural History, your first stop must be the Kenneth E. Behring Family Hall of Mammals. It boasts specimens of more than 250 species, with the diorama beautifully designed and presented in an uncluttered way befitting one of America's largest such collections.

Mr. Behring's beneficence is proclaimed on a sign above the hall's entrance. Left unexplained is the fact that Mr. Behring helped populate it—after traveling around the world to shoot these animals and have them stuffed.

Behring and the Smithsonian ran into trouble after an attempted collaboration to import the trophies of some of the rarest animals in the world and to display them at the museum. In 1997, Behring traveled to Kazakhstan to shoot two endangered Kara-Tau Argali bighorn sheep, of which there are fewer than one hundred in all the world. Mr. Behring, past owner of the Seattle Seahawks and a fixture on the Forbes 400 list of the wealthiest Americans, paid Kazakh government officials a handsome licensing fee for the privilege.

It turns out that Behring paid for privilege on both ends. Just before the Smithsonian requested the imports of the sheep, it received a $20 million gift from Behring to create the Hall of Mammals. The Kara-Tau Argali is on the federal government's list of endangered species, so Behring was forbidden from importing any into America. But because the law allows scientific institutions to seek import permits, provided they "enhance" the conservation of the species, there was a narrow loophole for the Smithsonian to seek permits for the Argali trophies.

I worked hard to draw press attention to the Behring-Smithsonian relationship and to argue that importing these endangered animals would not enhance their conservation—quite the opposite, in fact. In response to stories in the *New York Times* and *Los Angeles Times*, and later on ABC News, in 1999 the Smithsonian withdrew the import request. It didn't help matters that, while all this was going on, other details of Behring's hunting expeditions came to light. Using a helicopter, Behring and his hunting party (including two past presidents of Safari Club International) had shot three bull elephants in Mozambique in 1998, in addition to a lion, a leopard, and a buffalo. Mozambique had banned elephant hunting in 1990, after a wave of elephant slaughtering by poachers. And for all of Behring's munificence toward local government officials, bestowed in hope of legitimizing his slaughter, the nation's chief wildlife officer concluded that the elephant killing was, in fact, illegal. Behring had tried to get Mozambique to declare the slain creatures "problem elephants," as if he had bravely performed some kind of public service. But money didn't talk this time: the government concluded that Kenneth Behring himself was the problem and wanted him out of their country. He apparently got out just in time, as Interpol was looking into the case of the missing tusks, and in the end, he and his safari companions escaped punishment.

As a kid who loved natural history, I had always assumed that animals at museums had not been deliberately killed to put them on display, though the circumstances were always a bit murky. The specimens seemed a little too well preserved for animals who perished as roadkill or died from natural causes. As an adult, I realized that people like Ken Behring were responsible for many of the specimens, and the details were a closely held secret. For the hunters, donating the trophies wasn't just a way to lend an air of charity to their avocation—there were also some very practical benefits. Behring and other well-to-do hunters have sometimes looked to museums as pathways to get their endangered animals into the country, just as with the Argali sheep. Most of their scams went unnoticed. Other trophy hunters have taken it a step further, using a loophole in the tax code to confer nonprofit museum status on their living room or den. They would then "donate" the trophies to themselves, which allowed them to write off a good share of the costs of their last hunting excursion. That savings, in turn, would help finance the next big-game safari, in what one writer has described as a kind of "frequent slayer" program. At the urging of HSUS and Senator Charles Grassley (R–Iowa) in August 2005, Congress put an end to this enrichment scheme for the trophy-hunting lobby and closed the loophole. Grassley, as chairman of the Senate Finance Committee, said his aim was to take "the tax cheating out of taxidermy."

Trophy hunters are an especially bizarre case of the contradictory impulses many people bring to the treatment of animals. The people who do it obviously have a fascination with the animals; otherwise, they would not go to such lengths to pursue them and to keep and display what they have killed. But there could hardly be a more frivolous reason to kill an animal, and they are constantly searching for high-sounding reasons to justify their pastime. In Behring's case, his great wealth allowed him to pass off his bloodlust as a grand charitable enterprise. But strip away the pretenses, and all that's left is a fascination with animals that's been twisted into a desire to pursue, kill, and possess them.

Before taking full possession, of course, the trophy hunter must pay a call on his taxidermist—and not just any old practitioner of what in the hunting trade is considered a high, almost sacred art. Here, too, we find a distorted version of the human-animal bond, expressed in professions of high esteem for the animals who have just been so needlessly slain. "It takes dedication to the animal," as one taxidermist explains. "You are trying to honor the animal as much as you can." Meditations of this kind have even inspired a work of philosophy on the subject, *Taxidermy and Longing,* stuffed with every deep reason one could imagine for taking the life of an animal and then trying to make it as lifelike as possible.

Sometimes, old and new ways of viewing animals are gathered up in the character of one man. The best example of this is the figure who stands tallest in the founding history of the American conservation movement, Theodore Roosevelt—a big-game hunter fascinated by wildlife and bird watching; a cofounder, with other trophy hunters, of the Boone and Crockett Club; and, simultaneously, a staunch advocate of the bird protection movement. He berated market and pot hunters as a scourge, yet he shot and killed buffalo, bear, bighorn sheep, and other wild animals even then facing extinction in the United States. He railed against game butchery, but on his storied African safari he alone killed 269 animals, including nine lions, thirteen rhinos, eight elephants, seven hippos, and seven giraffes—principally for museum collections. He was a man with an extraordinary fascination and understanding of animals, but who nevertheless reveled in killing them. For these paradoxical reasons, his memory is honored by both hunters and conservationists.

Roosevelt fancied himself a naturalist in the tradition of John James Audubon, believing that shooting animals was a part of studying them and then putting them on display. He could not, or would not, readily transition to the "do no harm" form of nature study represented by John Muir, Enos Mills, and other contemporaries. Roosevelt sought to recast hunting, once a utilitarian and then a commercial pursuit, as a recreational pastime and a defining element of American masculinity and strength. For him, hunting was an ennobling pursuit that embodied the kind of Darwinian struggle that would keep the United States vigorous and forever free.

Many historians defend Roosevelt and his passion for hunting by saying that we should not judge him by modern standards. They overlook the fact that he had many prominent critics in his day, including Mark Twain and George Thorndike Angell, founder of the Massachusetts SPCA. John Muir told historian Robert Underwood Johnson that he had challenged the president on their well-publicized trip to Yosemite, asking, "Mr. Roosevelt, when are you going to get beyond the boyishness of killing things? It seems to me it is all very well for a young fellow who has not formed his standards to rush out in the heat of youth and slaughter animals, but are you not getting far enough along to leave that off?" If Roosevelt's reply was as Muir claimed—"I guess you are right"—then it must only have been to change the subject, because history records no change in TR's attitude toward hunting,

Yet for all that, Roosevelt left an unparalleled and enduring wildlife protection legacy. Presidents before him dabbled in land preservation, but Roosevelt surpassed them all, even if you add up the collective efforts of his twenty-five predecessors. In an era when forest lands were largely depleted, farmlands were under severe strain, and extractive industries were flourishing without serious restriction or regulation, TR turned the federal government into an instrument of conservation. He viewed the conservation of America's wild lands as an essential incubator of American character and wealth—one that would preserve and strengthen American democracy.

He created 150 national forests, quadrupling the forest reserves of the United States—from forty-three million acres to nearly two hundred million. He established fifty-one federal bird reservations in seventeen states and territories, mitigating the disastrous effects of the millinery trade. He was responsible for eighteen national monuments and four national game preserves. He doubled the number of national parks from five to ten, and he saved the Grand Canyon from zinc and copper mining interests with the stroke of a pen. He was responsible for the creation of the U.S. Fish and Wildlife Service, and he shepherded to passage the Antiquities Act of 1906, which would be used by later presidents to protect some of the most beautiful expanses of the United States.

Altogether, Roosevelt placed 230 million acres of land under permanent public protection—that's the equivalent of the landmass of the Atlantic coast states from Maine to Florida. Remarkably, Roosevelt accomplished most of this against a legion of opposition, including titans of industry, market hunters, and developers. The idea of national parks has also provided inspiration to the world, and now dozens of countries have established their own national park systems.

Today's animal advocates find Roosevelt's legacy challenging. He appears to have had a kind heart for domesticated animals, rescuing kittens, adopting stray dogs, and opposing the tail docking of horses. But his tenderness vanished when he was in the wild with a gun. Understanding his contradictions is no easier than trying to fathom how the nation's constitutional framers demanded human liberty at the same time that they were personally involved in chattel slavery. My own view is to recognize good where you find it and, with serious reservations noted, to honor Theodore Roosevelt's memory for the noble and farsighted things he achieved as president.

## The Era of Rational Slaughter

THE INTERVENTION BY ROOSEVELT and others came at the right time—except where it came too late for some species. Contrast the America of 1900 with the state of the nation a century earlier. In the first few decades of the nineteenth century, elk and bison roamed the forests of the Upper Midwest, as did wild cats, wolves, and grizzlies. Passenger pigeons numbered in the billions, with flocks often darkening the midday sky.

Farther west, tens of millions of bison pounded the rich topsoil of the Great Plains—from Sam Houston's Texas all the way up to the prairie provinces of Canada. They commingled with elk and pronghorn and deer, and they were all on alert for plains wolves, grizzlies, and Indian tribes. The bison sidestepped prairie dog burrows as they grazed, converting tall grasses into sinew and muscle on their two-thousand-pound frames. The prairie dogs retreated underground when they spotted hawks, golden eagles, or swift foxes but had no refuge from black-footed ferrets who could navigate their carefully designed tunnels.

We know what happened to change that picture forever. Past the midpoint of the nineteenth century, and as the settlement of the West quickened with the passage of the Homestead Act of 1862, few wild places were beyond the reach of our expanding republic. With the transcontinental railroad and a network of other railways linking the growing urban populations to the resource-rich hinterlands, the technology and transportation systems were in place for a liquidation of natural resources on a vast scale.

With demand for wood surging to build homes, buildings, and fences, the timber industry stripped forests, including the great forests of the Upper Midwest. The denuding of these habitats delivered a blow to passenger pigeons and every other species that depended on forest lands. But just as deadly to the passenger pigeons were market hunters who killed the birds to sell their meat on the cheap to the growing class of urban poor. Market hunters also killed for the elites—shooting herons, egrets, and other birds with elaborate plumage for the millinery trade, so wealthy ladies could adorn their hats with ornate feathers.

With the advent of the repeating firearm, no animal was more vulnerable than the bison, who massed in herds in open fields and whose hides could fetch up to $50 at market. One professional hunter killed more than twenty thousand bison, and casual hunters and travelers even shot the animals from the open windows of trains. The U.S. Army joined in slaughter, calculating that eliminating the buffalo would doom the resistance of the Comanche and other Plains Indians. In just three decades, from 1850 to 1880, these forces practically shot the species into oblivion, with the evidence piled up on the plains in the form of rotting carcasses and bleached bones.

The American conservation movement was born in response to this catastrophic destruction of wildlife, just as the humane movement was created in the same era in reaction to the abuse of horses and other working animals in cities. Prominent Americans—including John Muir, the artist George Catlin, and the author George Perkins Marsh—sounded the alarm at the mass slaughter of wildlife and pleaded for restraint. They lobbied to create national parks, which were some of the earliest tangible policy efforts to preserve what was left. The Lacey Act of 1900, which made transporting meat and other wildlife parts not allowed under state law a federal crime, marked the first major national wildlife conservation law.

To some, it seemed like madness to set aside these lands, to close off any space whatsoever to hunting, or to curtail the trade in wildlife parts. Yet the idea of protecting millions of acres and placing limits on their exploitation reflected a new sensibility and code of ethics: that other species have a place and purpose of their own, that there are limits not only to what is possible with nature but also to what is right, and that public access to these great places was a unique source of American vitality.

In the early decades of the twentieth century, the slaughter of wildlife had been checked, in part because so little wildlife had survived. America saw the first serious campaign to restore some of what had been, and along with that came a debate about the correct philosophical approach to conservation. One of the men who shaped that debate was Aldo Leopold—another figure who, like Roosevelt, was fraught with contradictions and whose legacy is still felt today. A forester, conservationist, and, like Roosevelt, a sport hunter, Leopold is recognized as one of the founders of the game-management model that has dominated wildlife policy in America in the post–World War II era.

A professor of game management at the University of Wisconsin, he argued that the hunting of animals could be rationalized—that government wildlife agencies could treat deer, ducks, and other game species as a crop and manage them scientifically to produce a maximum-sustained yield for the benefit of society. By blending the science of wildlife management with agricultural principles and theory, Leopold focused not on individual wild animals but on the maintenance of populations. State and federal fish and wildlife agencies embraced these principles and sought to rebuild animal populations, such as deer and elk, and to stock pheasants, turkeys, and other game animals to cater to hunters, setting up an interlocking relationship between hunters and government agencies that continues to this day.

But Leopold also advocated restraint. In his master work, *A Sand County Almanac,* he advanced the often-quoted principle: "A thing is right when it tends to preserve the integrity, stability, and beauty of the biotic community. It is wrong when it tends otherwise." Leopold opposed state killing of predator species, but despite his warnings, the war on predators never really stopped—it only changed form. Most hunters viewed them as competitors, and ranchers and farmers thought of them as vermin. In 1931, Congress passed the Animal Damage Control Act promoting the elimination of predators "injurious to agriculture." Leopold opposed this idea, pointing to the example of the Kaibab National Forest in Arizona. There, he claimed, after government hunters killed all the mountain lions, the deer population increased by tenfold and then crashed, causing starvation and suffering, as well as serious damage to forest health.

The scenario that Leopold laid out has been challenged over the years, but Kaibab became a defining case study in American wildlife-management theory, and it drove the argument that human hunters were needed to control deer populations, absent predation by wolves and mountain lions. This idea influenced the 1937 Pittman-Robertson Act, which levied a federal tax on gun and ammunition sales and directed the proceeds to state fish and wildlife agencies as an incentive to support hunting. Natural resource programs at major universities churned out staff for the state agencies and incorporated the thinking of foresters and agricultural economists in creating the new breed of wildlife managers, providing intellectual reinforcement for the game-management model.

As wildlife managers succeeded in growing larger game populations, states sold more hunting licenses, and the industry grew. The dogmas of wildlife management took hold and would not come under serious challenge until the last quarter of the twentieth century. That's when the animal-protection movement began to stand up to the hunting industry. And that's also when conservation biologists emerged to challenge wildlife policies focused almost exclusively on producing game for hunters instead of protecting wildlife for its own sake and for broader ecological purposes.

Unaccustomed to being questioned, hunters during this same period formed new lobbying groups, such as the U.S. Sportsmen's Alliance and Safari Club International, and existing groups like the

National Rifle Association took on a harder edge. These organizations can still be heard trading on the memories and reputations of men like Roosevelt and Leopold, even as they accept and promote practices considered contemptible and unsportsmanlike, such as shooting animals confined to pens, setting robotic ducks afloat to act as decoys, and using high-tech gadgetry like telescopic sights and global positioning systems that remove from the blood sports whatever modest measure of fairness they ever had.

## "Let's Kill Them All": The Attack on Alaska's Wolves

A FEW YEARS AFTER my winter 1989 trip to Yellowstone, I was dispatched to Alaska. It was the winter of 1993 and the Fund for Animals was dealing with a different wildlife controversy. The issue was Alaska's notorious wolf-control program—one of the intractable wildlife-management debates of our time.

The occasion for the trip was a so-called Wolf Summit, a gathering hastily called by Governor Walter Hickel to facilitate "the exchange of ideas" after he grudgingly suspended plans to launch an aerial wolf-hunting program. Hickel, elected in 1990 as the candidate of the Alaska's newly formed Independence Party, convened the three-day summit in mid-January, when the sun peeks above the horizon for about four and a half hours a day and the thermometer can sink to forty below zero.

He set the meeting at a hockey rink in Fairbanks, the state's second-largest city but its main stronghold for prohunting activism and political conservatism. Governor Hickel advertised my presence and the attendance of a small cadre of other animal advocates from Outside—as Alaskans term the Lower 48. The governor's antics had the desired effect. About fifteen hundred locals, some decked out in wolf fur, turned out for the summit to rally for the "Alaska way of life." The pro-wolf-slaying assembly was about as interested in the exchange of ideas as the beer-swilling crowds at the hockey games that usually filled this venue. It didn't take long for me or the other Outsiders to realize that the purpose of the whole exercise was to make us feel as unwelcome as possible, while the state "reeducated" the American public on why Alaska is different and wolf control desperately needed.

A few months earlier, Hickel had advocated the aerial-hunting plan by declaring, "You just can't let nature run wild." But wolves were not a safety threat to people or to livestock in the vast wilderness between Anchorage and Fairbanks. Hickel had called moose and caribou "Alaska's livestock," and the transparent purpose of the wolf-control program was to boost their numbers, so that hunters could shoot more of them.

Even so, the reeducation efforts were not taking—and most Americans asked only that the wolves be left alone. About one hundred thousand people wrote or called the governor's office, and many promised they'd boycott travel to Alaska if the state went ahead with its aerial killing. As national director of the Fund for Animals, I urged the public to take their wildlife-watching trips elsewhere if Alaska intended to slaughter wolves from the air. Alaska couldn't have it both ways—luring nature lovers to the state with glorious images of wildlife and stunning landscapes, while engaging in a deliberate program of wildlife extirpation.

Tourism is Alaska's third-largest industry, and business owners who relied on Outsiders made their worries known to the governor. But even after Hickel bowed to their concerns and suspended the program, he remained defiant, and I had no illusion that the state would relent for long. At the summit, Hickel had pandered to the raucous crowd, declaring "I will not be a part of Alaska giving away its sovereignty over the management of fish and game. We have a right to care for this land, according to our knowledge of the North." He said people from Outside knew about Alaska from the television show *Northern Exposure,* but that "people here have to deal with northern reality."

Since Alaska's statehood in 1959 successive governors have had the opportunity to put their signatures on wolf-management pro-grams, but none had ever given wolves a reprieve. It generally just got worse for the wolves. Hunters and trappers annually kill five hundred to a thousand wolves, and the only reason it's not more is the vastness of Alaska and the difficulty in accessing the bush— hence the interest in aerial wolf control. In terms of standard tools, trapping has been the cruelest method used. If wolves are snared around the neck, blood is trapped in the head and it swells to double its normal size—a gruesome phenomenon known as "jelly head." Moose, lynx, and other animals also get snared, and this "bycatch" is another awful consequence of the method. If caught in a leghold trap, the struggling wolves sometimes resort to chewing off a frozen limb. Scenes of just that agonizing spectacle were captured on film in the 1990s by Dr. Gordon Haber, an independent wolf biologist, and for millions of Americans these images were hard to forget.

Cooked up by political appointees at the Alaska Board of Game, the plan was to use fixed-winged aircraft and helicopters to reduce targeted populations by 80 percent—with the aerial gunners killing three hundred to five hundred wolves. This wasn't a novel program. In the 1960s, the aerial gunning of wolves and polar bears in Alaska had shocked the nation and prompted Congress to pass the Airborne Hunting Act of 1971. That law forbids killing wildlife from the air or shooting animals within twenty-four hours of land-ing the aircraft, with exceptions for the protection of public health and of wildlife populations. Land-and-shoot hunting was, in many respects, worse than airborne shooting. The pilots would harass and chase the wolves by aircraft, exhausting them to the point that they could run no longer. They'd then land the aircraft and shoot the ter-rified animals unable to flee because of lactic acid built up in their muscles. Under Hickel's plan, the state sidestepped the federal law

by arguing that aerial wolf hunting would boost depleted populations of moose and caribou.

The summit proved to be nothing but a chance for the wolf-
killing lobby to posture and vent. I had a few conversations with
state wildlife officials, but the leaders of the Alaska Outdoor
Council, the state's primary hunting lobby, had no interest in talking. Hickel took the press on a same-day land-and-lobby jaunt to
Minto, an Athabascan Indian village forty miles west of Fairbanks
that was strongly in favor of wolf killing. In this piece of sylvan the-
ater, a tribal member named Ronnie, one of a roster of speakers at
the local gym, complained, "We don't manage your cattle, so why
should you manage ours?" He added, "I'm all for wolf control. Let's
kill them all." His point, apparently, was that the state should not
manage wildlife, but actively hunt wolves to the point of extinction.
I was thankful that many tribal members disagreed, and some of
them spoke up at the Wolf Summit.

Having made their case at the summit and elsewhere, the Board
of Game convened and reauthorized a wolf-control plan for the next
year, only slightly scaling back its original plan. Hickel had packed
the board with the most notorious wolf-control advocates in the
state, many of whom were closely affiliated with the Alaska Outdoor Council. Still, it wasn't as if the Alaska Department of Fish and
Game had a starkly different viewpoint. The department's director
of conservation had himself joined aerial wolf-gunning expeditions
as a regional wildlife manager in Tok and earned the moniker David
"Machine Gun" Kelleyhouse.

Wolf-control programs went forward that year, but after Alaska
elected Democrat Tony Knowles as governor in 1994, he grounded
the aerial-gunning program. And two years after his election, voters
approved by a wide margin a citizen initiative on the November
1996 ballot to ban aerial wolf gunning except in "a biological emergency." This outcome demolished the wolf hunters' claim that this

was a clash between Alaskans and Outsiders. Alaskans wanted, at the very least, aerial wolf killing to stop, and polls revealed even a majority of hunters favored this position.

That should have settled the matter. Yet remarkably, state lawmakers overturned the voter-approved citizen initiative in the following year's legislative session. Unbowed by the actions in Juneau, wildlife advocates put a second measure on the ballot in 2000 to stop aerial gunning and voters approved that reform, too. And again lawmakers overturned it—defying their constituents for a second time on the same issue.

But the most destructive thing that lawmakers did was to pass, back in 1994, the Intensive Management Act, declaring that "the highest and best use of most big game populations is to provide for high levels of harvest for human use." Signed by Governor Hickel as one of his final acts, this legislation effectively committed the state, as a matter of law, to manipulating predator populations so that people could kill more moose and caribou for the freezer. Wolves and bears be damned.

After Knowles left office at the end of 2002, his successors Frank Murkowski and Sarah Palin lifted the restraints on wolf killing. Both had close ties to the Alaska Outdoor Council and an unqualified political enthusiasm for wolf killing. They launched aerial-gunning programs not only for wolves, but for bears, and Palin even proposed paying bounties for the left forelimbs of wolves. Seeing this harsh turn, voters put a third anti-aerial-gunning measure on the ballot in August 2009, but this time, in a low-turnout primary, it was defeated. The governor's office itself committed $400,000 to an in-state PR campaign promoting the value of "predator management" programs. Having twice voted to end a practice they considered cruel and unwarranted, clearly Alaskans themselves now needed to be "reeducated."

## *"Garbaging for Bears": Commercial Hunting and Contemptible Conduct*

AT HSUS, WE HAVE not campaigned against all forms of sport hunting in Alaska or any other part of the country. Despite the overblown claims of the NRA and other prohunting organizations that we are working to ban it all, we've largely asked hunters to hold themselves, at the very least, to their own professed standards of conduct. Retrieve and eat what you kill, treat wildlife as a public resource and do not kill for commerce, observe traditional norms of sportsmanship and fair chase, and do not subject animals to lingering deaths.

These are standards that no politically influential hunting organization advocates today—indeed, none of these groups asks its members to abide by any serious ethical code of conduct. On the contrary, the politically active hunting groups are the ones who promote the worst practices and resist any restraints at all. When it comes to the insanity of predator-control programs, *Audubon* magazine columnist and lifelong advocate of hunting ethics Ted Williams captured the mind-set long ago: "Wolves don't pay for hunting licenses and every moose or caribou lost to a wolf is one less hunting license fee paid to the state."

As unfair or unethical as wolf control in Alaska may seem, it is hardly unique in our era. Everywhere predators live in the United States, they are under assault from ranchers and sport hunters. In the northern Rockies and Great Lakes region, wildlife officials are doing their best to delist wolves and begin killing them for sport. Mountain lions inhabit a dozen western states, and they are hunted in nearly all of them, principally by hunters who use hounds fitted with radio telemetry collars to track down the animals and kill them. Only California bans trophy hunting, and that came through a ballot initiative in 1990. Even there, the state's hunting lobby persuaded lawmakers to put a repeal measure on the ballot just six years later, after two fatal lion attacks on people. On this second vote, the margin of victory against hunting was even greater—not because of any lack of sympathy for the victims, but because there was no evidence that trophy hunters killing lions would at all diminish the small possibility of a lion attack on a person. Voters in Oregon and Washington followed up, in 1994 and 1996, respectively, by banning the hunting of lions with packs of dogs, again scoring a victory for the principles of sportsmanship and the place of predators in the ecosystem.

In the interior West, however, thousands of mountain lions are still shot and killed each year. Montana trophy hunters kill hundreds annually, even though the state has no idea how many lions there are. In Utah, hunting guides have used "roping and choking" techniques, in which they tree a lion, rope and choke the creature to the point of suffocation, and then release him just ahead of the dogs of a fee-paying client. The table set for him, the so-called hunter then shoots a dazed, half-dead lion out of a tree. Such heroic exploits can be enjoyed for about $3,000.

Black bears are treated no better. In more than a half-dozen states, it is still legal to hunt bears during the spring, when females are nursing dependent cubs. When the females are shot just three or four months after giving birth, the cubs are doomed to starve. In the spring, or in the more typical fall hunting seasons, ten states allow bear baiting, in which food is set out as a lure and the bears are shot while feeding. Ted Williams calls it "garbaging for bears," with pizza, jelly doughnuts, grease, and rotting meats set up in fifty-five-gallon drums to draw the bear into shooting range. The U.S. Forest Service and the Fish and Wildlife Service admonish forest users to "never feed bears," but they make an exception for baiters who dump millions of pounds of food in the woods during the hunting season in order to get an easy shot. The bears regularly visit the bait sites—making them less wary of people and more inclined to raid other human trash sources. It's the very thing that wildlife agencies claim is detrimental to bears and people, but they apparently suspend that thinking when it comes to trophy hunters.

More than half of the thirty states with legal bear hunting allow hunting with hounds. The dogs may chase their quarry for miles, until the exhausted bear climbs a tree or turns to face a pack of frenzied hounds—either way dying a merciless death. The same method is used for lion hunting. Voters in Colorado, Massachusetts, Oregon, and Washington banned both baiting and hounding. But voters in Idaho, Maine, and Michigan rejected these restrictions, after the hunting lobby claimed that bear populations would increase dramatically if these methods were prohibited.

In Maine, hunters shoot four thousand bears every autumn, and nearly 80 percent of them kill the bears over bait. Four out of five of the hunters come from other states. It's a lucrative business, with hunters paying $2,000 to $4,000 for the opportunity to kill a trophy bear. Ironically, when we helped to wage an unsuccessful 2004 campaign to prohibit killing bears over bait, the Maine hunting lobby attacked HSUS as an out-of-state group, arguing that it's up to Mainers to decide these issues. Apparently, out-of-staters are welcome in Maine as long as they're coming to bait, stalk, and kill the state's wildlife. But if you're there to question the whole sorry business of baiting bears—a practice forbidden in hunting most other animals—the message is just go back where you came from.

Hunters are never more resourceful than in thinking up new justifications for the things they do. They know that for nonhunters, killing for recreation alone, just for the fun of it all, is suspect. So they're always quick to stress that they eat what they kill, and even elephant trophy hunters boast of distributing the meat to locals—presenting the kill as a kind of feed-the-villagers program. Another attempt at self-justification is the claim of hunters to control overpopulation of deer and elk. Yet in every state where natural predators like wolves, lions, coyotes, bears, bobcats, and lynx exist, hunters also target them. Add in the 89,710 coyotes, 395 bears, 373 mountain lions, and 1,883 bobcats killed by USDA's Wildlife Services program in 2008 alone, and then the tens of thousands killed by ranchers, and you have hundreds of thousands of natural predators eliminated for no better reason than bloodlust from the hunters and contempt from the ranchers.

Then there's the practice of shooting animals in fenced enclosures, known as captive or "canned" hunting. I've been to the Texas Hill Country, where many of the nation's thousand or so captive exotic hunting operations are located, and visited a number of these places. Operators provide a price list of native and exotic species, and hunters shoot the animals in a "no kill, no pay" arrangement, with the outcome certain. Somehow this doesn't offend the hunting lobby's "sportsmanship" sensibilities. The Safari Club International even bestows "hunting achievement" awards and "Grand Slam" prizes on hunters who shoot animals trapped inside fences. In fact, one such hunting achievement award, "Introduced Trophy Game Animals of North America," is designed to send customers 'a-runnin' to these fenced-in hunting ranches. Handling the introduction are owners who, by means of fencing, baiting, or drugging, make it impossible to miss.

Captive shoots are a pure form of commercial hunting, with few of the usual lofty pretenses to necessity. The animal is private property, the killing is certain, and the hunter does not typically eat the meat. Many captive hunting operations have their own on-site taxidermy services—to minimize those empty days of "longing" before the trophy's on your wall. And if the taxidermist does his job right, no one but he and the hunter will ever know that the animal was shot in the back at the foot of a fence—completing the whole contemptible farce.

Despite an unapologetic defense of these operations by powerful groups like the NRA and the Safari Club, many states have banned them, including Wyoming, which is hardly friendly territory to critics of hunting. Montana voters outlawed canned hunts in a November 2000 ballot measure, while states from Oregon to Tennessee have either banned or restricted the practice. Canned hunts continue in about two dozen states.

There are other captive shoots in the United States. At perhaps more than one thousand captive bird-shooting facilities, pheasants, quail, and mallard ducks are typically pen-reared and shot. You pay only for what you shoot—the birds are planted in the grass before the hunters arrive or are thrown from the top of tall structures in "tower shoots." In Pennsylvania, "hunters" shoot live pigeons just for target practice, as if clay pigeons or skeet were not sufficient.

The trapping of animals for sport and fur is also inconsistent with the hunting lobby's rhetoric. It's a form of market hunting—a vestige of the nineteenth-century era of killing of animals for commercial sale of their parts. Like predators killed by hunters, trapped animals are not consumed for meat, and the states have no management plans, no population estimates, and no kill targets. Trappers kill millions of animals annually, and the intensity of the activity is driven largely by the value of the pelts. When bobcat pelts, for instance, are commanding high returns at fur auctions, there will be a surge in the killing of that species.

Leghold traps and snares are also indiscriminate, catching any animal that triggers the device. Add to that the fact that the animals languish in the traps for twenty-four to ninety-six hours (state policies vary on how often the trapper has to check the traps), and you get just a glimpse of the extreme suffering that the animals endure, with most trapping occurring in winter months when trapped animals often freeze to death. Thankfully, the public agrees that this cruelty is unacceptable and has approved HSUS-backed ballot measures in Arizona, California, Colorado, Massachusetts, and Washington to restrict the use of steel-jawed leghold traps and other body-gripping traps.

According to the U.S. Fish and Wildlife Service, only 12.5 million Americans went hunting in the United States in 2006. That is down from nearly twenty million hunters in 1975, with modest but measurable declines in hunting license sales just about every year since. Even so, hunters hold nearly all of the seats on state fish and wildlife boards—allowing policy decisions to be dominated by a small group determined to guard its privileges. Throughout the country, the system is stacked in favor of wildlife killing and against wildlife watching, even as eighty million Americans now watch wildlife every year. But while they've rigged the agencies, they can't rig the demographics; the trends will soon overwhelm hunters, and they seem to know it.

A 1996 Massachusetts ballot initiative that restricted trapping also eliminated the requirement that five of seven members of the state Fisheries and Wildlife Board had to have purchased hunting licenses in each of the last five years, and that four of the seven members had to "represent hunting, trapping, and fishing interests." It was a victory for the popular will that we need to see repeated elsewhere, so that wild animals are no longer parceled out as in some entitlement program for hunters. For too long, the management of wildlife has been left to a shrinking minority of sport hunters, who act as if wild animals have no conceivable value except as targets and trophies. But there is a public interest in protecting wildlife for better and more benevolent purposes, and it's about time the public itself was heard from.

## The Face of Innocence and the Shame of Canada

As I PRIED MYSELF out of a small helicopter, it seemed like I was stepping onto another planet—frozen and barren yet wondrous and beautiful. Ice floes reached for miles, merging with the horizon. Small, rounded mounds of ice, built up by the snows and sculpted by the winds, were the only vertical features on a vast, level plain. Even with sunglasses, my eyes strained against the white light gleaming off the surface. There were no trees, no other vegetation, no evidence of any prior human visitors. In the foreground, though, there were seals—thousands of them, white-coated, wide-eyed, newly born, and nursing from the bellies of their silver-coated mothers.

It was March of 2008, and I had come to Charlottetown, Prince Edward Island, to see one of the greatest wildlife spectacles in the world—the gathering of hundreds of thousands of seals in Canada's Gulf of Saint Lawrence. In the winter, as many as six million seals migrate south from Greenland, swimming across the North Atlantic and down the jagged coasts of Quebec and Labrador to reach the immense sheets of ice that interlock with the landmass of Canada. The range of polar bears does not extend to this area, so seals see it as a safe place to bring their young into the world.

I had first come here more than fifteen years before at the invitation of my friend Brian Davies, the founder and then president of the International Fund for Animal Welfare (IFAW). Brian had been a cruelty investigator in the 1960s with the New Brunswick SPCA, and in that capacity had been assigned to monitor the hunt. He was so disgusted by what he saw that a few years later he left the SPCA and founded IFAW as a campaigning organization almost singularly devoted to stopping the hunt—no small task, since it had been going on for at least two centuries. For Brian, the campaign would become his life's work.

I was working at the time for the Fund for Animals and its president, Cleveland Amory, who in the 1970s had raised his powerful voice against the hunt. In 1979, the Fund purchased a ship, the *Sea Shepherd,* for marine-mammal campaigner Paul Watson, and he and his team took to the ice to spray paint the coats of the baby seals, rendering the pelts commercially worthless. In response, Canada's government threw Paul and his crew in jail, roughing them up in the process. Brian admired Paul but not his tactics. He regarded them as a distraction from the more methodical campaign that he had organized and that seemed to be working.

After all, the annual toll taken by the sealers was in decline by the 1980s, especially after the European Union banned the import of fur from white-coated seals. The annual kills declined to as low as eighteen thousand, and the markets were disappearing. Most observers of the conflict assumed that time had passed the hunt by. They believed it would not survive long in the modern era—an assumption made a few times before. Even well before then, Newfoundland's most prominent newspaperman had all but pronounced it dead. "A very colorful page has gone out of Newfoundland's history," said A. B. Perlin, editor of the St. John's *Daily News*. "Seal fishery was a wasteful industry. It was, in many ways, an unpleasant industry. I've heard many a sealer talk about the small whitecoats, two or three days old, almost looking up with tears in their eyes as they killed them. And frankly it's an industry that we could do without . . . and from the standpoint of humanitarianism alone, it's probably a good industry to be without."

That's the way it looked again in the mid-1980s, like a dying industry. But Brian was not overly confident and invited me to the ice because he wanted the Fund as an ally in finishing the job. I was deeply moved by what I saw on that first trip, but I doubted that we at the Fund could add much to the campaign. We hated the seal kill, but we were still focused on America's many wildlife problems, and the Fund just didn't have much capacity in Canada.

For my part, I had no illusions that the seal hunters were about to give up. If there's one lesson I've learned in animal protection, it is never to write an obituary for any form of animal abuse. The animals have only one life, but the abusers seem to have nine of them—and you never want to underestimate the pridefulness of sealers in particular, or of their friends in Ottawa. As it turned out, in the early 1990s, the sealers tinkered with their calendar, waiting for the babies' coats to shed their pure white fur, revealing a silvery gray fur underneath, before they killed them. The sale of the silver fur from the slightly older baby seals would not violate the import ban on whitecoats, and the sealers could thus reclaim lost markets in Europe. They could also take advantage of newly emerging fur markets in China and Russia. In one respect, too, time was on their side: the global economy was humming, and suddenly the fur industry was on the rebound.

Canadian politicians, moreover, had devised a new strategy to blame seals for the government's dreadful mismanagement of the cod fishery. Canada's Department of Fisheries and Oceans (DFO), which was supposed to safeguard the fishery for long-term sustainable yield, had ignored warnings from scientists and fishermen about overexploitation and placed virtually no limits on the industry. By 1990, the cod had finally reached the point of commercial extinction—at only 1 percent of their historic populations. In the annals of North American natural resources management, this was one of the most appalling breakdowns ever—and it was directed by the same people who had overseen the killing of the seals. By 1992, overfishing had inflicted so severe a population decline that the Canadian government declared a moratorium on cod fishing, taking away jobs in every province of Atlantic Canada.

Political leaders in Ottawa and in Newfoundland needed scapegoats, and seals were the easy targets. Politicians cited the presence of the few million seals off Canada as evidence enough that these creatures, and not the thousands of industrial fishing crews, had decimated the cod. Harp seals do eat cod, but the species makes up just 1 to 3 percent of seals' diets. Moreover, because seals also eat predators of cod, they might actually have a net positive impact on cod populations.

Of course, neither the bureaucrats nor the politicians mentioned these facts. In 1998, John Efford, minister of fisheries and aquaculture for Newfoundland and Labrador, told the provincial legislature: "Mr. Speaker, I would like to see the six million seals, or whatever number is out there, killed and sold, or destroyed and burned. I do not care what happens to them . . . the more they kill the better. I will love it." Lawrence O'Brien, a former member of parliament from Labrador, told the House of Commons in 2003, "It is not study that we want. . . . We want those seals taken out. I do not care how they are taken out. Every bloody one of them can be killed. I will go in there myself with a rifle and help shoot them." It wasn't exactly the voice of reason. But such talk was the closest thing to it among Canadian politicians blaming the seals for their own failures of foresight and management.

I see this attitude all the time among the adversaries of HSUS and its mission—the same habit of excuse making and angry, vengeful blame shifting. I saw it with the bison and wolves, and here it was playing out with the seals. If you strained hard enough, you could come up with a plausible-sounding theory to ascribe all of maritime Canada's troubles to the seals, but if you go by common sense, all of this was plainly a con job. And in Canada, they somehow pulled it off every year. Summoning the same indignation over the same phony claims and numbers, they did it again in 2003, when the national government ramped up its seal quotas. Officials announced that they'd allow the killing of a million seals over three years—more than three hundred thousand a year.

Just a year later, Dr. John Grandy, HSUS's longtime senior vice president of wildlife programs, came to me in my new role as CEO and made an impassioned plea that HSUS should gather its forces to oppose the hunt. Images of my 1993 visit to the ice came to mind. Cleveland Amory had passed away in 1998, and Brian Davies had left IFAW. John Grandy recommended that we hire Rebecca Aldworth, who had been one of IFAW's top seal campaigners, and put her in the forefront of a revitalized, global campaign to counter a hunt that was now bigger and bloodier than ever. I knew it would be a very difficult battle, but also that we couldn't allow a wildlife massacre of this magnitude to unfold just a few hundred miles to the north of us without a fierce counter-response. If sealers were going to declare war on the pups, then we were going to declare war on the hunt. It was now our fight, and we've been waging it ever since. And for me, it was time to go back up to the ice.

Spring was around the corner, but it was a long bend. Winter still had a tight grip on this part of the world. We flew out in two helicopters to begin our annual Seal Watch program. It was March 2008, and the hunt was still three weeks away. We were there, along with our camera crew and some journalists, to provide the "before" picture—a glimpse of the grandeur of this open-air nursery before the arrival of men with clubs and skinning knives.

Though there's considerable local support for the hunt, there is global opposition to its savagery. Since the pelts are sold around the world, the goal was to choke off those markets and render the hunt unprofitable. Rebecca Aldworth was with me on this trip, and the seals have no more capable advocate than her. She is not only incredibly articulate and knowledgeable about every facet of the hunt, she's also a native of a small town in Newfoundland, in the heart of sealing country, and not easily written off as someone who does not understand local traditions. Also with us was Nigel Barker, the handsome young television personality who is also a world-class photographer whom we had brought to record the scene and to help tell the world about it.

About forty minutes into our helicopter flight from Charlottetown, we began to see small specks on the ice. There they were—newborn seal pups bunched together in loosely knit herds. Even though I'd seen them before, the sight transfixed me. We pressed ahead, and then saw thousands more. We were witnessing the terminus of the biggest annual mass movement of wildlife in the world, bigger even than the storied migrations of wildebeest, zebra, Cape buffalo, and other animals every year across the Serengeti.

Rebecca surveyed the concentration of seals below and signaled to the pilot that this was a good place to land. He moved us closer to the ocean and set down the helicopter's floats on what we hoped was a thick patch of ice. We cautiously stepped out and felt sturdy ice beneath our feet and then the frigid winds on our face. We were dressed in survival suits, which provided a layer of defense against the biting air and against the water in case we fell through. It was below zero, and the windchill made it seem even colder. But the sun was shining, and, amazingly enough, the seals seemed unfazed by the conditions. Many mothers were still with their babies, and most simply stood their ground as they saw us, while a few fled to openings in the ice and slid into the water. The mothers typically venture off to feed and then return to nurse the pups. There were streaks of blood on the ice, but not because any seals had been killed already this year. This was placental material, still fresh and not yet covered up by the snows, and a marker of new life.

We approached some of the babies, but I was always sensitive about getting too close. In seeing these balls of fur, almost out of a toy store, it's tempting to want to pet them. But I did my best to resist the impulse. A few of the baby seals put on a brave display as we approached, trying to assume the defensive posture of a much older seal. But most others barely reacted, though they must have been puzzled as they saw, for the first time, a creature other than a seal.

I gave the mothers an even wider berth, knowing that they'd probably witnessed men killing seals in years past. If I kept my eyes focused on an opening in the ice, I'd occasionally see a dozen adult seal heads pop up in unison. Then, just as quickly, they'd all vanish, in the most elegant synchronized swim I'd ever seen. Their wariness is well justified, since shooting with a firearm is now a primary method of killing for the sealers.

We spent three cold hours on the ice, walking around, taking photos, and filming. We wanted to document this experience and kindle people's bond with wildlife and nature. The babies are born with thick white fur, and they drink up the fat-rich milk of their mothers for about twelve days. Then the mothers leave them, forever. For another two weeks, the babies still can't swim, and they move by clumsily dragging themselves along the ice with their developing flippers. It is during this most vulnerable phase of their development—after their mothers have left but before they're swimming proficiently—that the sealers come.

There could hardly be a more defenseless, unoffending animal on earth—a newborn marine creature still unable to swim, without legs to run away, and without maternal protection. There is something so obviously and grotesquely unfair about the killing. The men had ships, clubs, guns, and not an ounce of remorse and came charging like frenzied warriors into one of nature's nurseries. Even sport hunters, in pursuing bears, deer, or other animals, usually abide by an unwritten rule not to shoot baby animals. No such code of honor here: killing the newborns is the whole point of it.

I am an equal defender of all animals, regardless of how they look, whether they are imposing and fierce or harmless and endearing. It is not the size of their eyes, or the richness of their fur, that drives my instinct to shield them from abuse. But there's no denying that beautiful animals, especially babies, stir something in people, and not all animal abuse is equal in the eyes of the public. It's not just explained by species or cultural preferences for certain animals, but also by the needlessness of the abuse and the perceived motives of the abusers. That's partly why puppy mills elicit more of a reaction than the use of mice or rats in animal testing, or why dogfighting is more widely reviled than killing animals for food.

The beauty and vulnerability of these newborn seal pups, so defenseless against their slayers, was only the beginning of the case we had to make on their behalf. There was the sheer carelessness in the manner of execution, taking a scene that would be ugly and violent in the best case and adding an extra measure of viciousness and cruelty. Then there was the colossal wastefulness of the hunt, leaving hundreds of thousands of small carcasses behind after the skin was peeled off. Finally, there was the vanity that inspired the whole enterprise—all of this suffering and bloodshed for a product nobody needs.

Americans decided they didn't need seal fur long ago. The United States banned the killing of seals in 1972 when President Richard Nixon signed the Marine Mammal Protection Act. So Americans can see Canada's hunt not only as cruel and unnecessary, but also as beneath our own national standards. For me, one of the keys to animal advocacy is to identify existing moral and legal standards and then ask simply that they be applied in a consistent way. This appeal will carry you a long way in the case against the slaughter of seals, since if Americans were doing the killing, it would be a criminal offense.

They didn't have a law in Canada banning the sale of seal skins, but it turned out that Canadians weren't buying them either. That's left the industry entirely dependent on foreign markets and undercuts its own endless refrain that what they do to seals is nobody else's business. When you ask another country to import a product, you have literally made the matter that country's business and invited others to make judgments about all that went into it. They have based the whole enterprise on the willingness of foreign markets to accept seal skins, and still they're shocked and insulted when, for moral reasons, those countries decline the offer. The sealers just want the rest of the world to pay up and shut up, to buy baby seal fur and keep their opinions to themselves. But that's not how it works, and they cannot have it both ways.

The first blows in the seal hunt are struck in the area where I had been standing, near the Magdalen Islands. The ship's captain pulls up to an ice floe, like a boat pulling into a dock, and sealers jump off—clubs in hand and knives sheathed on their hips. The seals try to scurry away as the men run toward them, and sometimes rise a few inches from the ground to deliver a desperate, futile little snarl. The men swing at them, striking them with sharp blows to the head with a device called a hakapik, which at the top has a blunt metal face on one side, and a curved ice pick on the other. Once the sealer has struck a pup one or more times, he chases after the others, methodically attacking one after another. After all the seals in sight have been bludgeoned, he often goes back and starts the skinning process. He either flays the seal on the ice, cutting the pup down the middle and peeling off the fur, or he does a half spin of the hakapik and then gaffs the seal's throat before dragging the creature onto the boat. There, the fur is peeled off, and the carcass dumped in the water or hurled back onto the ice.

Often, the sealers do not deliver lethal blows. I've had to watch hours of seal bludgeoning, as recorded by HSUS staff who have braved the threats of sealers to document this horrific cruelty. The sealers are in a frenzy, running on the ice and trying to kill as many pups as possible before any can get away. Because the seals squirm around frantically, the blows that fall on them do not always land where intended. I've watched seals in a pool of red, gasping for air and trying to raise their heads out of the blood. In 2001, a team of veterinarians observed the hunt and then did postmortem examinations of the pups. Their report concluded that the seal hunt causes "considerable and unacceptable suffering," noting that in 42 percent of cases, the seals did not exhibit evidence of cranial injury sufficient even to guarantee unconsciousness when the knife was applied. In other words, many may have been skinned alive. In 2007, HSUS escorted an international team of veterinary experts to the seal hunt. Their report noted a widespread disregard for the regulations by sealers and a failure to enforce the law by authorities, concluding that both clubbing and shooting seals is inherently inhumane and should be prohibited.

All of which explains why Canadian authorities work so hard to keep the HSUS camera crews as far away as possible, and why the sealers menace and bully anyone who gets too close. But those scenes were recorded, and the cries of the seals have been heard across the world. The response of citizens has been to demand that their governments have no part in this cruelty. And that's exactly what's been playing out. In 2009, HSUS helped convince the European Union to ban the import of any seal parts into any of its twenty-seven member nations. That's far more consequential than the whitecoat ban of the 1970s, partly because the EU is so much larger now. Canada's other NAFTA partner, Mexico, has also banned seal imports. And the Russian government, in a decision that surely got the attention of Canadian sealers, has banned its own baby seal hunt, which had claimed the lives of up to thirty-five thousand seals a year—described by Vladimir Putin himself as a "bloody business."

All of this, combined with the long-standing prohibitions in place in the U.S. market, has left Canada with few options. In 2010, the allowable kill, known as their official quota, was 388,000 seals, but because of rock-bottom prices and surplus pelts piling up in warehouses, the sealers killed just a fifth of what they were allowed—some 66,509 pups. It was the same story the year before, and there is little prospect that it's going to turn around. And for all of their defiance, you won't find anyone who calls this an industry with a future. Just try to imagine any investor of any type anywhere in the world putting a dime into the sealing industry.

As it is, there are only about fourteen thousand licensed sealers in the whole country. And even when the pelt values are highest, only about five thousand or six thousand actually participate in the hunt. Although it's a commercial activity, for the guys who do it sealing offers a minor, off-season supplement to their regular income as fisherman. Sealing is tough on their boats because of all the heavy ice. And when you consider all the costs of fuel, repairs, and maintenance, you can understand why only one of every three licensees even bothers. When the furs drop in price, the whole thing turns from a source of extra income into a costly hobby.

If the free market alone were guiding events here, no one at all would head out to the ice floes for the killing. For Newfoundland sealers themselves, the industry generated about $1.2 million in gross revenues for 2010. That sounds like a good year only if you compare it with gross revenues in 2009, which came to less than $900,000. Over the last decade, pelt prices have gone down from $30 or $40 a skin to just $15 or $20 today—which includes both the fur and the oils extracted in the processing, which are used for commercial lubricants, animal feed, and omega-3 supplements. So on the financial plus side of the modern sealing industry, we're looking at total gross receipts of about $1.2 million a year. To put these numbers in perspective, mining in Labrador is a $2.5 billion industry, and the harvest for fish and shellfish approached $450 million.

If those are the pluses, what's the financial downside, apart from the operating costs borne by the sealers themselves? Well, for starters, there is the cost of deploying the Canadian Coast Guard for six or seven weeks, and these costs, of course, are borne by the taxpayers. It's the Coast Guard that breaks the ice and allows sealers to get into the nurseries. This federal agency is also there to "monitor" the hunt, which in practice seems to mean hindering efforts by the press and others to document the slaughter, and to conduct search and rescue for the sealers themselves when they get into trouble in these ice-filled waters. All of this costs the people of Canada from $3 million to $5 million a year—so right there, the public costs are already three to five times the business revenues in recent years.

Along with assistance from the Coast Guard come tens of millions of dollars that sealers have received over the years in public subsidies, which take many forms. The government withholds a full accounting, but the ones we do know about are considerable. There are direct subsidies for seal products, grants and loans to processing plants, and investments in product marketing and development, so that all those skins have somewhere to go. And then there's all that lobbying and diplomacy by Canadian government officials, who shuttle between Ottawa and the capitals of the world pleading the case of the sealing industry.

It adds up to millions upon millions of dollars every year in public money, all to preserve an industry with gross receipts of a million dollars and sometimes less. And when you factor in the sealers' own individual costs, their actual profit is just a fraction of receipts. For every baby seal they club to death, they're pocketing only a few bucks—if they clear any profit at all.

And this doesn't even count the spreading economic toll of the international boycott of Canadian seafood—provoked entirely by the furor over the seal hunt. Canada exports more than $3.5 billion a year in seafood products, with two-thirds of that going to America. It's hard to measure the precise economic impact of the boycott that HSUS launched in 2005, and which has now spread to Europe. But with thousands of food retailers, restaurants, and other food service providers having joined our campaign, there is no question that Canada's seafood sales abroad have been reduced by tens of millions of dollars, if not hundreds of millions. Since the boycott started, the value of snow crab exports—which in 2004 accounted for half of all Newfoundland's fish exports—has fallen by $900 million.

Any way you look at it, the hunt is a net loser for everyone involved, and above all for the fishing industry. Yet the hunt goes on year after year. And if it's not economics that drives it onward, then what's left but foolish pride and political pandering? The Atlantic provinces resemble what Americans call "swing states," with all three major parties—the Liberals, Conservatives, and New Democratic Party—competing for votes. And though national polls invariably show that most Canadians would gladly be rid of the hunt, in this locale protecting the seal hunt is the only acceptable position.

Even with lucrative shell fisheries in Newfoundland now, the sealers cling to the hunt as a traditional way of life, and a rational analysis of the economic situation is wasted on them. It was overfishing that caused a decline in a number of stocks, but settled opinion holds the seals responsible. In reality, the seal issue is just a proxy for a whole assortment of misfortunes, frustrations, grievances, and resentments—all of which the political class is skilled at exploiting and reinforcing. So the subsidies flow, and the politicians keep their act going with loud scorn for the opinion of outsiders, calculating that local votes count for more than global disapproval.

Yet, at some point, the angry posturing will have to give way to common sense—to an honest accounting of seal hunting and of the gathering strength of the seafood boycott. As one Canadian writer has put it, the sealers and their political patrons "have as much chance of stemming this tide as Germany did of stopping the Allies after D-day. The battle is lost. But because of ideological fanaticism they keep fighting, secure in the delusion that the Canadian taxpayer, like the cod, is an inexhaustible resource that will forever fund this foolishness."

As in other such clashes over the fate of animals, however, the sealers may find that what they now fear as a bitter defeat will actually bring a better day. The seal hunt is a relic of the old economy of North America, and a willful clinging to it is the surest way to remain stagnant and to shut off new opportunities. Times change, and so do people's hearts. And these communities, including the sons and daughters of today's sealers, may one day discover that the path to progress need not involve liquidating wildlife and destroying nature. More than ever, just the opposite is true. For every one customer who will buy a seal fur, there are at least a hundred people who would pay good money to witness the serene beauty of the nurseries of the North Atlantic. Seal watching and seal slaughter, however, can never go together. They are incompatible, both morally and economically. The people of Atlantic Canada, and ultimately all Canadians, must choose one or the other. The continued slaughter of seals is a shameful, sorrowful thing, and unworthy of an otherwise fine nation.

# Part III

*Building a Humane World*

# CHAPTER SEVEN

# Cruelty and Its Defenders

My CHILDHOOD HOME STOOD at the intersection of a busy street and a not-so-busy one—a place plenty dangerous enough for a dog off leash, and especially if no one's looking out for him. I'll always regret that as a family we didn't dutifully abide by that rule. That's how one summer day turned tragic for us when I was eight years old.

My first dog, half poodle and half beagle, was named by my older brother for the Athenian statesman Pericles. It was a big name to live up to and didn't quite fit our little guy. Pericles was an always happy, go-along sort of dog, not an alpha male as the august name would suggest. It was no surprise, then, that one afternoon Pericles was out and about, tagging along behind another dog. The lead dog stepped into the busy street by our home, and Pericles, always the loyal soldier, fell in right behind him. The big dog made it across, but our little Pericles didn't. A car hit him straight on, and he died there on the road.

My family shielded me from the details as they relayed the grim news about my best friend. I was devastated, perhaps because I was just old enough to grasp what had happened, but still too young to be prepared for such sorrow. I bawled for hours—sitting down with my back against the garage wall on our makeshift basketball court, head buried in my folded arms, chest heaving, and my shirtsleeves absorbing the tears. The feeling stayed for a long time.

I now know that grieving after losing a pet is not unusual. It would be odd to have any other reaction. We feel empathy and love for the dogs and cats with whom we share our lives. We treat them as members of our families, and that means dealing with all they bring into our lives—the good and the bad, including the inevitable loss that comes with their shorter life spans.

Even if their passing is not premature—through an accident or sudden illness—it is still wrenching and painful because we see their entire life cycle unfold. We gleefully watch their frolicking as puppies or kittens, see them mature into their prime, and then in a matter of years their bodies start to decline, and too soon they slow down and begin to fail. In many cases, we are ultimately confronted with an awful choice: we keep on loving them but watch them face the frailties of old age, or we consult with a veterinarian and decide to end their suffering.

That's a moment we never forget. When I was a teenager, my parents and I stroked and gently whispered to Brandy, our Labrador mix, as the vet injected a solution into her veins. She trusted us to the end, but her plaintive eyes and pinned-down ears told me she knew that something serious and very different was happening. She could see it in our eyes, despite our vain efforts to pretend this was just another routine visit to the vet. After it was all over and we left the vet's office, we silently got into the car. We took the longest ten-minute drive of our lives back to the house. When we got there, we felt emptiness—in the house and in our hearts. This little life, such a big presence in ours, was gone.

We all felt incredible loss, but it was more than just the grief of knowing we'd never see our friend again. Our sorrow was compounded by having seen her suffer so much. We could sense her discomfort and even her fear that her body was breaking down. As self-aware people, we put ourselves in her place and felt her pain. We felt empathy.

That empathy is the root of compassion, and it inspires some of the greatest works that are done in this world. We feel the greatest empathy for those we love, but empathy does not require an intimate connection. Our religious and secular traditions call on us to care for the unknown poor in our communities, and the hungry and ill everywhere. Many of us donate funds or volunteer to ease the hurts of others or to better their circumstances, no matter if the victims look different or live halfway around the world. Countless nonprofit organizations are alive with this spirit, and industrialized nations give billions of dollars in aid to answer human needs in every nation.

But do we extend this same generous spirit to animals? It's not hard for anyone to understand my distress at Pericles's sudden death, or our anguish at Brandy's painful decline; they were my loyal friends, eager to please me and a source of joy and companionship. But what of the animals we never name or know—the stray dogs and cats at the local shelter, the animal victims of disasters like Hurricane Katrina or the Gulf oil spill of 2010, the seal pups on the ice floes of Canada, the sows imprisoned in crates on an Ohio factory farm, the zebras confined on a Texas hunting ranch? That's a more complicated question, but still a relatively easy one to answer. Of course we should care about the suffering of these creatures, and all the more so when their afflictions are laid on them by human beings. Almost to a person, we would care and hurt for these creatures if we saw their suffering up close, just as we care for the animals we know and love as our pets. Often the only difference—the only thing that dulls or quiets our empathy—is the simple fact that they are far away and out of sight. And the people who profit from harsh and cruel practices want to keep it that way.

## The Moral Scale and Humane Living

FOR ALL THE VARIOUS theories of rights concerning animals, the cause of animal protection appeals ultimately to our sense of fairness, to our capacity for mercy, and to standards of conduct that are ancient and universal. When we read in Proverbs that "the righteous man regardeth the life of his beast, but the tender mercies of the wicked are cruel," that's the language of humanity speaking, and it's heard in every culture and every time. In fact, far from advancing novel theories and new agendas, often the greatest challenge of the animal-welfare movement is to remind people of things they already know to be true—that to mistreat any animal is beneath us, that cruelty of any kind is dishonorable and inexcusable, and that we all have duties of kindness and self-restraint in the treatment of our fellow creatures.

But believing in general principles is one thing; putting them into practice is another. Making choices to reduce our impact on other creatures' lives is not a simple proposition, given our traditional dependence on animals for food, clothing, and recreation. The use of animals was long ago built into our way of life. And making adjustments and changes, even when past necessities are now simply options, and the basic needs of former times are the minor conveniences of today, still seems like asking a lot. And for many people, the big step comes when they make the connection between the animals in their own lives and animals in general, and empathy and moral concern begin to extend outward.

As always, there are challenges in reconciling moral concern with our daily practices, and the most problematic is our attachment to a meat-based diet. Most of us are taught as children to eat animal products to maintain our health and strength. Meat, milk, and eggs are everywhere—they are staples stocked in our refrigerators, on the shelves of our supermarkets, and on the menus of restaurants. At almost every turn, it's presented as a given that you cannot get by without meat. And if a person is not willing to question that, or allow moral considerations to influence their choices in what to buy and eat, then often the progression is reversed: food preferences are allowed to override any further moral inquiry. Instead of thinking clearly and objectively about our real options in diet, and our real obligations to avert cruelty and suffering, many simply go along with their impulses, submitting in effect to a tyranny of whim and appetite.

In endless forms, advertising does its part to silence any serious thought about the treatment of animals, whether it's "Beef: It's What's for Dinner," "the Incredible Edible Egg," or California's "happy cows." So does the federal government. When I was a kid, the federal government posted a diagram of the "Basic Four Food" groups. The message was you'd be unhealthy if you did not liberally consume meat, dairy, and eggs, since they represented half the schematic. Marion Nestle, professor of nutrition at New York University, has documented how these guidelines were formulated after heavy lobbying by the industries themselves. And though subsequent refinements have called for more fruits and vegetables in the diet, the Food Pyramid still calls for generous heapings of meat, milk, and eggs.

At the same time, the American Dietetic Association tells us that meat, eggs, and dairy are not needed to maintain good health. On the contrary, nutritionists note that these foods, consumed at today's per capita rates, contribute to serious health problems from heart disease to cancer. In a sense, the science of nutrition, by removing its imprimatur from meat or other animal products, has opened the door to moral questions a lot of us would rather not entertain. If

these products are not vital to our development and health, and in
some cases are plainly harmful, then what justification is left except
for taste and pleasure? Eating meat may be a powerfully ingrained
and familiar habit, and it may provide a taste and texture that satis-
fies the palate. But when it comes at a cost of mass confinement and
endless misery, at what point does the moral scale tip in favor of the
animals?

In the same way, fur coats were for ages a necessity in some
regions. What are fur coats and trim today but a shallow status
symbol? Household products and cosmetics tested on animals are
everywhere, and the testing is not only painful but endlessly re-
dundant. At a time when alternatives to testing on live animals
are available and effective, can't we finally call an end to this needless abuse of millions of rabbits, dogs, and other animals? There are
so many other forms of violence against animals that long ago lost
whatever justification they might have claimed. Yet somehow they
persist, and whatever sustains these practices, it is not reason or conscience. The moral basis for these harsh uses of animals depended on
human need, and when that has passed away, all that usually remains
are excuses.

In a marketplace with more choices than ever, there are alternatives. But change is often inconvenient and threatening, and none
of us likes to feel as if we're being told what to do. Cruelty-free food
and clothing, moreover, can be harder to find or more expensive.
Eating animal products from more humanely raised animals or an
entirely plant-based diet requires a shake-up in our shopping habits,
along with a higher degree of discipline and personal commitment.

A new diet can be off-putting to others. It can make family
members and friends feel awkward, as if a moral judgment is being
made about their own choices. And it can make us feel awkward,
too, since we don't want to be seen as preachy or extreme. At that
point, we have two rational impulses in conflict—a social concern
for animals, and a social instinct for conformity. It's easy to oppose

animal cruelty in principle; wringing cruelty out of our lives is a much bigger challenge in practice.

I was in college when I decided to stop eating meat. I'd seen images of factory farms and gathered more details about the types of agriculture that had largely displaced traditional farming. It wasn't done as a fad or on a whim—this lifestyle change required some measure of sacrifice, and I took that step because I thought it was the right thing to do. I also hoped my decision might prompt serious-minded people to consider factory farming as an important moral question. Just as others had caused me to face the matter, I hoped to set an example of my own.

Whenever I explained my rationale, even when I did my best to be measured and inoffensive, the first instinct of my classmates was defensiveness, especially if the subject came up in the dining hall—where the food on our plates put the moral question literally right in front of us. "You aren't eating meat, but you are wearing leather shoes or a leather belt, aren't you?" Or "Yes, you are eating a veggie sandwich, but the combine that harvested those vegetables chews up rabbits, mice, and other small animals." And without fail, "You are wearing a cotton shirt—what about the pesticides that were applied to the fields, poisoning countless birds in the process"?

They were all arguments not to act, not to care, and people can be awfully sharp and resourceful debaters when that's the ground they're defending. At some level, however, their arguments had a certain logic. Almost every action we take has some downstream effect on animals, even if it's indirect and unintentional. If the aspiration is to do no harm through our diet and other consumer choices, then that's an impossibly high standard—it's unachievable and impractical. By leading a normal life—eating, hopping in a car or flying in a plane, buying clothes, or flipping on a light switch—we have an impact on animals, though it often goes unseen.

What struck me about my college classmates is that they often invoked these arguments not to enlighten me, or to help me sort through the most challenging moral circumstances. The strategy was simple: if they could poke a hole in my thinking, or find some inconsistency in my behavior, they could absolve themselves of any moral responsibility. In practice, it was a case for doing nothing, merely because one could not do everything. And if we followed that kind of reasoning in any altruistic enterprise, we would despair immediately and never even try to take that crucial first step. Even a lifetime's worth of compassionate work will never be enough, by this measure, and so what is the point of even trying?

This all-or-nothing attitude had another serious flaw, and the smartest of the skeptics understood as much. If our inability to be perfectly consistent absolved us from any moral responsibility toward animals, then by this logic we had a green light to do just about anything to them. Not only were factory farms and fur production okay, but so too were puppy mills, dogfights, and even horrific abuses of pets. A world with these reductionist rules had no limits. The sophistries tossed off in the Yale dining room are equally available to the worst abusers, and liberally employed to discourage reform of any kind. If it's all or nothing, nothing will always win. And by that logic, no objective standards are left to prohibit even the most extreme forms of cruelty.

Of course, all of my college friends, if the question were put to them, would concede that cruelty to animals was wrong and that perpetrators of malicious cruelty should be punished by law. They wouldn't really take their insistence on perfect moral consistency to its logical end—because it would require defending all sorts of appalling and vicious things. When it came to their own pets, they opposed cruelty strenuously, even if other such wrongs could not be eradicated entirely. Whatever arguments they made, they generally shared society's conviction that animal cruelty is a vice to be prevented and punished.

Once this dawned on them, their instinct was to pivot and attempt to distinguish one form of animal use from another. They tried hard to reconcile their basic intuition that animal cruelty is wrong with society's acceptance that eating meat, wearing leather, and hunting is right. "Maybe anticruelty principles apply only to pets, and not to wildlife or farm animals," they ventured. Or "perhaps it is not cruelty if it's done for a compelling social purpose— like controlling wildlife populations, producing cheap meat, or in finding a cure for a disease." In other words, the ends justify the means, and the material well-being of a society and our personal comfort depend upon these kinds of unpleasant trade-offs. In each case, they sought a comfortable moral resting place between two beliefs that seemed intuitively correct—objecting to cruelty and holding on to their habits involving more routine uses of animals.

Here is the greatest challenge in the cause of animal protection: the problem is not disagreement with the broad principle that animals deserve kindness and protection, but in the consistent application of that truth. Observance of anticruelty ideals demands something of us, especially in a world where the mistreatment of animals is rampant and, therefore, moral opportunities abound. It is challenging at times to live by the highest application of these principles, and we may continually struggle with the boundaries of ethical conduct. But those difficulties should not prevent us from doing anything at all. Here, as in other matters of conscience, we should be wary of arguments that try to smother doubts and advocate a course of action that requires no reflection, no inconvenience, and no sacrifice.

## Rhetorical Camouflage: Debating Opponents of Animal Protection

IN THE YEARS AFTER graduation and finding my way into full-time advocacy, the question of animal protection became more than Ivy League table talk. Now that I was campaigning nationally on behalf of animals, my actions had real-world consequences. My goal from the start was not to prescribe all personal behavior relating to animals, but instead to prevent and ultimately end the worst abuses.

In my new role, I encountered the many actors on all sides of the debate about animals. There were animal advocates loudly protesting outside fur stores, and others lobbying their elected officials. There were hunters and factory farmers defending everything about their industries, while a handful among them saw the need for some reform. And then there were those not nearly as engaged— members of the general public who had become increasingly familiar with the issues but was still sorting through their own conflicted views and habits.

Among animal advocates, there were more women than men and more whites than blacks or Latinos or Asians. There were more college graduates than working-class folk, and more people from the cities than from the country. More often than not, advocates started with a keen affection for their pets that led to a broader awareness of animal cruelty. But there were still animal advocates of every background, with diverse beliefs. Some were particularly passionate about companion dogs or feral cats, others about laboratory monkeys or farm animals, and still others about wild mustangs or endangered wolves. Some were vegetarians, while others minimized their meat consumption or tried to source it from more humane farms. But they shared common goals: to live with greater ethical consistency, to raise society's awareness, and to make life more bearable for animals.

Like advocacy groups of most every kind, their mission is to draw public attention to great wrongs and needs. The public, by definition, was not nearly as engaged as the activists, but neither were they hostile to the cause. If asked, most people would agree that cruelty was wrong and show no sympathy for animal fighting and other malicious acts toward animals. A majority, if given the chance, would vote for a ballot measure to ease the suffering of farm animals or to ban steel-jawed leghold traps or canned hunts. Some even boycotted fur, meat and eggs from factory farms, and cosmetics and household products tested on animals. Others contributed money to animal charities, especially during times of heightened awareness, as we saw after Hurricane Katrina or the revelations about Michael Vick's dogfighting.

Then there were the die-hard opponents. My work has brought me into contact with trophy hunters, bear baiters, seal clubbers, factory farmers, commercial breeders, and dogfighters. I debated them on television, on radio, and on college campuses. I appeared at legislative hearings, with the room packed full of cockfighters and trappers. I saw them out in the field—on the ice floes, in the hunting fields, or on their factory farms. I addressed hundreds of outdoor writers and farm broadcasters at their conventions. And I subscribed to the major hunting, cockfighting, and farm publications, acquainting myself with their thinking and their strategies. Typically, they were angry about the criticisms we leveled and they were overtly hostile to the goals of animal protection. Through the years I've heard almost every imaginable argument against the ethic of animal protection, or at least against reform when it came to their brand of cruelty.

I've also seen a lot of cruel practices grow from bad to worse. The agribusiness and hunting industries have strayed farther and farther from mainstream sensibilities, treating animals like a harvestable crop, and using technologies that destroy any claim they ever had to benign husbandry or sportsmanship. More hunters are using laser sights, radio tracking collars on dogs, robotic ducks, automated corn feeders, night-vision goggles, and high-fenced, "guaranteed kill" properties that make sport hunting ever more unfair. And factory farmers are using severe confinement systems, fast-growing breeds, hormones and antibiotics, genetic engineering and cloning, and other methods that show no respect for animals' dignity as living creatures. As the public learns of these routine practices, they're growing increasingly skeptical of industry leaders' claims to respect animal welfare, and increasingly supportive of laws to restrain their conduct.

Our opponents have sensed this shift in public opinion, and they've been trying to adapt. Two decades ago, these hardened political opponents took a disdainful attitude toward our movement. They viewed animal advocates as meddlesome outsiders and treated us with contempt, arguing that animals had no rights, that humans could act more or less as they pleased, and that, in any case, what they did with *their* animals was nobody else's damn business. They especially disliked what they perceived as our air of moral superiority. This new movement was, as they saw it, a threat to their lifestyles and livelihoods.

But as the years passed and the momentum for reform increased, the dismissive tone gave way to a more nuanced rebuttal, and brusque denials gave way to comforting euphemisms. Realizing that the general public believed animals should be treated humanely, the savvier spokesmen for industry began arguing that any business using animals had an interest in their welfare, because sick and maltreated animals are less productive and therefore less profitable. They drew a distinction between "animal welfare," which they professed to support, and "animal rights." Giving their rhetoric a makeover, factory farmers suddenly became advocates of "science-based animal production," hunters suddenly did less killing and more "harvesting," and trappers and sealers were now the exemplars of the "sustainable use" of wildlife.

Animal advocates, according to this reframing of the debate, might be well intentioned, but they were hopelessly naive and veering toward the extreme. Farmers, researchers, and hunters didn't hesitate to play the fear card. "Do you know what your life would be like if the animal advocates governed?" cautioned Wesley Smith, a full-time critic of animal protection. In the picture he paints, we'd all be eating tofu and wearing hemp clothing, and the nation would witness a decline in farm jobs, an increase in nuisance wildlife problems, and a halt to medical progress.

Frederick K. Goodwin, an outspoken defender of animal use, told Smith in 1998 that radical animal rights groups have "hijacked" the broader cause of animal welfare. "The cruel irony is that they have drained funds from traditional welfare activities, programs such as neutering pets and stopping cruelty," he said. "All of the things which have really helped animals historically are now run on shoe strings." Goodwin associated himself with "stopping cruelty," but like many others who serve up this bromide, he never lingered long on the subject or offered much in the way of specifics. He's against cruelty in general, but he declines to identify, much less to work against, any particular cruelty. We're left to assume that cruelty relates only to the mistreatment of pets, and even then only in random, egregious cases such as clubbing or starving a dog or cat to death. The one step that industry representatives never take is to consider cruelty problems of a more systematic and institutional kind.

Cruelty, as they use the terms, is defined entirely by motive, rather than by actual conduct and the objective result. If a high-minded motive is offered for any practice—done in the name of science, progress, or even profit—this is all-absolving, and the matter is considered closed to further moral inquiry. The incoherence of this position is especially clear in the case of the very animals Goodwin and others say we should focus our attention and concern upon—dogs and cats. The people who supply dogs to research facilities—known to the USDA as Class B dealers—pick up "random source" dogs and respond to "free-to-good home ads," selling off people's pets to laboratories and into painful and often lethal experiments. Same animals, same fear and pain, but because it's done at the behest of a laboratory, then somehow the dogs' fate is not supposed to concern us anymore. This is a terrible and deceitful abuse of companion animals, and industry apologists like Goodwin simply refuse to open it up to moral examination.

We have learned so much about animals in our lifetimes, and it confirms what common sense told us all along: that animals have the capacity to suffer, and motives and settings do not alter that reality. What is the objective result of relegating a chimpanzee to solitary confinement for decades, after deliberately infecting the animal with a crippling disease? How about cramming a breeding sow into a cage barely larger than her body—condemning her to a life of unrelieved confinement above manure and urine pits that fill the air with ammonia and sear her eyes and lungs? Or how about an animal caught in a leghold trap and languishing for hours or days before the trapper checks his sets and then stomps or clubs the victim to death? By any reasonable measure, these actions constitute the intentional and knowing infliction of pain and emotional torment for no good reason—and isn't that the very definition of cruelty?

It's not that animal advocates have broadened their claims of what constitutes cruelty, but rather, the detractors' views have become too narrow, selective, and arbitrary. A classic example of this capriciousness is found in the federal Animal Welfare Act, in which the research industry succeeded in defining "animal" to exclude birds, mice, and rats, who are surely animals by any other definition and who happen to make up 95 percent of all animals used in research. If they're not animals, as the experimenters see it, then what are they? And what kind of scientific mind disregards such a fundamental truth out of sheer convenience, just to spare themselves the responsibility of compliance with even minimal standards of care?

Consistency, to say nothing of scientific integrity, requires that we address cruelty of every kind—both the random abuses that everyone sees for what they are, as well as those still cloaked in the false respectability of business and industry. Viewed in this light, so many of today's controversies over the treatment of animals are a rearguard action by animal-use industries to prevent the public from making increasingly obvious moral connections—between those individual acts of cruelty and the systematic ones. The industries and trade groups all have their various arguments, displaying what Cleveland Amory once called man's "infinite capacity to rationalize his cruelty," but their whole cause ultimately amounts to a campaign of denial. And it's part of our mission to keep making those connections, overcoming their denials with an appeal to objective standards and universal values.

One thing you have to give the opponents of animal protection is that they know how to organize. No institutional abuse of animals is without its allies and lobbying apparatus. You expect the trophy hunters and the factory farmers to have their Washington offices and lobbyists, but even cockfighters have a national trade group. The animal-use industries have also sought in recent years to invent a new public persona, financing front groups with sober and civic-minded names like Americans for Medical Progress or the Center for Consumer Freedom. These groups have staffs, offices, stationary, websites, money, and everything else that a real public-interest organization has—except members. In most cases, they are entirely the creation of animal-use industries, and they exist mainly to present the illusion of mainstream opposition to animal protection.

I've noticed something else about the names our opponents take for themselves. It used to be they formed groups with unambiguous names like Putting People First, which really meant putting people first, second, and third, and relegating animals to the absolute bottom of the heap after people had their way with them. But names like that went out with the blunt, aggressive rhetoric we used to hear. Nowadays, the casual observer of animal-welfare controversies has a little more trouble figuring out just who the opponents of reform are. The rodeo folks and horse slaughter people are the

Animal Welfare Council. The bear baiters we went up against in one ballot measure called themselves the Maine Fish and Wildlife Conservation Council. And in the fight over Proposition 2 in California, the factory farmers were Californians for SAFE Food. In their continuing quest for respectability and decriminalization of their hobby, the cockfighters go by the name of the United Gamefowl Breeders Association, and in Michigan hunters in the debate over shooting doves preferred to be called the Citizens for Wildlife Conservation Committee.

When the whole mission of your group is to trap, bait, shoot, fight, confine, or otherwise torment animals, it's hard to find a name with just the right ring. And when a group cannot bring itself to state its real purpose in honest and plain language, that tells us most everything we need to know about the merits of its cause. For us at HSUS, this rhetorical camouflage can make things a bit complicated in defining the issues and the parties involved. On the other hand, when your opponents' strategy is to look and sound as much like you as possible, and to wrap themselves in the language of animal protection, you know you're making progress.

## Collateral Damage and the NRA

OPPONENTS OF ANIMAL PROTECTION are never angrier than when someone they assumed to be one of their own turns against them, usually by simply upholding the very rules and standards that animal-use industries profess to support. This is how a California wildlife official ran into trouble a few years ago with the NRA— the group that talks of liberty but defends license.

At the time that Governor Arnold Schwarzenegger appointed him to the state Fish and Game Commission in early 2007, R. Judd Hanna seemed like a safe, down-the-middle choice. Hanna was a lifelong Republican, a rice farmer, and a real estate developer—

hardly the credentials of a radical environmentalist. And yes, perhaps most important for any appointee to the Fish and Game Commission (the state's five-member panel that sets the rules for fish and wildlife policy), he was an avid hunter. Even though just 1 percent of Californians are hunters, it's been almost an unbroken political tradition for governors to appoint only hunters to the commission.

The NRA and other hunting groups were accustomed to calling the shots at the commission. Perhaps because he was a newcomer, Hanna still operated under the assumption that he was there to serve only the public interest, and that this had something to do with protecting wildlife. Early in his term, he started showing a high degree of independence, and the hunting groups began to take notice.

When the commission began to examine the effect of lead shot on the endangered California condor, it didn't take long for Hanna to stir controversy. He circulated 160 pages of documentation to his fellow commissioners, including scientific studies and even information from the Audubon Society, that provided compelling evidence that lead shot discharged from hunters' firearms was poisoning and killing condors. He underlined passages in the packet and even made notes, "much like I did when I was in college," he told me.

At the time, there were only about 150 condors in the wild—about half of them in California, most in the central part of the state. Another 150 lived in accredited zoos, which had been running a captive breeding program that started with a few founder animals removed from the wild in the early 1980s in a desperate effort to save the species.

Hunters proved the greatest threat to the condors, though not by taking potshots at these prehistoric-looking birds. There is a wounding rate for hunted deer, wild pigs, and other big game—perhaps 10 or 20 percent—shot but not retrieved. They are left instead to face a prolonged death, lasting hours or days. Condors eat carrion, so they'd ingest lead pellets as they fed on the carcasses. A 2007 scientific study by the Peregrine Fund found that lead poisoning was "the most frequently-diagnosed cause of death among free-ranging California condors." Cynthia Stringfield, a Southern California veterinarian who advises the U.S. Fish and Wildlife Service's Condor Recovery Program, said twelve of the fifty-one condor deaths that have occurred since the endangered birds were re-released were definitely the result of lead poisoning. She believes the actual death totals are much higher. "I'm sure the mortality rate from lead poisoning is more than half."

There was an easy fix: require hunters to use a nontoxic shot, like copper or bismuth. The move had precedent because the U.S. Fish and Wildlife Service, after years of equivocation, had banned lead shot for hunting ducks, geese, and other waterfowl. The spent shot had accumulated in wetlands, and bottom-feeding birds would incidentally pick up the shot in the normal course of feeding. The NRA had bitterly fought the imposition of a lead-shot ban, claiming it would all but eliminate duck and goose hunting, but the evidence was so overwhelming that even the federal wildlife agency knew it had to act. After all, the Service protected waterfowl, in part, so that hunters could shoot them, not so they could poison them with spent lead.

As it turned out, hunters easily made the switch to nontoxic shot. Yes, it was slightly more expensive, but even from their standpoint, there would be an advantage to using it: reducing the lead in wetlands and other waterfowl hunting areas would save millions of birds, leaving more to shoot.

But the federal lead-shot ban still riled the NRA. And when it became a state issue before the commission, the group rallied its members—against the ban in general and Judd Hanna in particular.

The NRA viewed Hanna as an apostate, who had betrayed the hunting fraternity by openly advocating for the ban. The NRA arranged for a letter to the governor from thirty-seven Republican state legislators—a good share of the party caucus—expressing "grave concerns" that Hanna was not "impartial relative to his participation in the commission's decision-making." The day after the governor received the letter, the director of the Department of Fish and Game requested Judd Hanna's resignation. Sensing this was more than a "request," he left the commission after just nine months of service.

His offense was simple: as a member of the commission, he worked to protect wildlife and he dared to defy the NRA. His forced departure sent a signal to the other sitting commissioners that they, too, had better know their place. As far the NRA saw it, the condors were nothing but collateral damage, and the same would go for any commissioner who got out of line.

As it turned out, the controversy over Hanna's ousting backfired on the NRA. The attack on the independence of the commission led the Democratic-controlled legislature to push through the legislation—over the objections of every Republican lawmaker—to ban lead shot in condor habitat. Feeling the public pressure and recognizing the scientific consensus, Schwarzenegger reluctantly signed the bill into law. The NRA had won the battle to oust Hanna, but lost the war. In the fight over policy on lead shot, they were now zero for two.

"The science is irrefutable," Hanna told the *Washington Post*. "There isn't a shadow of a doubt that lead from ammunition is the leading cause of death and illness in the California condor. Lead was already identified as a problem in 1987 when the last of the wild condors were captured. It should have been outlawed then, but with the birds out of the wild they had stopped dying."

In his letter of resignation, Hanna admonished the hunting lobbyists that they should have been the ones to lead the fight to eliminate lead shot. "We may be missing what could, perhaps, be our last opportunity to salvage not only the reputation of our hunting community, but also hunting itself in California," wrote Hanna. The

99 percent of Californians who do not hunt "will tolerate hunting providing it is done ethically and honorably. Poisoning the California Condor is neither honorable nor ethical."

IT WASN'T EXACTLY THE first time I'd seen the NRA defend an unconscionable practice. In fact, the very first protest I ever attended was aimed at ending something the NRA was defending as innocent fun. It was a big pigeon-shooting event, and along with some bad memories the experience left me with a much clearer picture of the opposition.

I had just returned to Yale from the far reaches of Isle Royale National Park and heard from a friend that a protest was planned on Labor Day in Pennsylvania against the nation's largest pigeon shoot.

The shoot was a fund-raiser for the local fire department and a cultural marker in the thickly forested, undulating Appalachian hills in the heart of coal country. It had been a tradition in Schuylkill County for decades—a big family event complete with beer and hot-dog vendors. They called it a pigeon shoot, though that hardly begins to describe the spectacle. To the delight of a cheering throng of local residents, shooters readied themselves in an open field. Dozens of crates were packed tight with terrified pigeons, and ropes were used to pull off the lids one at a time. As each lid was removed, two or three pigeons would flutter upward, and then just as quickly fall from the sky amid bursts of gunfire.

I couldn't fathom the cruelty and callousness of it all. Making matters worse was the crowd milling about and watching the maiming of these animals without pity. The spectators weren't directly shooting the birds; they were just there to laugh and drink, as if the killing of thousands of animals was a good excuse for a party.

My first thought was that there must be a better way to raise money for the fire department. My second thought was that the cause of animals was very much about people. The behavior of the adults was appalling enough. But the spectacle also featured "trapper boys," who would scramble out and stomp on the wounded birds, or twist their heads off—encouraged by more cheering from their proud parents and the other spectators. To see ten- and twelve-year-old kids conscripted in the crushing and killing was beyond belief. I was watching not just a massacre, but also a kind of moral numbing and indoctrination. Doubtless some of the children had felt a fondness for animals similar to what I'd felt as a boy, but this ritual seemed designed to squeeze every ounce of that feeling out of them. It was easy to imagine, fifteen or twenty years into the future, that these same kids would be the adults in the community—either down in the field shooting the birds, or reveling in the bleachers as others did the killing, perhaps with their own children dashing onto the field to do their part. I don't know which affected me more—the massacre right in front of me, or the cruelty-training-camp feel of it all.

The organizers were quick to claim that the shoot controlled the pigeon population. But given that they purchased the birds pre-captured from dealers throughout the Northeast, it was hard to take that claim seriously. On occasion, a shooter would misfire, and a hungry pigeon would escape and fly off, actually augmenting the local pigeon population. Nor did anyone eat the birds—the estimated five thousand killed birds were just taken to the local dump, some still alive. No shooter required a license, so they couldn't even claim that their activity contributed to the state's wildlife-management programs. You had to search high and low for a single good reason for any of this, and you'd still come up empty. The truth was the event was conducted because it was a tradition and the organizers and participants thought it was fun, and besides "they were just pigeons." It brought to mind Thoreau's line about people he saw shooting squirrels just for fun—that "the squirrel that you kill in jest, dies in earnest."

I naively thought that such a thing as I'd seen in Pennsylvania could not persist long in a civilized country. Adults in the local community would step in to stop this nonsense. It turned out that those adults were complicit, and I found myself returning year after year to protests that were never any match for the zeal of the participants. Each year, there were more protesters, but also more fans of the shoot. It had turned into a flashpoint in a larger culture war, with locals and gun enthusiasts defending their home turf and "way of life," and others just showing up because they found the ruckus entertaining.

With the organizers not budging, we turned to the state legislature. There, surely, lawmakers not beholden to this community would outlaw such barbarism. That, too, proved an optimistic assumption. It was during this process that I learned my earliest political lessons. It was hard for me to believe that a majority of elected officials in Pennsylvania, or in any state for that matter, could conclude that pigeon shooting was acceptable. But I soon figured out how they really looked at these issues: that it was not worth antagonizing the NRA.

Lawmakers, especially from rural areas, wanted to show off perfect scores from the NRA and to associate themselves with the right to bear arms and the right to hunt. Though target shooting could certainly continue uninterrupted if pigeon killing were outlawed, the NRA treated any limitation on shooting sports or hunting as a first step toward banning them all. "Pigeon shooting is an historic and legitimate activity steeped in tradition with many participants throughout the Commonwealth and around the world," read an NRA alert to its members. "For over one hundred years, shoots have been held in Pennsylvania by law-abiding, ethical shooting enthusiasts, hunters, and sportsmen who would not tolerate an activity that would constitute cruelty to animals."

Here again was an anticruelty bromide—just like Dr. Goodwin's above—without any rational application of the principle. These shoots bloodied and killed real animals—with their bodies pierced by hot lead. It wasn't done for any purpose but target shooting, and alternatives were aplenty. You couldn't even call this conduct hunting, since there was no licensing and the birds were captive. It was no more than a hollow imitation of hunting. In this case, the NRA was saying it was against cruelty, but it was doing everything in its power to protect the cruelty and keep it legal.

For a decade, the issue stalled in the legislature—with anti-shooting and proshooting lawmakers in a standoff, and the resulting inaction favoring the NRA. But in 1999, the Pennsylvania Supreme Court delivered good news. In a case brought by a Pennsylvania anticruelty officer and backed by the Fund for Animals, the court ruled that state humane agents could arrest pigeon shooters under the state's anticruelty law. That was enough to convince organizers of the Hegins event to call it off that year. There was a hitch, though. Other shoots, defying the threat of prosecution, quietly persisted—without the big crowds or beer and hot-dog vendors. HSUS and other animal advocates have continued to gather intelligence and provide the information to law enforcement, but a few shoots still survive to this day. Pennsylvania needs an unambiguous state law banning shoots, but the NRA continues to stand in the way.

If you were to take a survey of NRA members, most would probably favor an end to pigeon shoots. The group's leaders, though, rally members by playing the antigun, antihunting card. "National 'animal rights' extremist groups, led by The Humane Society of the United States, have organized and funded efforts in Pennsylvania and around the country to ban this longstanding traditional shooting sport," wrote the NRA to its members. "Make no mistake, this isn't just about banning pigeon shooting, but banning all hunting species by species."

By stirring up the hard-core sportsmen, hunting groups attract more donations, and lawmakers score with gun-rights activists. I've seen this same political dynamic again and again. It's one reason why HSUS has pursued ballot initiatives on hunting and trapping issues. Pennsylvania doesn't allow such initiatives, but about half the states do. By bypassing lawmakers so closely aligned with the gun lobby and placing reforms directly before the voters, we've helped to pass restrictions on abusive practices like the use of steel-jawed leghold traps and bear baiting. But there is no national initiative or referendum process, and the political dynamics I saw in Pennsylvania, and in California with Judd Hanna, can be seen as well in the Congress of the United States. The NRA has pretty much captured Republicans, with the exception of some suburban members of the caucus, and many rural Democrats, too. With this combination, the NRA has warded off efforts to halt the most reckless practices, such as commercial trapping on national wildlife refuges and even the shooting of animals in fenced enclosures in canned hunts.

No case better demonstrated the political challenge in dealing with the NRA than the debates in Congress over protecting bears. The organization has never been too fond of bears in general. For example, it fought off an effort in Congress to crack down on the trade in bear gallbladders, which poachers sell into the traditional Chinese medicine market for thousands of dollars apiece—a truly vile practice that has nothing to do with hunting. But the NRA really put its stake in the ground in the case of bear baiting.

The National Park Service and the Fish and Wildlife Service have strict regulations barring the feeding of any wildlife, including bears, because it is ultimately a threat to them and to visitors who might encounter a habituated bear. But the Forest Service and the Bureau of Land Management, America's two largest land managers, allow bear baiters to set out food piles on federal lands in states that allow the practice. The bait piles set by hunting guides virtually guarantee their fee-paying clients an opportunity to shoot a bear. "In a normal season we will go through 10 tons of pastries and about 8 tons of meat," boasts one hunting guide on his website.

Some guides will even burn honey to attract their targets from miles around. Others use "walk-in" baits, in which they load up an old horse or mule with food, walk the animal into a forest, shoot him, and then add his carcass to the bait pile.

In the fall, bears feed for up to fifteen hours a day to build fat reserves for a long period of dormancy. In the spring, bears emerge from their dens hungry, and berries and other favored foods often don't ripen or develop until the late spring or early summer. The bait piles are a welcome buffet for the bears. Somehow, under the current policy, it is wrong to lure bears with food if you want to shoot them with a camera, but just fine if you want to shoot them with a gun.

The logic behind this policy was a mystery to Congressmen Elton Gallegly, a Republican of California, and Jim Moran, a Democrat of Virginia, who introduced a bill to ban baiting on federal lands. It seemed a pretty straightforward idea to their colleagues, and the measure quickly gathered 190 cosponsors. Then the NRA, perhaps seeing the momentum of the bill, announced its opposition and added the issue to its annual congressional scorecard. Just like that, twenty-six cosponsors withdrew their support—something you almost never hear of on Capitol Hill. When the matter came before the House as an amendment to an annual spending bill, it got only 170 votes, well short of the 218 needed to win. Some of the initial cosponsors of the legislation even voted against it.

The way those lawmakers folded would not have surprised Judd Hanna, who had seen the entire Republican membership of the California Assembly obediently fall into line at the summons of the NRA. These days, Hanna leads the California Game Wardens Foundation, where his efforts to enforce laws against the poaching of bears and other wildlife will doubtless run him into more trouble with the NRA. This time, however, he knows who he's dealing with. After his forced resignation for the offense of trying to save an endangered animal, Hanna put it all in perspective. "This is not about me," he said. "It's about the condor. It's about the NRA hijacking the system."

## "The Agro-Industrial Complex": Harvesting Subsidies

IT TAKES A MAN of real backbone to stand up to the NRA and other powerful animal-use groups, and Kevin Fulton fits that description. He's a guy who stands out in any crowd, but especially the company in which we found ourselves in the spring of 2010. We were in Ohio for an organizing meeting against factory farms. A fair number of other men attended, but mostly the room was filled with young and middle-aged women who are such an important part of the animal-protection movement. Kevin is, of all things, a Nebraska cattle rancher and a big one at that—tipping the scales at 315 pounds.

Now fifty years old, but looking a bit younger, Kevin is mostly muscle, with thick shoulders that come to a V at his neck. I had heard from a colleague that Kevin said he could take a standard frying pan and roll it up like a newspaper. I was skeptical, so we got a frying pan to see if he could make good on his word. I tried it first. After I was done straining for several minutes, the pan was still in perfectly good shape for the stovetop. Kevin grabbed hold of it, grimaced a bit, and proceeded to do just what I had heard he was capable of. When it was all said and done, the frying pan looked more like a rolling pin. I thought to myself, *Here's a man made for the farm, not the kitchen*.

Kevin had done big-time weightlifting some years back, and at one point, he made a dead lift of more than 650 pounds. I asked how he maintained his strength these days. "Building fences and knocking in posts," he said, "and just the stuff you do on the farm." He said he keeps very busy running his twenty-eight-hundred-acre organic farm, raising cattle, pigs, and sheep, and looking after three children of his own.

He and I connected for a whistle-stop tour through Ohio, joined by a half-dozen other farmers from the state united by a concern over the direction of industrialized farming. Ohio is one of the biggest factory-farming states, with twenty-seven million laying hens and nearly two hundred thousand sows in harsh confinement.

"We don't run our farms like that," Kevin told the crowd. "We let our animals outside, so they can be animals.

"In my small town in central Nebraska, there used to be a thousand people, and now there are three hundred. It's happening all over farm country. When they become too industrialized, there's hardly any need for labor. I employ several people, and these are good jobs, and that's why my state senator called and recently said 'Hey, I'd like to come over and see what kind of stuff you are doing there.'"

Kevin didn't think much of American Farm Bureau president Bob Stallman's recent statement that critics of industrialized farms want to return to the days of "40 acres and a mule," an applause line he tossed out to the crowd at the farm lobby group's spring 2010 national convention. "My father was a veterinarian," says Kevin, "and I got a degree in animal science from Kansas State. Stallman couldn't be more wrong. I use modern technology, but I also use the practices of my father and grandfather and they work. Last year, my wheat crop doubled the yield per acre of any of my neighbors and I didn't apply any chemicals at all. It's good for me and my family, it's good for my community, and it's good for consumers."

Kevin Fulton does not like the way the Farm Bureau operates, and it's probably safe to say the Farm Bureau doesn't appreciate an outspoken critic with his background. But Kevin is just the sort of farmer that Robert Martin, of the Pew Charitable Trusts, considers the best hope for agriculture in America. In 2006, Martin and his colleague Josh Reichert assembled a panel of veterinarians, animal scientists, farmers, ethicists, and former public officials to study major issues in livestock agriculture. The twenty-member commission, including former agriculture secretary Dan Glickman, conducted hearings throughout the nation, and deliberated for more than two years before issuing a damning analysis of industrialized agriculture and a far-reaching set of recommendations. Martin invoked President Dwight Eisenhower's admonition, at the end of his second term, about the enormous power of the military-industrial complex. Now, Martin said, there is a new worry in America: "the agro-industrial complex—an alliance of agriculture commodity groups, scientists at academic institutions who are paid by the industry, and their friends on Capitol Hill."

Today, an interlocking network of private and public institutions controls our food production system. From one perspective, it has been a raging success, consistently generating enormous yields of grains and animal products at low prices for consumers. But these prices mask an array of social costs—animal cruelty, environmental pollution, dangerous new pathogens and other food-safety concerns, and the bankrupting of family farmers and the unraveling of rural communities. It is the tension between these two contrasting pictures of agriculture that dominates the debate today.

During the twentieth century, as the U.S. population quadrupled, the federal government developed a public support system for the private food-production sector, as a hedge against food scarcity and dramatic fluctuations in production. The foundation of this system was laid with the creation of the USDA in 1862 by President Abraham Lincoln to support agriculture. The federal government also helped to establish, through the Morrill Acts of 1862 and 1890, a network of land-grant colleges and universities, which included departments concentrating on production agriculture. And it built an infrastructure to promote agriculture—a system of aqueducts to move water in the arid West for irrigation and a network of railroads to carry food and other commodities across the nation.

Massive farm bankruptcies in the 1920s had convinced Franklin Roosevelt, by the time he took office in 1933, that there had to be strong state controls to protect farmers from the self-inflicted wound of overproduction. In the 1930s, the Great Depression and the Dust Bowl dealt two unforgiving blows to the nation, causing a mass migration of beleaguered farmers from the Midwest to California. The nation had to buffer farmers from dramatic downturns in the economy and the twin threats of drought and overproduction. Massive inputs of federal dollars and advances in the science of agriculture provided an answer.

In the coming decades, the "green revolution" sparked a remarkable surge in crop outputs, with tools like genetic selection for higher-yield crops and faster-growing animals, irrigation, and chemical fertilizers and pesticides. Traditionally, farmers used manure and crop rotation to replenish the soil. But in this new era, they could keep growing the same crop on the same land each year, thanks to the rejuvenating effects of fertilizers.

During the postwar years, some scientists and industry leaders realized that keeping animals indoors could produce cheaper meat and eggs. They got a boost from private companies that specialized in breeding and genetics, reengineering the animals to grow larger and faster. The results have been startling. Since 1960 milk production has doubled, meat production has tripled, and egg production has increased fourfold. The broiler chicken of our day is twice the size he used to be and grows to that size in half the time, due in large part to the work of geneticists and commercial breeding scientists at companies like Aviagen and Cobb-Vantress.

In the 1950s, Ezra Taft Benson, President Eisenhower's secretary of agriculture, famously said that farmers had two choices: "get big or get out." The goal was to rationalize production, even if that meant bigger farms and fewer farmers. With hefty government subsidies favoring larger operations, production increased, and the number of farms declined at an inverse ratio. There are now about

2 million farmers, down from 3 million in 1970 and 5.3 million in 1950, and just 10 percent of the largest and richest farmers get three-quarters of all subsidies.

The USDA, a vast bureaucracy with tens of thousands of employees and a budget now approaching $150 billion, pushes this support out to the agricultural sector in endless ways. According to Ken Cook of the Environmental Working Group, "the federal government paid out a quarter of a trillion dollars in federal farm subsidies between 1995 and 2009," principally through "an interlocking maze of subsidies that, taken together, force taxpayers to spend billions of dollars no matter what the condition of the farm economy." These subsidies come in the form of cash payments, market loans, revenue assurance programs, federal crop insurance, and disaster payments, together adding up to tens of billions of dollars a year, mainly for the largest and most profitable farms in America. Today, federal subsidies account for about 20 percent of all farm income.

This accounting of federal subsidies does not include technical support staff, research dollars for intramural and extramural projects, surplus commodity purchases, market promotion efforts, or a range of alphabet soup programs authorized by Congress and implemented by the USDA. Nor does it include multibillion-dollar subsidies to the corn industry through the 2007 energy bill for ethanol production. The lawmakers who push for these massive payouts to farmers mostly populate the agriculture committees in the House and Senate. These lawmakers typically represent rural, farm-dominated regions, and in return for these subsidies, farm trade associations, their political action committees, and their members provide campaign contributions and endorsements.

When the Obama administration and Congress bailed out the carmakers and the banks in 2009 in hopes of preventing a deeper recession, it not only required full repayment of the loans, but also a series of reforms within these industries—high fuel-efficiency standards, for example, from Detroit's car manufacturers, and no overleveraged financial instruments from Wall Street traders. By contrast, Congress and the USDA ask for nothing in the way of reform in return for their annual payouts of billions of dollars to agribusiness—no repayments and no standards on animal care, sustainable practices, manure management, or other environmental protections. In return for all of this government bounty, you would think that the industry might at least feel a mild sense of duty to consumers—that it would want to make things right with taxpayers who, largely unaware, fund all of these subsidies. But it is quite the opposite. The American Farm Bureau Federation and other leading agribusiness organizations lobby for deregulation. They have opposed reform efforts to ban the slaughtering of downer cows, to restrict the extreme confinement of breeding sows, and even to make them account for their own pollution or greenhouse gas emissions. Even as they collect billions every year in government subsidies, agribusiness leaders and corporate farmers dismiss talk of all such reforms, and lecture their critics on how the free market should work.

The federal government funnels 70 percent of farm subsidies—about $5 billion a year—to producers of just five crops: corn, cotton, rice, soybeans, and wheat. Given that more than 97 percent of soy meal and more than 60 percent of corn and barley are fed to farm animals, these subsidies are an enormous benefit to the livestock sector, because feed is their biggest cost and they get it at artificially low prices. According to a 2007 study by Tufts University, subsidized corn and soybeans provide a savings of $288 million a year to Tyson Foods. The same study found that the government pays a hidden subsidy of almost $10 to pork producers for every pig raised and killed.

In 2010, as it does every year, the USDA went on a spending spree with Americans' tax dollars in buying up surplus animal products. Consumers obviously did not want these products, but the farm lobby pleaded. In response, the USDA doled out hundreds of millions more to buy up surplus pork, chicken, and eggs for the National School Lunch Program and other federal food-service plans. This was in addition to the billions of dollars' worth of agricultural commodities that the USDA is obligated to buy for these programs each year.

Few people want to eat the meat from spent hens no longer needed by egg producers. Weakened by the enormous outputs needed to produce an egg almost every day of the week, the hens have fragile bones, which break easily and splinter into their flesh. The old cliché was that spent hens become Campbell's soup. However, it's been many years since major soup companies wanted any of this meat, apparently finding it unfit for their customers. As many as 24 percent of hens suffer broken bones following removal from their cages, and as many as 98 percent of carcasses have broken bones by the time they reach the end of the evisceration line.

Yet the U.S. government has "become the largest single buyer of spent hen meat," according to the United Egg Producers (UEP). Based on the trade group's estimates, the USDA purchases about 10 percent of all spent hens, and the UEP proudly notes that many of the carcasses are sent to the school lunch program. One study found that these hen carcasses are several times more likely to be contaminated with salmonella than the carcasses of chickens bred for meat production.

For all of their unwitting generosity, Americans don't seem to get much of a break from the egg industry, which is now the target of a series of class-action lawsuits for illegal price-fixing. In 2008, it came to light that the UEP had set up an anemic animal-welfare program that seemed to serve to limit production and keep prices high. It also manipulated egg exports to the same end—to fix prices and increase profits. In June 2010, Land O'Lakes became the second major egg producer to settle the case, agreeing to pay $25 million to the plaintiffs.

The dairy and hog industries also rake in public support while resisting reforms at every turn. The Hallmark/Westland slaughter plant that closed in 2008 and that specialized in slaughtering spent dairy cows was the second-largest supplier of ground beef to the National School Lunch Program—providing cheap, possibly unsafe beef to kids in school districts in nearly every state. In 2009, the National Pork Producers Council (NPPC) asked Congress to provide more than a quarter of a billion dollars in subsidies to the industry, including tens of millions for direct purchases of surplus pork that consumers didn't want. Governors of nine leading pork-producing states, presumably at the industry's request, also wrote to Agriculture Secretary Tom Vilsack urging him to make the payout. When the House Agriculture Committee held a hearing on the subject, it did not invite a single witness to testify who did not represent industry.

In return for government payouts, the hog industry has degraded rural communities by releasing huge volumes of liquefied manure that foul waterways and kill aquatic life. It got so bad in North Carolina, the nation's second-largest hog producer, that the state imposed a moratorium on new manure lagoons, essentially halting construction of new hog farms. The NPPC has also led efforts to block Congress from phasing out the use of antibiotics for nontherapeutic purposes—a practice that has hastened the spread of antibiotic-resistant bacteria that threaten public health. When the H1N1 swine flu pandemic hit, the industry claimed it was a victim of media hype, even in the face of overwhelming evidence that factory farms are the most dangerous incubators and mixing bowls for virulent strains of the flu that threaten the safety of tens of millions of people across the globe. A hog factory run by an American corporation in Mexico may have even played a central role in starting the pandemic in the first place. Now American factory farms threaten to create a new superbug through their routine crowding of thousands of pigs on giant farms.

With all of the public wealth directed at propping up industrial agriculture, you would expect at the very least that the federal government would make some provision for the animals caught up in this vast and often ruthless production system. But you would be mistaken. Laws to protect these billions of creatures during production or transport—the sum of their lives until almost the very moment of death at the slaughterhouse—are weak or nonexistent. Those final moments are regulated under the terms of the federal Humane Methods of Slaughter Act, and even there, the industry has managed to keep poultry excluded from any protection—and birds account for more than 95 percent of all slaughtered farm animals. As for the Act's minimal protections for mammals, the Hallmark case showed how much those are worth.

Corporate welfare has a lot more pleaders in Washington than animal welfare, and lobbyists for agribusiness are unrivaled in their ability to extract public money for the advantage and enrichment of relatively few. But something has gone seriously wrong when the government of the United States basically contracts out agricultural policy, not only to the industry it is supposed to regulate, but to the worst elements in the industry. What they have built, literally at our expense, is a system that operates without the least respect or concern for the animals they exploit, and little regard for the people they claim to serve.

## "The Rising Plague" and the Veterinary Oath

THE PEW COMMISSION'S 2008 report on industrial farm animal production was a small victory all by itself, reflecting an understanding by serious thinkers in agriculture, including a few people close to the industry, that reform of some kind is urgently needed. Among those thinkers was the commission's vice chairman, Dr. Michael Blackwell, who hardly fits the industry's caricature of its critics.

Dr. Blackwell was raised in Idabel, a small town in the far southeast corner of Oklahoma, wedged between Arkansas and Texas. Farming was a way of life there, as was hunting. Rodeo was a form of family entertainment, and the area was for decades a hotbed of cockfighting, until the practice was finally banned in the state in 2002. Growing up in the 1950s, his father was one of the few rural veterinarians in the tristate area, and he covered a huge expanse of territory to treat animals large and small. The younger Blackwell grew up dealing with farmers and farm animals, and he inherited his father's deep disapproval of animal abuse of any kind.

After obtaining a degree in veterinary medicine from Tuskegee and then a master's in public health, he ran two private veterinary clinics. He was then called to public service, eventually becoming the chief veterinarian for the U.S. Public Health Service, deputy director of the FDA's Center for Veterinary Medicine, and chief of staff to the U.S. surgeon general. He left government in 2000 to become dean of the College of Veterinary Medicine at the University of Tennessee, overseeing an expansion of a range of veterinary programs at the school until his departure in 2008.

When we spoke in my office in downtown Washington in June 2010, nearly two years after the Pew Commission had completed its report, Dr. Blackwell explained why he agreed to sit on the commission. He told me that agriculture is as fundamental an enterprise as there is, with implications for animal welfare, public health, and rural communities—subjects that he's examined and seen from many angles throughout his life. He's seen things take a bad turn, and he accepted the assignment as a chance to help.

Reflecting two years of debate, study, and public hearings, held all across the country, the Pew report was filled with informed analysis about modern agriculture and its troubles. In its conclusions, the report offered sweeping recommendations to restore rational limits and humanity to livestock agriculture. When the industry rejected the report out of hand, Dr. Blackwell was not surprised. Harder to understand was the reaction within the leadership ranks of his own profession—of fellow veterinarians who had taken the same oath he did.

The American Veterinary Medical Association, the largest professional association in the field, dismissed the report, saying it "contains significant flaws" and major departures "from both science and reality." In particular, the AVMA attacked the commission's recommendations to phase out the nontherapeutic use of antibiotics on farm animals and slammed the recommendation that the intensive confinement of laying hens, breeding sows, and other animals be phased out. They were, said the AVMA, "dangerous and under-informed recommendations about the nature of our food system."

Some years ago, I assumed, along with most Americans, that veterinarians and their professional associations would be strong advocates for animal protection. But through the years, I've seen AVMA align itself with big agribusiness on issue after issue. The harsh and defensive reaction to the Pew report was hardly an aberration, but just the latest evidence that agribusiness and the AVMA have intersecting economic interests and nearly indistinguishable positions on many animal-welfare issues.

People rightly look to their individual veterinarian as an expert on the care of their companion animals, and this faith is generally well placed. Veterinarians invest years to learn their art, and they take a professional oath to alleviate animal suffering. Veterinary medicine is generally not a lucrative profession, and most veterinarians are in the field because they believe in their vocation and honor its purposes. Collectively, veterinarians brandish a remarkable set of tools and skills, treating cancer and bone disorders and extending the lives of animals who would have suffered or been euthanized in years past.

Medical assistance for dogs and cats accounts for three-quarters of veterinary income, against a little more than 10 percent for horses and another 10 percent for farm animals. But it is the specialists in agricultural animals—equine, swine, bovine, and avian—who continue to exert influence beyond their numbers within the profession. The vets in these fields typically work for farms, government, or agricultural research centers and share the industry's mind-set. They are generally not focused on healing individual animals, as are companion-animal veterinarians, but rather on maintaining the collective health of the population, even if that means leaving particular animals to suffer, and even if 5 or 10 percent of all the animals die before reaching the slaughterhouse. In this view, it doesn't matter if the individual animals are in misery, just so long as they are sufficiently productive and the whole operation is profitable.

The American Association of Swine Veterinarians publishes a professional journal, and its sponsors are primarily pharmaceutical companies, such as Alpharma, Bayer Animal Health, and Pfizer Animal Health. It's no surprise then that swine vets generally favor the routine dosing of pigs with antibiotics on large, industrial farms. The American Society of Laboratory Animal Practitioners objects to the political efforts of animal-welfare groups to ban the use of random-source dogs in experiments. The American Association of Bovine Practitioners still won't endorse a legislative ban on tail-docking dairy cows, expressing a reluctance to impose its views on dairy producers.

Where there is no large, organized industry with a professional veterinary presence, as on the issue of animal fighting, the AVMA often takes the proanimal position. But if vets are in the employ of a legal and established animal-use industry, they can usually be counted on to endorse the corporate line. Whatever degrees might hang on their office walls, these vets are in the same business as the factory farmers, with a powerful financial interest in keeping things just the way they are. This unnatural alignment of interests helps explain the long list of positions shared by the AVMA and agribusiness, and why the veterinary association is so deeply estranged from widely accepted animal-welfare principles.

To give just a few examples: the AVMA sat on the sidelines during the debate over the slaughter of lame or sick downer cows until the industry itself finally conceded the point. The veterinary association has refused to take a stand against foie gras production, which involves forcing a pipe down the throat of a duck or goose as a standard feeding practice until the liver is as much as ten times its normal size. The AVMA admits that this practice induces lipidosis, a painful disease of the liver, but neither that fact, nor the other obvious abuses of foie gras production has been enough to warrant the group's disapproval. And until just a few years ago, the AVMA supported the egg industry's routine practice of starving hens for days to extend their laying cycle—a practice known as forced molting. It disavowed this cruelty only after the industry had stopped doing it anyway. Likewise, for years the AVMA supported confining calves in narrow veal crates—once again changing its position only to conform to changes already afoot in the industry.

On most issues of animal welfare, the industry will not relent at all, and the AVMA won't either. One of the support systems of hyperintensive animal agriculture is the routine feeding of antibiotics to farm animals. Dosing the animals with these drugs helps factory farmers keep them alive and productive in overcrowded and profoundly unnatural conditions. It is a classic case of the whole mind-set that drives the industry: the antibiotics are introduced as a solution to serious problems of their own making—the confinement of thousands of animals in small buildings. But rather than address those underlying problems, by giving the animals more space, the industry layers on a supposed "corrective"—the massive, nontherapeutic use of antibiotics—that only compounds the original error and creates a whole new set of problems that now gravely threaten public health.

If anyone should recognize that this is the wrong prescription for a fundamentally rotten system of agriculture, it's the men and women at the AVMA. But no, not even here, are they willing to contradict the factory farmers. The AVMA defends the mass use of antibiotics as perfectly acceptable and as "science based." But it's not the science of animal welfare or of public health that now shapes their thinking; it's the science of economics, and it's never been more dismal.

As columnist Nicholas Kristof related in the *New York Times,* because of the overuse of antibiotics in livestock, "now we're seeing increasing numbers of superbugs that survive antibiotics. One of the best-known—MRSA, a kind of staph infection—kills about 18,000 Americans annually. That's more than die of AIDS." This is just one example of the threat that superbugs pose, and serious scientists were warning of it long ago. Donald Kennedy, a former FDA commissioner, recalls how the agency tried to control the use of antibiotics on farm animals back in the early 1980. "Even back then," he writes, "this nontherapeutic use of antibiotics was being linked to the evolution of antibiotic resistance in bacteria that infect humans. To the leading microbiologists on the F.D.A.'s advisory committee, it was clearly a very bad idea to fatten animals with the same antibiotics used to treat people. But the American Meat Institute and its lobbyists in Washington blocked the F.D.A. proposal."

Today, serious scientists are more alarmed than ever about what one author calls "the rising plague." The best-known public-health organizations in the country—including the American Medical Association, the American Academy of Pediatrics, the Infectious Diseases Society of America, and the American Public Health Association—have urged the Congress to act before it's too late by banning nontherapeutic uses of antibiotics on farms. Only one major medical group opposes this public-health reform, and that's the AVMA.

As widespread as the mass confinement of farm animals has become, there are still vets in practice who can remember a day when it wasn't so—a time when animals not only knew the feel of soil and sunshine, but also the caring touch of a veterinarian. Michael Blackwell saw that spirit at work when as a boy he tagged

along with his father, who brought the healing arts to animals because they were in need or distress, and not just because he viewed
them as broken-down production units. With his colleagues on the
Pew Commission, he tried to call fellow veterinarians back to that
noble standard of care and compassion for all farm animals. "Federal
standards for farm animal welfare," said the Pew report, "should be
developed immediately based on a fair, ethical, and evidence-based
understanding of normal animal behavior."

Having lost all sense of what is normal in the treatment of farm
animals, much less what's fair, the AVMA has endorsed the mass-
confinement system. AVMA leaders are so caught up in the industry
and its peculiar logic that "normal," for them, is a confined animal
who can scarcely move and never sees the light of day. And this
doctrinaire adherence to the mechanics of mass confinement has
put them at odds, not just with traditional animal-welfare organizations, but also with the growing ranks of veterinarians not steeped
in the ways of the livestock industry. In the 2008 fight over Proposition 2, for example, we actually had the support of the California
Veterinary Medical Association, whose president, Jeff Smith, put
the matter very simply in a piece for the *Modesto Bee:* confinement
systems are "clearly not defensible from a welfare perspective." The
group hung its decision on its eight principles of animal care, including the principle that "animals are sentient beings with wants
and needs that may differ from those of humans and are worthy of
respect from individuals and society." The AVMA operates without these detailed animal-welfare principles and leaned hard on the
CVMA to reverse its stance in favor of reform—but to no avail.

One of the hardest AVMA positions to explain is its vehement
opposition to federal legislation to stop the slaughter of American
horses for human consumption. Over the last decade, horse slaughter has been repeatedly debated in Congress, generating more con-
stituent mail than any other animal-welfare topic. The HSUS and
allies have led the fight to shut down horse slaughter plants within

the United States and also to stop the live export of horses destined for abattoirs in Canada and Mexico. In 2007, legislative and court action shut down the last three domestic horse slaughterhouses, but the industry simply began moving larger numbers of horses to plants over the borders. According to an HSUS investigation, the horses are typically jammed into cattle trucks, with mares and stallions bunched together. They are transported as far as fifteen hundred miles, sometimes into central Mexico, and then off-loaded, shot with a captive-bolt gun, a rifle, or even stabbed with a short knife before they are killed and then processed for meat.

Agriculture groups, led by the Farm Bureau and the Cattle-men's Association, have opposed the legislation. They see horses as livestock and would rather have their members pocket a couple of hundred dollars for a horse they don't want rather than pay a couple of hundred dollars for humane euthanasia and carcass disposal. It comes down to that simple calculation—owners subjecting horses to misery, fright, and a terrible death, just to save themselves a couple of hundred bucks. And their most outspoken political ally has been the AVMA, whose own defense of slaughter is presented in terms of concern for the horses. If slaughter plants aren't allowed to acquire and kill American horses, according to the AVMA, then American horse owners will starve and abandon their animals. The AVMA says there are tens of thousands of abandoned, unwanted horses, and unless there's an outlet for them, people will just turn them loose. It's a cynical, deeply pessimistic view of American horse owners, and a backhanded acceptance of illegal behavior, since abandon-ment and neglect of horses is a crime in just about every state. Are farmers really so willing to commit criminal cruelty just to avoid the cost of sheltering or adopting out an animal who has served them or even granting the horse a decent death? The AVMA thinks so. Just as the AVMA's leadership has internalized the mind-set of factory farmers, they have accepted all of the assumptions of the negligent horse owner and made it their official policy.

Dr. Doug Corey, an AVMA spokesman and a past president of the American Association of Equine Practitioners, objected to the antislaughter legislation, telling a congressional committee that "processing at a U.S.D.A. regulated facility provides humane euthanasia." To Dr. Corey, apparently there's no difference between a professional veterinarian gently administering a euthanizing solution at someone's home, and transporting forty horses packed together in a cattle trailer, arriving at the place where they will watch other horses slaughtered in terror before it happens to them.

As with all of these issues, many rank-and-file veterinarians break with the AVMA on horse slaughter. Dr. Nicholas Dodman, a veterinarian at Tufts University, told Congress that the slaughter industry is "predatory and brutal" and is "sucking healthy horses out of owners' hands to support the industry." In a remark clearly directed at the AVMA, Dr. Dodman added that "Any group or organization that supports it really needs to evaluate what they are all about." His view is that horse owners must be held to a standard—to provide proper care for their animals—and not given the option of betraying a creature who had trusted them and depended upon them.

The AVMA position on slaughter assumes that horse owners will be cruel to their animals if they are not allowed to slaughter them. Yet, in other food-production debates, the AVMA tells us to trust farmers to never mistreat their animals on factory farms, and it resists every attempt to impose legal protections. It's an incoherent position, held together only by the convenience of the moment. These conflicting arguments lead to precisely the same place: deregulation, and no standard of care at all for animals. What's really behind it is subservience to agribusiness, a betrayal both of the animals and of the veterinary vocation.

Veterinarians should be the best advocates animals have. For now at least, we'll have to rely on individual veterinarians like Michael Blackwell to carry the torch, because the profession's leading trade association is usually going to be found on the other side of the fence.

All of this is just a glimpse of the corrupt and widespread practices promoted and explained away by opponents of animal protection. They have money, they have connections, they have power, and they know how to use them all. For these industries, their trade groups, and their apologists, the suffering of animals is the most incidental of details. They prefer not to think about it, and they try very hard to make sure that you don't either.

# CHAPTER EIGHT

## The Humane Economy

We stepped aboard one of the whale-watching boats that launches from the north shore of Cape Cod, heading toward Stellwagen Bay in hope of glimpsing a fin or fluke, or with luck even a breaching whale. It was spring, the best time to see whales in the Northeast. The seasonal currents and the bay's curvature together produce an upwelling of cold water, rich with plankton and schooling fish such as herring and mackerel. Whales and dolphins have learned that there's many a meal to be had in these waters, and they congregate here in unusual numbers. That, in turn, attracts the whale-watching boats, which for $45 a head give their customers a chance to see creatures bigger than any dinosaur, yet whose languid movements leave only a modest ripple in the water and a gentle wake.

To draw customers, the vessels have to get close to whales but not too close to scare them off. Approaching them carefully and showing the proper respect produces the right results. On this day, if the whales had any inkling of who was aboard our ship, they might have felt especially at ease, since it included a half-dozen leaders of the world's top animal-welfare charities—all of whom had campaigned at one time or another to protect whales from modern-day threats. This was our recreational outing at the end of a three-day meeting in nearby Chatham to define long-term objectives for our cause, producing a kind of bill of particulars to advance the "great republic of the future" that Henry Salt imagined in his own time.

At that World Animal Forum conference in May 2010, we renewed our commitment to protecting whales and other wildlife from commercial killing and slaughter with the impatience of an effort that should have prevailed long ago. We resolved to end, by 2020, the euthanasia of healthy and treatable dogs in the West and the inhumane culling of companion-animal populations in developing nations; we resolved to spread the use of alternatives to animal testing in science, setting the goal of universal acceptance by 2025; by 2030, we would help to secure the adoption and proper enforcement of anticruelty and animal-welfare legislation in every nation—including a worldwide ban on factory-farming practices by 2050.

These are aspirations that many of us will not live to see fulfilled, in the long arc that great reforms usually follow until their completion. But they are among the worthiest goals of the animal-protection cause, promising a better day where the need is greatest in our own time and the abuses most severe. In each of the industries affected, talk of long-term reform and change on this order seems wildly impractical and untenable. Yet such transformations do happen, and over time what once seemed radical and unattainable becomes an obvious and welcome change for the good. What better place to reflect on such possibilities than the New England waters once filled with whaling ships?

There was a time when few could imagine venturing out in a boat merely to catch a glimpse of whales, instead of to catch and kill the whales. Back then, and really up until our own lifetimes, most everything that human beings knew about these grand and beautiful creatures we learned from the accounts of people who hunted and killed them. And there was a time, at the height of commercial killing, when industrial whaling fleets slaughtered more than sixty-five thousand great whales in a single year. It wasn't until the arrival of underwater photography and sound recording in the 1960s that we saw them in their own world, instead of being seized and dragged dead into ours. We have now heard the "songs" by which they communicate, seen mothers nursing and guiding their calves, and in other ways come to observe and appreciate these animals who are not only the most powerful and formidable on earth, but somehow among the most peaceful. It took ages to understand whales on their own terms, and by then they were almost gone—even the great leviathan, celebrated in the Psalms, nearly killed off by the recklessness of a few. But just in time, we have awakened to the glory of these animals and have all but stayed the hand of those who hunt them. And except for those hunters, who look at these miraculous creatures and see only meat, who isn't glad that the whaling industry is all but gone?

Today's international agreements protecting whales mark a fine achievement for humanity—one of those moments when we stood back and said "no more." It was a moment of reprieve and clemency for creatures we realized deserved better. And what has happened since shows how a violent and exploitative commercial interest can be supplanted by a more benevolent one that is just as profitable and far more enduring. There is money in whale watching, and at last men and whales can thrive together.

## "Creative Destruction" and Living Capital

AT THE FORUM IN Chatham, we asked which other industries could wring the cruelty from their methods, for their own good and the good of animals. We understood, as all of us in the animal-protection cause must, that when old ways are let go something has to stand in their place. It's not enough to argue, however persuasively, against wrongful practices—we must point the way to alternatives. Where local economies rely upon the needless abuse and killing of animals, it's for us to show that there are profits and jobs in worthier enterprises. The appreciation of animals presents a vast market of its own, and often, as in the case of the sealers in Newfoundland, it's these old practices and mind-sets that stand in the way of new economic opportunity. The same is true in places all across the world, as people and industries cling to habits and customs that have blinded them to the much greater possibilities of a humane economy.

What are the other business models that could achieve similar outcomes for other animals, while adding economic value and sustaining jobs? It's not an abstract question. For our cause to succeed, we have to carry the day with our moral and legal arguments, but also every workday show that business can succeed and also be good to animals. Corporations must provide opportunities for consumers to be humane, and consumers have to shop and otherwise act with animals in mind—with all parties demonstrating that social responsibility is more than just a hollow slogan.

In this way, the marketplace puts everyone to something of a test. We will be closer to success when consumers see themselves as more than just consumers—more than passive receivers of all that the market puts in front of them. The market responds to morally informed choices too, and when we decline to buy the products of cruelty, it will get to work on products made humanely. Other than the force of law, nothing persuades industries that operate by cruel methods more than the sight of unsold goods, vacant parking lots, and empty seats. The first step in building a humane economy is always to take our business away from the inhumane economy.

None of this, however, will be achieved without considerable resistance, and the biggest critics will be the businesses that now use animals in harmful ways and don't readily see the alternatives. People have a way of convincing themselves that whatever they're doing now is the right thing, and it only seems more right when they contemplate the difficulty and cost of changing course. Take just about any practice described in this book, and you'll find someone who can explain away every sorry detail, all the while insisting that no other way could be workable, much less profitable. They view themselves as the pragmatists in a world full of do-gooders with silly and often dangerous ideas. Somebody's got to face the hard truth that we need to use animals, as they see it. It's dirty, bloody work, and somebody's got to do it.

These folks typically view attempts to impose new standards or regulations on them as a form of economic interference. Thus, a ban on battery cages in the egg industry is an infringement on individual liberty, and an end to whaling, as they see it in Japan and Norway, is not just an assault on national sovereignty, but also an attack on free enterprise. If reforms prevail, they insist, people will be put out of work. A sentimental or weepy concern for animals just isn't that important or practical, we're told, and animal advocates should stop meddling in the affairs of commercial enterprises that produce good economic outcomes for communities and nations.

There's no doubt that a ban on battery cages or whaling would affect individual producers and industries, and at some level it's understandable that the people involved see the debate from a highly personal and self-interested perspective. But their reaction generally reflects a serious misreading of the broader economic principles at work. And least of all are corporate farmers in any position to complain about things changing too fast. They talk as if their own production practices are static, as if they cannot adjust to changing circumstances in the marketplace and adopt more humane practices. What they overlook is the incredible speed with which they changed to the new and harsher ways of industrialized production—to the "new agriculture" with mass confinement and all of its other merciless innovations. The changes in that sector would have been almost unthinkable to farmers fifty years ago, but now they are the standard. If the industry can demonstrate that kind of movement in one direction in so little time, they can turn in the other direction too. Factory farming is the creation of human resourcefulness detached from conscience. What innovations in agriculture might come about by human resourcefulness guided by conscience?

The market itself is constantly adapting, changing, and transforming, caught in the tug and pull between producers and consumers. In *Capitalism, Socialism and Democracy,* the famed Harvard economist Joseph Schumpeter celebrated this essential quality of capitalism as "creative destruction," the process by which entrepreneurs and innovators introduce new goals, new means of production, and new products in support of their visions. The critics who represent the established ways of business make apocalyptic assumptions, fearing that they and everyone else in their industry will be driven out of work if forced to change in any way. But this sort of change in business practices, as Schumpeter noted, drives new growth and is the lifeblood of the economy: businesses that do not adapt will be left behind, while innovators claim an ever larger share of the market.

Factory farms and the fur trade, to take two examples, operate within larger industries—the food and garment sectors. Food and clothing are obvious essentials, but they are, in our time, much more than that—they give us pleasure, they provide comfort, and they enhance our quality of life. Yet vital as they are, there's nothing essential about the *particular products* that factory farmers or furriers sell. Food aisles and coat racks brim with options for consumers, who are free to choose among many products of comparable quality—with no *net* economic loss for the broader industry. The loss experienced by the furrier is the gain of the seller of cloth or synthetic coats. The same is true for the factory-farming corporation, which stands to lose market share to more humane producers

or to the makers of plant-based alternatives as consumers put their
money where their mouth is and start eating with conscience. In
the end, everybody is still buying food and clothing; it's just that the
economic benefits flow to business in different directions based on
the elastic decision making of consumers.

Nothing obligates you or me to purchase low-quality products
or morally questionable ones in order to preserve the economic livelihood of one class of producers. That's a decidedly upside-down
way of looking at the relationship between consumers and suppliers.
If a person has a choice between an elegant cloth coat and a luxuri-
ous fur coat—both comparable in terms of style and warmth, but
only one requires killing animals—there's no question that the cloth
coat is the one to choose, based on a balancing of basic needs and
moral interests.

If the coat maker is to succeed in business, she's got to produce
goods not only of interest to consumers, but also in sync with their
sensibilities and values. We should not shed a tear for her if she fails
to account for their demands. If the furrier closes a sale, conversely,
he does not shed a tear for the cloth coat seller who loses out on a
purchase. Nor does the factory farmer need to offer an apology to
the pasture-based farmer when he succeeds in his operations. One
seller's loss is always another's gain.

Consumer prerogative and the value of competition are among
the most widely accepted precepts of Western economic philosophy. Competition drives producers to excel and to innovate, but it
also means that there are winners and losers. It's true that the state,
by providing a wide range of subsidies and other safety nets, can
guarantee that certain businesses won't fail. As for the rest, they
take their chances in the marketplace, where every day there are
casualties, with companies bankrupted or otherwise driven out of
business, sometimes by poor management, but just as frequently by
the changing wants of customers and the agile marketing and production strategies of competitors.

No one in the food industry has been more agile than John Mackey, who started a grocery store out of his garage in Austin, Texas, some years ago called Safer Way that has since become Whole Foods Market. An outlet that once catered to a small group of alternative shoppers now has nearly 300 stores worldwide, 54,000 employees, and $9 billion in annual sales. He refused to go along with the business mind-set that assumes that shoppers, when they grab the cart, leave their values behind in the parking lot. He knew they would consider and balance a larger array of factors including price, brand, quality, availability, and, increasingly, social responsibility. Whole Foods' competitors assumed that people cared only about cheap food, regardless of any other consideration. But Mackey believed if better options were offered, people would respond.

The success of his chain proves he's right. Consumers have their own bottom line, once they are alert to broader concerns. They are interested in fresh and local foods, they want foods without pathogens or dangerous bacteria, and they want to see animals treated humanely. More and more consumers understand that supposedly cheap food comes with far-reaching environmental and moral costs, even if the scanner at the checkout counter doesn't pick them up. Good businesses, like Whole Foods Market, are showing the way—and consumers and competitors are following their lead. Within the sector, foods labeled as "organic" and "humane" are suddenly big sellers, and even the "superstores" are getting the message and reserving shelf space for products with those labels. This is just a single example of how much one man and his team, by basing a business model on doing the right thing, can begin to change the standards for an entire industry.

Values matter, especially in business. Americans generally don't favor driving down cost by abusing workers or despoiling the environment. In a talk I had with evangelical minister Matthew Sleeth, who himself grew up on a dairy farm in Kentucky and now speaks out against factory farming, he underscored that considerations about a social concern for others, including animal cruelty, make us who we are as a people and a nation. "When you have the mind-set to have the cheapest price, you end up with slavery and animal cruelty," he told me. "We choose to make the society we want."

After all, just about any product can be made cheaper if you don't mind taking moral shortcuts—if you simply subordinate every other consideration to cost. But there are natural costs to producing goods—costs that assume the basic moral boundaries will be observed—and a proper accounting includes them all. In every system of democratic capitalism, it is understood that even with all its virtues, the free market has its dark underside, where the weak are left at the mercy of the strong unless the law stands in their defense. This is where farm animals have been left, and why legal protections are so desperately needed.

Realizing just how bad things have gotten in modern farming, many people are searching for companies with a conscience, and for industries with a sense of social responsibility. And this presents a growing opportunity. There's plenty of evidence to show that a new, humane economy has greater earning potential for businesses, generating more economic activity and more jobs, as well as a safer environment and healthier and better lives for consumers. Commerce built around appreciation or respect for animals has a far larger consumer base than the businesses that disregard these principles. Cruelty and destruction of nature have, for good reason, become deeply unpopular, and no successful business can survive in the long term by relying on them.

WELL BEYOND THE BAYS of Cape Cod, the progress of the global whale-watching industry shows us how a humane economy can work. Coastal towns all over the world are gaining both jobs and profits from whale watching. The prospect of seeing whales draws a hundred thousand visitors a year to the town of Kaikoura in New Zealand, which as a nation generates more than $80 million directly from this line of business. In western Scotland, on the Isle of Mull and in other coastal communities, whale watching generates $12 million a year. Worldwide, there are now more than three thousand whale-watching operations, employing about thirteen thousand people. The global business has grown bigger every decade since it all began around 1950 at Cabrillo National Monument in San Diego, where visitors could spot gray whales from an observation point. According to the International Fund for Animal Welfare, more than thirteen million people took whale-watching tours in 119 countries worldwide in 2008, generating ticket fees and tourism expenditures of more than $2.1 billion.

The whale-watching industry is a product of Schumpeter's "creative destruction." For two centuries, America had been at the center of a global whaling industry, as boats filled with "iron men in wooden ships" launched from several dozen Atlantic seaboard towns to ply the world's oceans for months at a time in pursuit of whales. Even by the standards of that time, it was horrible and dangerous work. After a whale was spotted, the crew would lower smaller, more maneuverable boats into the water, and the men on these boats would close in on a whale and throw their deadly harpoons by hand. The harpoons had ropes tied to them, tethered to the front end of the hunting boats, often smaller than the whales themselves. After the whale was struck, they'd be pulled along as the whale fled. Once the whale succumbed, typically to blood loss, and was hoisted on board the main vessel, the men cut up the carcass to extract the commercially valuable products, mainly the oil. It was a precious cargo, which, once delivered to port and sold, would fuel the lamps and grease the machines of a growing nation.

Then, in 1859, a gusher in Titusville, Pennsylvania, signaled the beginning of a new era of energy production. The extraction of fossil fuels from the earth—first oil and then coal—over time made the slaughter of whales for such purposes obsolete. Oil and coal were abundant and seemingly inexhaustible. Whaling by other nations would continue to thrive well into the twentieth century, providing oil for the manufacture of the nitroglycerine-based explosives of both world wars, the lubrication of the heavy machinery of the old Soviet Union, and reduced friction in the finely calibrated instruments of space exploration. But it would never again be at the center of any nation's economy.

Although innovation and entrepreneurialism in the energy sector triggered the decline of whaling in America and other nations, it was love of whales that created a thriving global watching industry, whose revenues now far exceed the money made by the few nations that still kill the animals—if they make any profit at all. It is successful not only because it is humane, but also sustainable. There is no harm in watching animals over and over again. They keep on living, and, economically speaking, keep on giving. That contrasts with the harpooning of whales or clubbing of seals, which, by definition, involve onetime uses of individual animals. The exploiters of these animals hope that survivors will replenish the population, but in so many cases, whether it is with whales or with elephants slaughtered for their ivory, there are often no limits observed and the killing results in depletion.

The extractive practices are built around taking—akin to grabbing fistfuls of cash and running away with it—while the sustainable approach is grounded on the strategy of keeping all the living capital in place and profiting from these investments in perpetuity. And on the demand side, there is no comparison. Few people anywhere in the world desire whale meat or sealskin coats, and that market is constricting further as younger generations of Norwegians and Japanese take a pass on these products. At the same time, the market built around appreciating animals is vastly larger, and by all indicators is expanding rapidly.

For my part on that spring day, I was happy to be just another customer of an industry of the new, humane economy. There were perhaps a hundred of us on a 130-foot boat, with no clouds in the sky and the water reflecting a bright sun. The air temperature was a seasonable fifty-four degrees, but it was chilly, with the wind blowing from the north and the boat moving at twenty-five knots. Most of us huddled inside to escape the cold, though we were all poised to race to bow or stern when the captain cried, "There she blows!" It's no small thing to lay your eyes on a living creature fifty feet in length and weighing fifty tons, and all of us were primed for it.

After thirty minutes, we were in the mouth of Stellwagen Bay, where the ship's small crew spotted our first group of white-sided dolphins, who were pacing the boat on the starboard side in small pods and rising in an arc out of the water and then back in again. They may have been looking for fish, but it sure seemed like they were having a rollicking good time in the open ocean. Maybe they knew they were part of the draw for us, a kind of opening act, and they did their part to hold our attention and lift our spirits. As we studied the dolphins, all of us with smiles on our faces, we seemed to forget about the chill in the air. Standing outside in the ocean air and peering over the rails was the only place to be.

Then, from his perch on the top deck, the captain announced his sighting of small plumes of misty water up ahead, signaling the blowhole of a whale. He tacked to the right and headed in that direction, slowing down to avoid any collisions with whales—one of the few hazards that this industry can pose to the animals. It wasn't long before two humpbacks—a mother and her baby, about 150 feet apart—came into view.

The baby was splashing around at the surface, but the mother was more purposeful. The captain explained that she was blowing clouds or rings of bubbles around schools of fish, and as the bubbles rose, the fish got trapped within the water column she created. She then quickly rose up and took in an enormous volume of water through the baleen scales in her mouth, filtering out the water and ingesting the fish—a deft, efficient, and learned maneuver.

We watched for a while, and then mother and calf moved on, and so did we. Shortly thereafter, on the opposite side of the boat, off in the distance, we saw minke whales breaking the surface. In relative terms, they are "small" whales, with adults at twenty-five to thirty feet long but still twice the weight of an elephant. Minkes are among the most abundant whale species, and they also are the ones principally targeted by the handful of nations that still engage in commercial whaling, which left me wondering if they had faced threats during their transatlantic movements and knew that humans posed a danger. Certainly, they stayed a good distance from the boat, seemingly more wary than the humpbacks who had hardly seemed to notice our presence. I was unsure if they kept their distance out of instinct or as a learned behavior. Those who have studied whale communication have always marveled at its complexity, and a part of me has to believe that word gets around about human hunting and the ways of men.

It's always struck me that the hunters' worst offense was taking the lives of innocent animals. But a second serious offense was that their attacks on animals made the surviving members of their pod, or herd, or population more skittish and fearful of humans—thereby threatening the features of a humane economy inspired by an appreciation of animals. Fearful animals don't stick around to gauge the intent of humans, so that diminishes the experience of seeing wildlife for people who pursue them with cameras and binoculars. When I was in Kenya, which has banned sport hunting since 1977, I didn't have any trouble seeing animals—the great herds of game don't scatter at the first sight or smell of a person because they've become accustomed to a nonthreatening presence. The Kenyan government has affirmed this no-hunting policy time and again, rebuffing lobbying by the Safari Club International to overturn it, knowing that while hunters may generate some commerce, they will deprive many others from generating a far greater amount. The market for ecotourists dwarfs the market for killing wildlife, and if present trends hold, that gap will widen in the years to come. As for whales, just two or three governments engage in commercial killing, with government elites defending a practice that long ago lost its economic utility and cultural significance for coastal peoples.

The captain of our boat then sped farther north, and this allowed us to see our first fin whale a few hundred yards away. This is the second-largest whale in the world, with adults measuring up to eighty feet and weighing in at more than seventy tons. Rogue whaling nations target these animals, too, even though there's hardly any demand for their meat. Whale meat is stockpiled in government freezers in both Japan and Norway—a symbol of waste and stubborn allegiance to the abstract principle of cultural sovereignty and killing, rather than the necessity for any of it.

As we traversed the bay, we saw more white-sided dolphins, and then a few more humpbacks. One of the humpbacks flipped on his side and raised an eight- or ten-foot-long fin fully out of the water, slapping it down to create a magnificent splash. Other humpbacks rose almost entirely out of the water and crashed down on the surface, in the signature move seen so many times in television commercials and wildlife films.

While the morality and economics of commercial whaling have been a continuing matter of international debate, due to the intransigence of whaling nations, there's been consensus on the issue in the United States for some time. Here we were in Stellwagen Bay near the north shore of Cape Cod, which, along with the nearby island of Nantucket, once served as the hub of the global whaling industry. Now the place had been transformed, and it was just another coastal region in the world generating revenue from whale watching. And it was not just the businesses that supported this new economy, but also the political leaders. Senator John Kerry and the congressmen who've represented this area have long been among the strongest supporters of whale watching, as well as the fiercest opponents of commercial whaling.

To my mind, there's no better example of the path to progress for animals than the issue of whaling. A century earlier, ships that set out in search of whales had a far different purpose than the ones that launch today. America in the nineteenth century was an animal-based economy, with whales killed to make candles and horses yoked to punishing loads that wore them out in the nation's rapidly developing cities. As new sources of energy were discovered and new machines developed to run them, the human reliance on whales and horses fell away. New bonds could form between people and these animals. And with practical realities and economics no longer driving the moral decision making, Americans could begin to see them as something more than instrumentalities, and more as individuals.

This is the pathway for other species, too. With new economic opportunities and options, we can imagine a new and better relationship with animals. Everybody wins. We just have to be open to the possibility of change, and creative in the designs of new industries that would drive prosperity and growth, putting an end to behavior that causes harm and exploitation, and substituting what is both right and moral and profitable in every sense of the word.

## Wild Neighbors

THESE DAYS, WE DON'T have to go on a whale-watching boat or even hike in a national park like Isle Royale or Yellowstone to see animals in the wild. The venues for watching wildlife are often close by—at a local state park, a hiking trail, an Audubon nature center, or even in a determined patch of forest contained within a jumble of our homes and commercial buildings. Bird feeders hang in our backyards, brimming with seeds and grains that draw in birds, squirrels, and other creatures. Today, wildlife watching is officially America's favorite form of wildlife recreation, with estimated expenditures at about $50 billion—about the same amount as total spending on our pets. The number of people engaged in this pastime far eclipses the combined totals for participation in hunting, trapping, and fishing, with more than seventy million people, or one in four of us, considering it a hobby.

I count myself among this class of enthusiasts. And while it's on the upswing in popularity, it can hardly be considered a modern invention. Watching other creatures is built into our DNA, since for 99 percent of human history, our survival depended on studying other animals, so we could obtain food, protect ourselves from them, and even learn from them. Whether it was Native Americans before European settlement, or the explorers who gazed in wonder at the splendor and variety of wild animals in the New World, or city dwellers who trek into field and forest, people in this country have always drawn pleasure and inspiration from simply seeing and sharing surroundings with other creatures. And while our relationships with wildlife are conflicted, and sometimes contradictory, there can be no question that it's a basic expression of the human-animal bond.

For those of us who love watching wildlife, we typically have to search for them, going into natural areas to find them going about their business and straining to be quiet enough not to cause them to scatter. Sometimes, though, it's wild creatures who find us, showing up in the most unexpected places. In the spring of 2009, two wild creatures found me, and I soon realized life would change for me in a small way. They swooped down from above and made themselves at home on the small deck of my apartment in a high-rise in Washington, D.C.

My visitors were pigeons who decided to make a nest in a pot that itself houses a four-foot-tall plant. The plant looked beautiful when I got it at Home Depot, but I must confess it did not get the best stewardship from its caretaker. Horticulture has been an interest of mine but never a strong suit, and as a consequence the plant did not get properly trimmed, and it began to dominate an entire corner of my deck. The mother pigeon must have identified it as a pretty good hideout, and one also that would shield her soon-to-be hatchlings from the sun and predators.

She made a smart bet I wouldn't evict her. The HSUS has among its many operations an active urban wildlife program. That program is grounded on the conviction that people should not think of wildlife in our midst as intruders or trespassers, but as "wild neighbors." The unexpected arrival of pigeons is just the sort of wildlife experience we encourage homeowners to welcome. Like us, animals explore and search and find nice places to live, seeking comfort, protection, and pleasant surroundings. So when I realized she was going to stay for a while, I tried to be understanding. I asked Dr. John Hadidian, the director of our program, about the nesting habits of pigeons. John told me they'd be around for at least six weeks, with both mom and dad attending to the nest.

Despite both parents pitching in, they didn't make much of a nest—just a thin cluster of twigs. But after a short while, mom produced two fine-looking eggs, and she and dad started sharing duties incubating the clutch. Before long, there were two hatchlings who looked more scraggly than fuzzy. They would have been vulnerable to a predator or to a person not fond of pigeons, but with the cat inside, the deck five floors up, and a friendly mortgage holder, this was a bit of a sanctuary in itself.

Pigeons are also known as rock pigeons or rock doves, and before there were cities, they nested on cliffs. If this couple thought this was a cliff, then who was I to tell them differently? I decided I'd cede the deck to the family and do my best to minimize disturbance—choosing for the time being the inside of the apartment to read the paper and do my work. We humans hadn't been too good to pigeons—whether poisoning them in cities, or shooting them for targets, as in Pennsylvania—and I was going to try for a small makeup action for our species and give this little family the space they wanted.

Mom and dad were with the chicks almost all of the time for the first few weeks, taking turns to venture out periodically to collect food for the family. They'd come back and regurgitate the bounty—drawing a milky substance from their crops and delivering it right into the mouths of the hungry chicks. They grew fast, and within about five weeks, they had shed their scraggly, cottony covering and in its place came dark, smooth, iridescent feathers. They now looked like "regular" pigeons, only smaller. They stayed put on the soil of the pot for a couple of weeks. But one day, I came home and saw them up and about.

Each night I was in town, I'd come through my door and go straight to see them. They'd be walking around, with their stick-figure legs and bobbing heads. Occasionally, they'd stop and peck at some bug or speck on the deck and then resume their new exercise regimen. I was anxious to see their physical development, their fast progression from helpless hatchlings to self-sufficient juveniles.

Then one night, I came home, and they were gone. I wondered how these neophytes could muster the bravery and take to wing. But I knew that day would come, and here it was. They'd now start a much more challenging and exciting phase of their lives— and I could use my deck again. All birds must leave the nest and learn to fend for themselves. We regard this maturation process as routine, but in truth it is a miracle, especially on the time frame it happens for so many species. Life is so fragile, but so resourceful, too, whether it's a case of grasses or weeds reaching out of the cracks of a sidewalk, a squirrel building a life around just a tree or two on a city block, or a pigeon finding a plant to nest in on the deck of an apartment.

The mother and her two babies messed up my deck a bit, but that was no matter to me. They're welcome back anytime. Recent studies on the concept of "cognitive mapping" tell us that pigeons and other animals have an incredible memory for sites and other physical settings. So I wonder if my place will hold significance for them. I wonder if any of them will come back and stop in. I know that when I go back home to New Haven, where I was born and grew up, I sometimes drive by my childhood home to see what it now looks like. We think of animals as too practical for such reminiscing, but I am not so sure.

Pigeons are just one of many species we seek to help through our urban wildlife program. We provide information on the breeding habits and behavior of so many adaptable creatures. We also offer tips and guidance to people on how to deal with raccoons in the chimney, skunks under the deck, woodchucks in the backyard, birds in air vents, or advice on dealing with other wild creatures who've decided that some part of our home or landscape makes a good nesting site, resting spot, or playground. In dealing with these cases, it generally comes down to a matter of problem solving and tolerance, since, mostly, the worst the animals do is cause a minor inconvenience. If a conflict erupts, here, too, human ingenuity should be put to work. These conflicts can typically be solved by nonlethal means, such as by capping your chimney to exclude wild animals, so that smoke can come up but a raccoon cannot come down.

In fact, we have a staff of professionals, working mainly in the Washington, D.C., area, who run a program called Humane Wildlife Services, and they do nothing but respond to calls from people concerned about wildlife in or around their home. They're developing a new business model for wildlife conflicts, providing a happy contrast to the ways of commercial "nuisance wildlife control" trappers. These commercial operators, who often came out of the world of trapping animals for fur and who rarely have the animals' interest at heart, don't always inform homeowners about the lethal methods they use. By calling in these trappers, found online or in the phone book, residents often wrongly assume that their problem will be solved and the animal will be released. If they don't want any harm to come to the creatures causing a bit of trouble, they've got to ask

the right questions, and also find the right company, assuming outside intervention is needed at all.

Most of the "problems" that urban wildlife cause can be solved,
with some good common sense or a few technical fixes. But so
much of it comes down to perception and tolerance. For one person,
a deer in the backyard is a joyous sight—a treat for everyone in the
household. For others, that very same behavior and circumstance
on their property is akin to an unwelcome invasion, with the residents fretting about the landscape or ornamental shrubbery or even
alarmed about the threat of disease. Some people consider a deer
nibbling on azaleas to be a capital offense, though the creature only
is doing what comes instinctively to him or her. If it's the beauty of
nature they fear losing in those azaleas, they should stop and take
a longer look at the deer, and they'll see something more beautiful
than anything that grows in a garden.

Some people seek to wall themselves off from wild animals.
That's an odd view, given our settlement patterns these days. About
half of Americans live in the suburbs, and that typically means that
these homeowners have more land than city dwellers. Their homes
are often framed by careful landscaping, manicured lawns, and
young or old trees. The whole setting is designed to blend comfortable living within a natural setting, even if it's manufactured. The
growth of big-box stores like Home Depot and Lowe's cater to in-
dustrious people who take pride in their homes and invest in every
inch of their property.

For so many people, it's an emotionally satisfying place, con-
structed on the ideals of green and open space. But trees and plants
and open space attract other living things, and these creatures graze
on grass or plants, nest in trees, take refuge in window wells, or just
pass through. You cannot have one without the other. Many people
invite these animals in, with bird feeders or other setups built for
wildlife, welcoming these creatures. But there are others who want
to turn them away and don't much care what happens to them. I

am always perplexed by this selective attitude that appreciates natural surroundings, but not the animals who live there. After all, the entire landscape, especially if it's a new development, was probably the recent home of some bear or woodchuck, so the original tenants have claims of their own, too.

"We are about at the point with urban wildlife where Henry Bergh was in the 1860s with domesticated animals," Dr. Hadidian tells me, explaining that people are still adjusting to the idea of wildlife within their midst. He reminds me that we have little in the way of an infrastructure to help people sort through the conflicts. That's why some take matters into their own hands or call out a nuisance trapper, who is often just an exterminator by another name. "People want the right answers, but they don't know where to get them." Hadidian says the nation has a pretty remarkable network of wildlife rehabilitation centers, but they are dealing with just some of the victims of human impacts on urban wildlife. We need institutions to help prevent problems, or to solve them once they occur, much as the local Humane Society or SPCA helps with cats and dogs. "We have to recognize the need for humane treatment of these animals, too, and for understanding that they deserve their own place."

All of our communities are environments shared with other animals, and it's our duty more than ever to be tolerant and even generous, especially given all of our intrusions into their traditional habitats. A red-tailed hawk named Pale Male nested on a building on Fifth Avenue in Manhattan in the 1990s, just across from Central Park, which itself was one of the first urban parks and a template for other such parks in other major cities. Pale Male and his mate attracted an enormous following, successfully breeding in the city for years before a co-op board abruptly decided to destroy the nest. The whole episode was a reminder that the divide between urban living and wild areas is itself artificial. "Wildlife" and "wilderness" are no longer "far away"—a romantic ideal removed from daily human existence. Like the pigeons who came down from above to roost on my deck, or the hawks who made a home in New York, they are adaptable and resourceful animals, and they are here with us now. That's a wonderful thing, and something to appreciate, even if it requires some forbearance and maybe even a little effort.

The presence of wild, free-roaming creatures results in some conflicts, but also opportunities—both moral and economic. New businesses can emerge to solve conflicts, to sell goods to appeal to our instinct to feed or attract wildlife, and to nurse animals when they are injured or ill or orphaned. Community leaders can plan for green causeways and open space, to give residents the beauty and comfort that come with sharing our surroundings with fellow creatures. Some planned communities, such as Harmony in central Florida, have been designed with the idea of attracting wildlife and having animals live among them.

The new economy, and the new communities within it, show that we can be far more attentive to the land and life around us, reflecting that concern in the way we build our homes, design our transportation systems, zone our communities, and preserve open space. We can be developers without being destroyers, and kindly neighbors to the creatures in our midst.

## Chemistry and Compassion

WHEREVER HUMANS LIVE, WE affect the lives of so many animals and the fortunes of so many species. There is trouble on many fronts, but also a growing consensus on the side of the animals, to spare them from extinction while there is still time. The particular solutions are still being debated, and proportional responses are still resisted in some circles. But the good news is that the basic moral framework is established. As the naturalist Aldo Leopold noted more than a half century ago, the first rule of intelligent tinkering is to save the parts.

At the same time, there are still those who insist that there are simply too many animals—be they wild, feral, or domesticated. They regard wild animals as a nuisance or worse, and faced with any conflict or even any inconvenience their first response is always the harshest—whether it's culling seals or wolves in the most remote wilderness areas, shooting deer or bears in the suburbs, or even culling dogs or cats seen as pests in cities across the world. Such people view every animal-related problem as a call for raw force, and they themselves sometimes need to be corrected by force of law. In other cases, human intervention is indeed required, but it needn't be violent and indiscriminate. There are plenty of nonlethal means of avoiding or solving conflicts with wildlife, and plenty of people know how to apply them.

Not everybody is neighborly about pigeons, for example. Many people wouldn't think of ceding their deck to the birds for even a few weeks. And truth be told, those kindly people who dispense bread and seeds from their park benches, and are then encircled by a mob of birds, seem to be in the minority in many communities. If not contempt, familiarity with pigeons seems to breed at least an active dislike on the part of many citizens. There are people who protest to city hall about these "rats with wings," irritated by the droppings and sometimes even the mere sight of the birds. Typically, city officials oblige these people too readily, dispatching nighttime details to trap and kill the birds, or simply laying out poison-laced foods that produce the worst of deaths.

Pigeon culling is a not-so-closely-held secret in many cities— one of the lesser-known "municipal services," like the killing of strays in the dog pound. It's a matter best left unspoken, at least as the folks doing the trapping and poisoning see it. But a new attitude is taking hold in urban wildlife management, and even the most hardened opponents of pigeons and other "nuisance wildlife" can hardly quarrel with the fix: chemical birth control for the birds. Today, more cities, including Las Vegas, Los Angeles, St. Paul, and

Tucson, are opting in, and their product of choice is OvoControl, a kibble bait that, after it is ingested, reduces egg hatching rates.

Technology and innovation are making it easier than ever for us to handle abundant populations of animals in nonlethal ways. Innolytics, the maker of OvoControl, is also promoting similar products for use on other birds, including Muscovy ducks in Florida, feral chickens in Hawaii, and resident Canada geese in the Mid-Atlantic states.

Nonchemical strategies are also a part of the solution. In Michigan, HSUS has for years worked with the state's Department of Natural Resources on an annual egg hunt, finding eggs laid by resident geese and oiling them to arrest their development. The oiled eggs don't hatch, but because they're left in the nest, mom doesn't produce a new clutch—keeping the population in check. It may seem unfair to the mothers to prevent the eggs from hatching, but the alternative is far worse—roundups and gassing of entire flocks, including the mothers.

Wildlife scientists working in concert with the HSUS have also developed contraceptives for mammals—using the vaccine to control elephant populations in South Africa, as an alternative to culling, and to limit the growth of deer populations in the United States, as on Fire Island in New York and Fripp Island in South Carolina. Researchers Jay Kirkpatrick and John Turner first proved the efficacy of a vaccine known as PZP in 1989 on Assateague Island National Seashore to manage the herd of wild horses on the ecologically sensitive barrier island.

When it comes to wild horses and burros, however, the most important use for contraceptive technology is for herds that roam public lands in the West. In 1971, Congress passed the Wild Horse and Free-Roaming Burro Act, declaring these animals "living symbols of the historic and pioneer spirit of the West" that "contribute to the diversity of life forms within the Nation and enrich the lives of the American people." Even though Congress noted that the ani-

mals were "fast disappearing from the American scene," the federal Bureau of Land Management has from the outset implemented
the law through roughshod removal of horses from the range as a
population control tool. The National Park Service was even less
forgiving toward the wild burros of the Grand Canyon—who were
rescued by Cleveland Amory from a planned massacre and became
the first inhabitants of the Black Beauty Ranch.

The BLM says it puts the captured horses up for adoption,
but the agency has swamped the system by gathering too many
animals—usually at the prodding of ranchers who graze their cattle
on public lands at below-market rates and mistake that privilege
for the right to evict wild horses and burros. The result has been a
swelling captive horse population, which BLM maintains and feeds
through contractors. Within the last two decades, the captive population of wild horses and burros has climbed past thirty thousand,
and the government is spending more than $25 million a year—
75 percent of total funding for the program—just to care for the
captive animals.

This is a ridiculous and shameful outcome, far from the intent of
Congress, and the BLM needs to scale back its roundups and remov-
als, putting contraception to work broadly as an on-the-range management program. With support from the Annenberg Foundation,
HSUS has been working with BLM to this end. This approach is
the only way out of a decades-long problem that BLM has handled
so amateurishly.

Chemical contraception for animals will someday extend even
to dogs and cats. In the spring of 2010, I was one of a few nonscien-
tists to attend the latest conference of the Alliance for Contraception
in Cats and Dogs (ACCD), the primary organization devoted to
finding nonsurgical methods of sterilization for companion animals.
If we are to reach the goal of ending euthanasia of healthy animals
in America by 2020, and the particularly cruel killing methods used
to control the population of dogs and cats throughout the world, we

can't just keep doing things the way we have for decades. One sign that the tide is turning was the recent news that billionaire philanthropist and animal advocate Gary Michelson has pledged $75 million, through his Found Animals Foundation, in prizes and grants "for promising research in the pursuit of a safe, effective, and practical non-surgical sterilant for use in cats and dogs."

The standard method for controlling reproduction of dogs and cats, surgical sterilization, and the message "to spay and neuter your dogs and cats" have been hallmarks of the humane movement's approach to overpopulation for decades. Along with promoting adoption, these approaches have lowered annual euthanasia totals in the United States from perhaps twenty million in the mid-1970s to four million today. But surgical sterilization costs money, and the procedure can only be done by a licensed veterinarian. It's common and effective, but surveys of pet owners show that cost and access to this service are impediments. If the scientists working with ACCD succeed, it will give the cause a remarkable new tool—allowing public and private agencies to limit reproduction, and thereby to prevent the needless and often cruel killing of millions of dogs and cats. As I write, Iraq is now engaging in a mass killing of tens of thousands of street animals, and China has being doing the same in recent years. A contraceptive vaccine, administered through feed or other means, will spell the end of these barbaric methods, sparing dogs and cats from being poisoned, or being bludgeoned and shot on streets and sidewalks.

This new approach of contraception is just one feature of the humane economy, in which human inventiveness is deployed in the service of compassionate solutions. And as today's innovations become tomorrow's standard practice, there is just no rational argument for animal "cleansing" programs of any kind. Those who insist on lethal depopulation plans do so out of stubborn denial or worse—an actual loathing of animals and indifference to suffering. And when it comes to domestic animal populations like dogs or

cats, whether it's here or abroad, we'll soon be able to deploy superior methods of population control. The day will come when people
who prefer poisoning and killings will at least know enough to keep
their thoughts to themselves.

## Twenty-First-Century Science

TECHNOLOGY AND INNOVATION OF a different type can solve the
most complicated issues of animal welfare, and nowhere are the
stakes higher than in animal testing. For more than a half century,
chemicals have been routinely and deliberately put into the eyes,
skin, stomach, and lungs of rodents, rabbits, dogs, and other animals
in a misguided attempt to assess the safety of drugs, cosmetics, industrial chemicals, and other products.

Estimates are that ten to twenty million animals are used
globally in toxicity testing alone, and this number could increase
dramatically given legislative mandates for ever more chemicals to
be tested, all of it ostensibly for the public good. Partly because of
the suffering they endure—the animals are rarely given the benefit
of pain relief—testing of this kind has been a defining issue for
the modern animal-protection movement. The cruelty of toxicity
testing has inspired legislative campaigns and consumer boycotts,
prompting progressive companies to forswear animal testing and to
use "No Animal Testing" branding on their products.

Animal testing is a classic case of doing the wrong thing for the
right reasons. On the one hand, toxicity information is needed to
judge the safety of chemicals and products. However, while poisoning animals might tell us a lot about how large doses of single
chemicals affect small animals with short life spans, it is of little use
in determining how low levels of exposure to mixtures of chemicals
affect larger, longer-living human beings. A rat force-fed a chemical for his or her entire life—often causing painful symptoms such

as tumors and organ failure—does not reliably predict the effects of chemicals ingested by you and me.

As one might expect, the relevance of animal test results is often challenged (remember the legal battles over whether cigarette smoking causes cancer?), leading to endless disputes over the data and to unyielding misery for animals in laboratories. Even under optimum conditions, regulating chemicals on the basis of animal data takes years and relies on guesswork. A report by the U.S. Food and Drug Administration estimates that new drug candidates have only an 8 percent chance of reaching the market, in large part because animal studies so often "fail to predict the specific safety problem that ultimately halts development." Even Francis Collins, now head of the National Institutes of Health, told reporters that animal testing can be problematic for a number of reasons: "It's slow. It's expensive. We are not rats and we are not even other primates."

We are left to wonder why a practice so suspect is nevertheless treated by the scientific establishment as the sine qua non of safety testing. And often the answer may be simple inertia—the settled ways of people locked in a routine and going along with the crowd. Though one would expect scientists to be inquisitive and skeptical, few who conduct animal testing ever get around to asking themselves the most basic of questions: *Why are we really doing this, and is it really doing any good?*

Sometimes it's the person at the top who thinks to ask it. Not long ago, I spoke to my friend Bill Nicholson, who for years was the man at the top of Amway, the sales and manufacturing giant known for its multilevel marketing programs of cosmetics and other household products. Like so many other companies, Amway conducted toxicity testing for its products.

Two decades ago, Nicholson, a brilliant businessman and now a devoted animal advocate, decided he no longer wanted his company's products to be dripped into rabbits' eyes or force-fed to mice and rats. He called Greg Grochowski, the company's director of research and development, and asked him why the company conducted animal tests. It was done to protect the company from legal claims against it, Greg told him, and not because the company wanted to do it or because the government required it. "I want to be out of the animal-testing business within 40 days," Bill declared in response.

Greg heeded the message and wasted no time in hiring additional staff to pull reports and studies showing that the ingredients in Amway's products had already been safety tested. They had to hunt high and low for the information, but the exercise proved that the safety tests had already been conducted. The Amway products would be safe, Greg reasoned, if each of the ingredients had already been deemed safe. Greg put an end to animal testing within the forty-day period—a time frame he and others would have thought impossible, until he was told to do it.

Amway was hardly the first, but it was one of the biggest companies at the time to halt the animal testing of its products and to seek out other ways of ensuring product safety. John Paul Mitchell Systems, The Body Shop, and Tom's of Maine had already sworn off animal testing, as hundreds of other companies have done since (all of them listed in the LeapingBunny.org directory of cruelty-free cosmetics and products). Even now, however, many of the largest cosmetic and consumer product companies in the world still have not kicked their animal-testing habit.

For nearly twenty-five years, European law has required the use of alternative, nonanimal methods when available. Yet here in the United States, federal lawmakers have so far refused to follow suit, or even to take such basic steps as prohibiting duplicative testing or requiring companies to share existing test data. The Animal Welfare Act calls for basic care standards for animals used in testing, which offers some comfort until you recall that this law does not even cover the use of lab-bred mice, rats, or birds, even though they comprise more than 95 percent of warm-blooded animals used in experiments.

The HSUS policy on animal testing is centered on the "Three Rs"—refining techniques to eliminate pain and distress, reducing the number of animals in protocols, and ultimately replacing the use of animals altogether—an approach first laid out in 1959 by the English scientists William Russell and Rex Burch. This struck me as the right way to address the issue, rejecting complacency about animal use while striking a practical note about getting companies steeped in the world of animal testing to themselves commit to a plan to find their way out of it.

For years, our HSUS board of directors was led by a medical doctor, David Wiebers, a neurologist then at the Mayo Clinic, and our current vice chair is also a medical doctor, Jennifer Leaning, who has long taught at the Harvard School of Public Health. Their presence has blunted the argument thrown at animal advocates that we were hostile to science or public health and safety. Our line of argument is clear: there are moral, financial, and public health costs to the use of animals, and we should all agree that phasing out and ultimately eliminating such use is in everybody's interest.

Some companies, including consumer product giants Procter & Gamble and Unilever, have invested tens of millions in alternatives development, and other companies are now doing the same. In 2008, in response to a campaign by Humane Society International, the European Union committed to an aggressive timetable for adoption of scientifically proven nonanimal methods as they become available. In fact, EU member countries overwhelmingly adopted the nonanimal skin irritation tests as full and complete replacements to the use of animals. It's now just a matter of time before the animal-based skin test becomes a thing of the past in Europe. Since that requirement affects the practices of so many multinational corporations, operating in both Europe and America, it's safe to say they will adopt a uniform policy modeled on Europe's reforms.

The large companies that account for most testing can also see the stirrings of change in America, if they are paying attention. The most hopeful sign came with a 2007 report by the U.S. National Academy of Sciences (NAS), *Toxicity Testing in the 21st Century: A Vision and Strategy*. That report laid out a long-term proposal for shifting largely, if not entirely, to nonanimal testing methods for chemicals, drugs, and consumer products. The new approach, drawn from the deliberations of an expert committee, relies on modern advances in biology and technology and emphasizes human—rather than animal—biology. Some experts believe the proposed research and development can be completed within a decade, though funding from Congress and industry will be needed to meet this objective.

Instead of observing signs of severe sickness or death in chemically overdosed animals, the strategy proposed by the NAS is to study how chemicals interact with cellular "pathways" in the human body at environmentally relevant doses. The strategy involves looking individually at the various cell types in the human body (brain, skin, lung, liver, etc.). Each type is individually tested in a cell culture for different kinds of toxic response. To reconstruct the whole-body scenario again, highly sophisticated computer-based approaches, called "systems biology" and "pharmacokinetic models," are used. These relate toxicity information at the cellular level to expected real-world conditions for a living, breathing human being.

To evaluate the cancer-causing potential of a single chemical in a conventional animal test takes up to five years, eight hundred animals, and upwards of $4 million. For the same price and without any use of animals, as many as 350 chemicals could be tested in less than one week in two hundred different cell or gene tests using a high throughput, robot-automated approach. If we are ever to clear the current backlog of literally thousands of chemicals with little or no safety data in order to properly protect human health and the environment, we need credible tests that can deliver results in days, not years.

Cutting-edge scientific work is already under way worldwide to develop the next generation of nonanimal testing methods. More than $250 million in alternatives research funding has already been provided by the EU, with an additional $30–50 million being invested per year by multinational companies. The EU has also provided funding for an ambitious Humane Society–led initiative called "AXLR8," which aims to coordinate worldwide research in this area to help accelerate the transition to a future in which safety testing is animal-free.

In the United States, we have been successful in helping to secure federal funding for alternatives, and "Tox 21"—an interagency collaboration among the Environmental Protection Agency, the National Toxicology Program, the Chemical Genomics Center of the National Institutes of Health (NIH), and the Food and Drug Administration. Through this endeavor, scientists are identifying key cellular pathways in humans, and using high-throughput cell tests to rate hazardous chemicals. We are also pursuing an international strategy for broad, regulatory acceptance of these new approaches as soon as possible. At the same time, we've helped persuade a few states to adopt reforms modeled on the EU's requirement to use alternatives wherever they exist—which is not only good in itself, but also shows the way for Congress.

So, with respect to toxicity testing, the key question is: Do we have the will to achieve the goal of total replacement in ten to fifteen years or will it take decades? To achieve this goal without more years of delay, and millions more animals sacrificed, we need something akin to the Human Genome Project, with a total budget of several hundred million dollars per year and a focused and coordinated research-and-development program.

We should work together—government, industry, and

advocates—to find a way out of this morass of animal testing. Continuing to treat animals as "guinea pigs," even in the face of evidence that the results are of limited value to human safety and even knowing the enormous costs in dollars and innocent lives, is a fool's errand. It is a twentieth-century approach to public health, discredited by serious science as much as by clear moral reasoning, and human creativity and ingenuity can show us a better way in the twenty-first century. If those qualities of genius belong anywhere, they belong in the realm of science. Applied to the great mission of freeing us from the use of animals, they can make modern science a powerful force for the humane economy, instead of just a drag upon it.

## Innovation and the Spark of Life

PEOPLE CAN SURPRISE YOU, and I've learned over the years that sometimes the best allies can be your former adversaries. It's a rare thing to meet someone who doesn't feel some measure of concern or sympathy for animals, or at least the occasional moment of doubt about the cruelties committed at their expense. We're a cause made up of converts; the doors of this movement are open to anyone whose heart and mind lead them inside.

And sometimes the least likely people take you up on the invitation. Such is the pull of our bond with animals that now and then you come across someone who has every reason to turn away from the cause, yet somehow still answers the call. I think of Chuck Anderson, a fellow you wouldn't expect to care much for wildlife and for sharks in particular. If anyone is entitled to despise these creatures, or even to wish them misfortune, it would be Chuck, who encountered a shark in June of 2000.

He was swimming in the warm waters off the coast of Alabama in the Gulf of Mexico when a bull shark suddenly attacked. No dolphins around this time: it was just Chuck and the shark. He tried to fend off the animal, but its first bite took off four of his fingers. The shark then spun around and aimed for Chuck's midsection— this time doing little damage. The third and final bite was the worst, severing Chuck's arm below the elbow. The shark then retreated and disappeared, and Chuck struggled to shore, grievously wounded. Bleeding profusely, he went into shock. He was saved only by some fast-working medics, and in the end he felt lucky that he lost only a limb.

Nine years after the attack, in July 2009, Chuck and eight other shark attack victims made the unlikeliest trek. They went to Capitol Hill to ask Congress to halt the needless killing of sharks. Specifically, they made the case for a ban on shark finning—the practice of catching sharks, cutting off their dorsal fins, and then dumping their bleeding bodies in the ocean. All of this to make shark fin soup, considered a delicacy by tens of millions of Chinese. Every year close to a hundred million sharks are killed for their fins, causing untold cruelty and devastation to the many species worldwide.

If it takes a big man to lose an arm to a shark and still rise in opposition to the wrongs done to these creatures, what are we to say of people who perpetuate such cruelties, all for a bowl of soup? I wouldn't necessarily quarrel with people harboring ill will toward animals who had attacked them. But the sight of these men and women championing the cause of animals, in spite of bitter experiences, is an exceptional expression of human altruism. They didn't blame sharks for the harm they suffered; they traveled the path from forgiveness to advocacy. And their example is more than a witness to Congress, it is a witness to all of us—showing the human capacity to look beyond ourselves and find goodness in all creatures.

This story is unusual in degree, but I've seen the same spirit in action all my adult life, in the company of people who take up the cause of the defenseless. And though the issues and arguments have changed over the years, the basic convictions have endured. At its best, the cause of animal protection is one of the more altruistic concerns you'll find. It's a cause that arises from some of the best instincts of humanity. It reminds us that animals have claims of their own in the world. They are not just here to be used and killed. They are not just things, or resources, or commodities, or targets, or economic opportunities in the waiting. Animals have the same spark of life that we have, formed from the same dust of the earth. They want to live just as badly as we do. Often, they experience life as we do. They can feel playful or angry, affectionate or afraid, sad or joyful.

In the end, the case for animals arises from the recognition of these qualities, and from the sense of kinship they instill. The case stands on its merits and needs no other concerns or connections to give it importance. Yet today, more than ever, there is indeed a close connection between cruelty and other pressing social concerns, and that reinforces the case for animal protection in the modern era and makes it relevant to every one of us.

No longer are the consequences of cruelty confined to its intended victims. So many troubles in the world can be traced in part to the abuse of animals. If climate change is a matter that concerns you, then the environmental costs of raising sixty billion animals for food—more and more of them on factory farms—also warrant your urgent attention. If the spread of disease and the danger of pandemics are a threat, then we have to get serious about stemming the exotic animal trade, the global cockfighting subculture, and mass-confinement methods of producing poultry—since these industries are incubators of diseases that can jump from animals to people. If you wonder about domestic abuse or violent crime in our communities, then look no further than how people get started. Often it begins with the abuse of animals and the loss of empathy.

Civilized societies have always known that to abuse any animal is wrong. Yet civilized societies have always tolerated various forms of animal abuse. In the slow and halting progress of humanity, we too often resist change and defend what we've become accustomed to. Every age has its terrible blind spots, its massive moral lapses, and this is one that we must overcome in our age. We have exhibited such incredible ingenuity in building systems of exploitation—now we must use that same gift to replace them with the new systems of a humane economy. We need laws, and standards, and clear, bright lines to halt cruelty and abuse. It is for us to exercise restraint, because in our dealings with other animals, there is an asymmetry in power. We are entirely in control, and how we use our power is an important measure of our character and our own lives.

No matter what the reform up for debate, we will always hear from those skeptics who warn of the radical implications of change—and in a way, they are right. There is always something a little radical when we dare to live up to our own beliefs. We deal with those skeptics all the time in our work at HSUS. And apart from their cynicism, what always strikes me the most is the static nature of the arguments that our opponents make. They cling to old ways and old traditions as if nothing is ever allowed to change, no matter the costs or the harm.

The sealers in Atlantic Canada talk as if life could not go on without forever clubbing and killing newborn seal pups. And as soon as one reason for the slaughter is disproved, they'll quickly pick up another. They don't care that the world is moving on. They just want to keep things exactly as they are. It's the same with the whalers in Japan and Norway. They know that whale watching is a potentially more lucrative and sustainable source of revenue, but they don't want to hear it and they don't want to change. They just want to go on killing, and without international action they would stay at it and haul in and carve up thousands of whales, even the rarest. The same is true of so many of the other interests and industries we contend with at the HSUS—the fur farmers and the trappers, the horse slaughter plants and the killer buyers, the trophy hunters, the animal-testing lobby, the cockfighters, and on and on.

They all want to live in a static world, doing the same thing, in the same way, forever.

In an odd way, however, this is good news for the cause of animal protection, because it means that human innovation is our ally. We don't just have human conscience working for us, though that will always take the lead. We have all the resourcefulness of the human mind—the boundless capacity of this nation and others to change, to question, to improve, and to shake off old ways. The cause of animal protection speaks not only to the conscience of America. It speaks as well to the ingenuity of America. And that is always a force you want on your side.

Every so often you come across one story that captures the big picture in animal welfare. And I found one not long ago about a whale who was killed by native hunters off the Alaskan coast a few years back. These self-described subsistence hunters hauled in a bowhead whale only to find that they were not the first to hunt that particular creature. Embedded in the flesh of the whale were the fragments of a bomb lance, traceable to a type of shoulder gun last used before 1890.

Only in recent years have we learned how long whales live, and this creature killed in the year 2007 was at least 130 years old. The lance carried a small metal cylinder fitted with a time-delay fuse, but it had failed to kill the whale, and he survived the span of the entire twentieth century without further harm. When Edison was working on the phonograph, this whale was feeding on plankton and diving in Arctic waters. Before Theodore Roosevelt was president, before he had even charged up San Juan Hill, this whale was learning his migration routes.

He lived all that time, dodging the orcas that are the only natural predators he would have to face, only to be slaughtered by men with harpoons. But the hopeful side of the story is how much the world can change, in a hundred years, even in the life of a whale. It was a century that began with the old economy of hunting and killing whales and ended with a new economy of appreciating whales and watching whales. It was a century that began with a lonely few animal-welfare groups, a scarcity of laws to constrain human greed, and a worldview that animals were there for the taking. But by the end of that century, there were hundreds of new groups, thousands of new laws to shield animals from cruelty and abuse, and a deeper understanding that we are custodians of the other creatures—called to defend them and to be their voice. A lot can happen in a hundred years, or in much less time. The life and fate of this single creature tells us that change for the better is not just a matter of chance. It is, in the way of all great causes, a matter of choice, and it begins with each one of us.

# Fifty Ways to Help Animals

THE CAUSE OF ANIMAL protection cannot succeed without you. Here's a list of fifty potential action items for you and others who want to become more involved and to help animals in practical ways.

## Personal Behavior

- Follow the 3 Rs of eating: reducing your consumption of meat and other animal-based foods, refining your diet by avoiding animal products derived from factory farming, and replacing meat and other animal-based foods with vegetarian foods as you are comfortable doing so.
- Purchase cruelty-free cosmetics and household products. Look for labels and messaging on the products that indicate that the product came to the market without animal testing.
- Avoid wearing fur. HSUS maintains a list of fur-free retailers, designers, and brands and our guide on how to tell real fur from fake.
- Adopt a friend for life from a local animal shelter or foster an animal waiting for a home. Support your local shelter.
- Learn about nonlethal methods of managing urban wildlife and about creating sanctuary for wildlife around your home.

- Microchip your pet and place I.D. tags on your animals and encourage others to do the same. And keep your cats safe indoors.
- Provide for your animals' future in case you can't care for them by creating a pet trust. Many states have specific laws to allow for these trust instruments.
- Prepare a disaster kit for your animals, in case of a hurricane, tornado, earthquake, or other natural disaster.
- Sponsor a "Stop Puppy Mills" billboard in your community.
- Ask your local restaurants and grocery stores to switch to cage-free eggs. It's an easy way for retailers to help combat one of the most extreme confinement methods in industrial agriculture.
- Join us in applauding pet stores that have taken a stand against puppy mills by joining our Puppy Friendly Pet Stores initiative. Encourage local stores that do sell puppies to stop, and to work with local humane organizations to adopt out homeless animals, as PETCO and PetSmart do.
- Help spread the word about animal protection efforts on Facebook, MySpace, Twitter, YouTube, and other online networks.
- Add an HSUS video (humanesociety.org/video) to your website, blog, or social networking page.
- Download HumaneTV, the HSUS's video app for the iPhone and iPod touch, for an easy way to keep up to date on the organization's far-reaching work to protect all animals.

## *In Your Community*

- Write letters to the editor on animal protection issues and encourage radio and television talk shows to present animal issues (humanesociety.org is a great resource for information).
- Ask your local radio and television stations to air the Shelter

Pet Project PSAs, an unprecedented media campaign aimed at boosting pet adoptions nationwide.

- Distribute animal welfare literature at events and stores.
- Work to engage your church or place of worship with animal-protection issues. The HSUS's Faith Outreach program offers many resources.
- Encourage your office to implement dog-friendly policies. Our book *Dogs at Work: A Practical Guide to Creating Dog-Friendly Workplaces* provides step-by-step advice.
- Help feral cats in your neighborhood with our Trap-Neuter-Return resources.

## Get Training—Get Activated

- Participate in Disaster Animal Response Team (DART) training and sign up to be an HSUS disaster responder. Crises strike all parts of the nation, and our volunteers are a critical component in a successful response.
- Attend or help organize an HSUS Lobby 101 training program in your community, in order to build a more effective grass-roots movement to help animals.
- Train to become a humane educator, develop outreach programs in your community.
- Attend the HSUS's Animal Care Expo (location varies), Taking Action for Animals conference (Washington, D.C.), or the Genesis Awards (Los Angeles).
- Take a workshop or online course through Humane Society University.
- Organize a Spay Day USA event in your community.

## Volunteer to Help Animals

- Volunteer to monitor property with the HSUS's Wildlife Land Trust.
- Volunteer for egg addling or other humane wildlife control programs for Canada geese or other species targeted for killing by government officials.
- Volunteer time and skills with your local animal shelter or rescue group.
- Explore volunteer opportunities with our Humane Society Veterinary Medical Association Field Services program.

## Civic Activities

- Determine which elected officials represent you at local, state, and federal levels.
- Contact your federal and state legislators about animal protection issues. There is a raft of pending legislation before Congress.
- Sign up for free HSUS e-mail alerts at humanesociety.org/join to get involved in helping to pass state and federal legislation and to have existing laws enforced.
- Study our legislative priorities and attend lawmakers' town meetings to urge them to support these issues.
- Work for the passage of local ordinances in your community, for example to protect chained dogs or improve the lives of dogs in puppy mills.
- Register to vote.
- Gather signatures for and help pass animal protection ballot initiatives in your state.

## Work with Local Schools

- Sponsor a local classroom and give the gift of *KIND News,* our award-winning newspaper for children in grades K–6, to elementary students or a young animal lover you may know.
- Engage kids, tweens, and teens with humane education activities and lesson plans.
- Help students start animal clubs at schools.
- Work to get your local universities or your children's schools to join our "Cage-Free Campus" campaign or to add vegetarian options to their menu.
- Book and sponsor animal-friendly staff for lectures at schools and universities.

## *Shopping*

- Shop at Humane Domain (humanesociety.org/humane domain)—the HSUS's online store—for pet products, Cause Gear that supports our campaigns and programs, and unique gifts for animal lovers.
- Purchase pet health insurance from Petplan and receive a 5 percent discount. Use code SPD20002.
- Shop with HSUS Corporate Supporters at humanesociety.org/ shop and help animals with every purchase—magazine subscriptions, personal checks, coffee, flowers, wine, jewelry, skins for electronic devices, custom gifts including stamps and cards, and more.

## Fund-Raising and Networking

- Make a personal annual gift to the HSUS or sign up for an automatic monthly pledge, make a memorial gift in honor of a friend or animal companion, or give gift memberships to friends or family members. Members receive *All Animals* magazine.
- Arrange coffee or lunch dates to introduce the HSUS and our programs to people who care about animals or host a house party through the HSUS's philanthropy department, and have one of our executives or subject experts speak at the event.
- Ensure that the HSUS is eligible for giving programs at your workplace.
- Donate your used vehicle to benefit the HSUS, and include the HSUS as a beneficiary in your will.
- Support other animal-welfare charities you care about.

# ACKNOWLEDGMENTS

In more than a quarter century in the field of animal protection, I've learned from so many people along the way. I regret that I cannot thank them all here.

I do wish, however, to thank some individuals who helped very directly with this work. Meg Olmert helped me on the biochemistry of the human-animal bond, and Steve Kellert was generous in talking to me about the broader connections between humans and nature. James Serpell strengthened my understanding of pet-keeping through the ages. My friend Ed Duvin thoughtfully encouraged my work at many levels, as did another friend, Bill Nicholson. Laura Hobgood-Oster helped me to develop my ideas on animal sacrifice and religion by discussing her scholarship with me.

On the issue of animal intelligence, I want to thank Barrett Duke of the Southern Baptist Convention for sharing his insights at just the right time about the emotional intelligence of animals. Marc Bekoff and Jonathan Balcombe, through their writings, are helping to change the way the world views animal consciousness and cognition, and I am grateful to them both. Jane Goodall, Alexandra Horowitz, Jeffrey Masson, and Peter Singer have been pioneers in reaching a mass audience and helping to open up so many people to new possibilities in our relationship with animals.

Ron Kagan shared his experiences, good and bad, in the zoological profession. And Matt Smith of the Central Virginia Parrot Sanctuary reminded me of the commitment that so many animal rescuers bring to this work. Judd Hanna gave me a disturbing case example of the NRA's heavy hand with state fish and wildlife agencies and commissions. As a longtime dog breeder, Ted Paul drew the wrath of some within the dog-fancy world when he spoke out against puppy mills, yet he never strayed from doing the right thing for dogs.

Ken Cook of the Environmental Working Group helped me sort through the tangle of agricultural subsidies in the United States, while Josh Reichert, Robert Martin, and Michael Blackwell, especially through their work on the Pew Commission on Industrialized Farm Animal Production, gave me and all of us a landmark and critical analysis on the problems associated with factory farming. I developed a friendship with Kevin Fulton, a Nebraska cattle rancher, during the writing of the book, and he reminded me that there are many people involved in agriculture who are willing to speak out about problems within the profession. I so enjoyed visiting his farm and seeing the animals.

In writing this book, I had the best faculty in animal protection behind me: the staff of the HSUS. On wildlife issues, I am grateful to Rebecca Aldworth, Stephanie Boyles, John Grandy, John Hadidian, Andy Page, Heidi Prescott, Naomi Rose, and Teresa Telecky. Sara Amundsen, Troy Seidel, and Martin Stephens shared their extraordinary knowledge of the use of animals in research, testing, and education.

John Goodwin and Ann Chynoweth helped by sharing their detailed knowledge of animal fighting. Jon Lovvorn and other HSUS in-house attorneys, including Peter Brandt, Peter Petersan, and Leana Stormont, provided support on many chapters, particularly those dealing with wildlife and farm animals. Jon, who spearheads our animal protection litigation strategy, is one of the most talented and hardworking lawyers I've ever known.

I am grateful to HSUS's incredible team of advocates and scientists working on farm animal welfare, including Paul Petersan, Josh Balk, and Sara Shields. I am indebted to Michael Greger, a medical doctor on our staff, whose scholarship on the intersection of animal agriculture and public health is incomparable. Paul Shapiro, senior director of our Factory Farming Campaign, is one of the most gifted animal advocates in the United States, and I so appreciated his guidance and suggestions.

On companion animal issues and on Hurricane Katrina, I am grateful to Carrie Allan, Adam Goldfarb, Betsy McFarland, Nancy Lawson, Laura Maloney, Cory Smith, Stephanie Shain, John Snyder, and Kathleen Summers, who provided information and helped me sharpen my understanding. Paula Kislak, a veterinarian and HSUS board member, focused my attention on hereditary and genetic problems among purebreds.

Christine Gutleben was my guide on the topic of animals and religion. I received much-needed help from Jennifer Fearing, Nancy Perry, and others in our public policy operation. Roger Kindler, our esteemed general counsel, gave me great advice, as always, and applied his exemplary care and scrutiny to the manuscript. Michael Markarian, our chief operating officer, is a tireless and brilliant colleague, and provided steadfast support and encouragement throughout this project. Andrew Rowan, our chief scientific and international officer, gave me the benefit of his unparalleled knowledge of scientific and technical matters relating to animal welfare.

I've worked for almost seventeen years at the HSUS, and it is truly a one-of-a-kind organization—the most influential and important group that the field of animal protection has ever known. I was appointed president and CEO by the board of directors in 2004, an event that marked a major moment in my life and, I hope, in the path of this organization. I have never forgotten the faith that two dozen elected board members placed in me at that time, and I have vowed never to give them cause to regret it. The trust we share, to steward a great organization and to make it greater, is a sacred one.

Whatever my debt to the whole board—and it is an incredible

group of giving, selfless people—I would be remiss in not expressing my deepest gratitude to our then board chairman, Dr. David Wiebers. David is a visionary man who has helped the organization to navigate its way through many critical moments, including an election process with many outstanding candidates for the position of CEO.

Anita Coupe was vice chair at the time of my election, and no one was a more steadfast supporter. I'll be forever grateful to her for that confidence, as well as for her active support of this book project, and it is truly my pleasure to work with her on a daily basis now that she's been elected chair of the board. I must also express particular thanks to Joe Stewart and Patrick McDonnell for their incredible trust and support of me in my professional pathway and in this work; two finer friends I'll never know. Two others in our current board leadership, Jennifer Leaning and Rick Bernthal, give so much of their time and energy to the organization, and we are so much the better for it. I also have an incalculable debt to Marian Probst, who is an HSUS board member, but has always been much more than that to me. She and Cleveland Amory founded the Fund, where I once worked. She was Cleveland's close collaborator on his books, and I benefited from her advice and support in my own journey with this book.

There are a few folks who really dug in to help me with the research, editing, and framing of the whole work. Lewis Bollard was an HSUS intern from Harvard, who evinced an incredible understanding of our work, and he didn't hesitate to sign on when I asked him to help. He was the first to see each chapter I turned out, and his edits and criticism immeasurably improved the book. I am grateful for his research concerning animal intelligence. I know he will do great things for the cause of animal protection once professors at the Yale Law School are done with him in a couple of years.

John Balzar is a remarkable journalist who shared a Pulitzer

Prize at the *Los Angeles Times* and wrote an engaging and beautifully written book on the Yukon Quest before joining the HSUS staff as our head of communications. My book would be poorer but for his advice and guidance, especially his insistence that I infuse it with my voice and personal experiences. I could not be more grateful to him as a friend, colleague, and mentor.

My colleague Bernard Unti and I are about the same age, and we both came to the field of animal protection in our late teens. We've been friends from the beginning, and I've watched with pride as Bernie earned his PhD in history and distinguished himself as a leading historian of the humane movement. He has been unfailingly generous at every turn, and his imprint and perspective extend far beyond his own scholarship in the field of history. I am immensely grateful to him for his support and friendship.

John McConnell and Matthew Scully have been speechwriters for presidents, vice presidents, and presidential candidates for the last two decades, and they are gifted writers, editors, and thinkers. Matthew and I spent countless hours on Skype reading, debating, and honing the ideas in this work. Matthew is the finest writer I've known, and his own book, *Dominion*, is an incredible literary achievement and an inspiration to me. Not only do I consider him one of my dearest friends, but I am profoundly grateful every day that he's used his special gifts to benefit the cause of animals.

I am also tremendously indebted to my fiancée, Lisa Fletcher, without whom this book would never have been completed. Writing a book in a year's time was no small undertaking, especially since I maintained all of my duties as CEO of HSUS. But Lisa was unfailingly patient with me and gave me the support and space I needed to research and write. She too urged me at every turn to put my voice as well as my heart into it, and on long walks together, we talked about every detail, and the structure, narrative, and flow of this book emerged. Little did she realize that in encouraging me to use my voice, she would then have to suffer my reading the book aloud. She helped me immensely with my final revisions and notes.

Finally, I offer thanks to my family. My mother, Pat, taught me about kindness and compassion, and my father, Richard, who was a football coach, about competition and fair play. My life's work to help animals has been propelled by the lessons they've taught me. My brother, Richard, eleven years my senior, has been my greatest mentor throughout life. My sisters, Kim and Wendy, are among the most giving and generous people I know. As a son and a brother, I could not be more proud.

Much credit is due to the entire publishing team at William Morrow/HarperCollins for having the confidence in me to take up this project. My editor, Peter Hubbard, had great enthusiasm for the original idea. Throughout the process, he was a pleasure to work with, and his suggestions and editorial acumen sharpened this book greatly. My agent, Gail Ross, pushed me hard to construct the book proposal, knowing it was in me somewhere. During our discussions, Gail and her team locked in on the concept of "the bond," and I am so grateful to them for their insight.

To each of the people named above, and to the others not properly acknowledged here, thank you. It's been a decades-long ambition to pull together my thoughts in book form. I hope it helps advance a great and noble cause.

# NOTES

Page                                           *Preface*

ix  In the famous phrase of Edmund Burke: Burke, Edmund, *Reflections on the Revolution in France* (London: Penguin Classics, 1986), 135.

x  I came across a story by Kate Murphy: Murphy, Kate, "Birdhouses Designed for Repeat Visitors," *New York Times,* August 11, 2010.

xi  Abby Sewell of the *Los Angeles Times*: Sewell, Abby, "Rehabbers' Perform a Grueling Labor of Love," *Los Angeles Times,* August 1, 2010.

### Introduction: Sanctuary

2  Babe's entire family was killed: Olson, Deborah, *North American Region Stud Book for the African Elephant*, Indianapolis Zoo, 2008, p. 19; The Fund for Animals, "Statement About Babe," July 5, 2006, http://www.fundforanimals.org/ranch/residents/babe_statement.html.

4  the price tag just for day-to-day operations: Moxley, Angela, "Animals Find Quiet Refuge at the Ranch," August 28, 2009, http://www.blackbeautyranch.org/about/animals-find-quiet-refuge-at.html.

5  Nim had bounced around various research facilities: Terrace, Herb S., *Nim: A Chimpanzee Who Learned Sign Language* (New York: Columbia University Press, 1987); Hess, Elizabeth, *Nim Chimpsky: The Chimp Who Would Be Human* (New York: Bantam Press, 2008).

5  he learned 125 American Sign Language (ASL) signs: Terrace, H. S. "A Report to an Academy, 1980," *Annals of the New York Academy of Sciences* (1981): 98.

5  with the lead researcher, Dr. Terrace, concluding: Terrace, H. S., L. A. Petitto, R. J. Sanders, and T. G. Bever, "Can an Ape Create a Sentence?" *Science* 206 (1979): 891–902; Terrace, H. S., "A Report," 94–114.

5  scientists have decoded the genome: National Institutes of Health; "New Genome Comparison Finds Chimps, Humans Very Similar at the DNA Level," August 31, 2005, http://www.genome.gov/15515096; The Chimpanzee Sequencing and Analysis Consortium, "Initial Sequence of the Chimpanzee Genome and Comparison with the Human Genome," *Nature* 437 (2005): 69–87.

7  exposing in March 2009 the abuses: Alpert, Bruce, "New Iberia Research Center Routinely Mistreats Chimps, Humane Society Says," *New Orleans Times Picayune,* March 4, 2009; ABC News, "EXCLUSIVE: Ex-Employees Claim 'Horrific' Treatment of Primates at Lab," March 4, 2009, http://abcnews.go.com/Nightline/story?id=6997869&page=1.

8  About a thousand chimps: Humane Society of the United States, "Frequently Asked Questions about Chimpanzees in Research," http://www.hsus.org/ animals_in_research/chimps_deserve_better/chimpanzees_in_research_fact .html, January 27, 2009.

8  he and his hired cowboys: Hoffman Marshall, Julie, *Making Burros Fly: Cleveland Amory, Animal Rescue Pioneer* (Boulder, CO: Johnson Books, 2006), 65.

9  "to induce kindness, sympathy, . . .": Merriam-Webster, "Black Beauty," *Merriam-Webster's Encyclopedia of Literature* (Springfield, MA: Merriam-Webster, Inc., 1995).

9  exotic deer from Africa and Asia: The Fund for Animals; "Blue Boy's Broken Horn Is Pitch Perfect," August 31, 2005, http://www.fundforanimals.org/ ranch/residents/blue_boy.html.

9  twenty-one prairie dogs, who faced gassing: The Fund4r Animals; "Groups Sue Federal Government Over Massive Prairie Dog Poisoning at Colorado Prison," November 24, 1999, http://www.thefreelibrary.com/Groups+Sue+Federal+ Government+Over+Massive+Prairie+Dog+Poisoning+At . . . -a057790510.

10  boom in raising ostriches: The Fund for Animals, "Rescued Ostriches Find Life and Love at the Ranch," September 25, 2007, http://www.fundforanimals.org/ ranch/residents/ostriches_donjuan_yvette_yolanda_yesenia.html.

10  kangaroo named Roo-Roo: The Fund for Animals; "All Boxed Out: Roo-Roo's Injury Results in Retirement from Performing," October 6, 2006, http://www.fundforanimals.org/ranch/residents/rooroo.html.

10  made famous in three of his best-selling books: Amory, Cleveland, *Cleveland Amory's Compleat Cat: Three Volumes in One* (New York: Black Dog & Leventhal Publishers, 1995).

10  Omar the dromedary camel: The Fund for Animals, "Over the Fence and Through the Barn: Babe Overcomes the Species Barrier," July 5, 2006, http://www .fundforanimals.org/ranch/whats_going_on/babe_omar_friendly_scar.html.

11  Mari and Josie were on the slaughterhouse floor: The Fund for Animals, "The Long Road from DeKalb to Murchison: Horses Spared from Slaughterhouse Floor," April 18, 2007, http://www.fundforanimals.org/ranch/whats_going_ on/miracle_horses.html.

11  a federal judge ordered the plant: *Humane Soc. of the U.S. v. Johanns*, 2007 WL 1201610 (D.D.C., March 28, 2007).

12  often acts as an agent of agribusiness: American Horse Defense Fund, "USDA Defies Congressional Ban on Horse Slaughter," February 11, 2006, http://www .thefreelibrary.com/USDA+Defies+Congressional+Ban+on+Horse+ Slaughter.-a0141936481.

13  "animals constitute more than . . .": Steven Kellert is paraphrased in Olmert, Meg Daley, *Made for Each Other: The Biology of the Human-Animal Bond* (Philadelphia, PA: Da Capo Press, 2009), 10.

15  I also learned as an adult: Lonati, Staci, "Animal Shows Charged with Fakery ('Wild America' and 'Wild Kingdom' Under Attack for Alleged Cruelty to Animals)," *St. Louis Journalism Review* 26 (1996): 7. For a recent work on this problem, see Chris Palmer, *Shooting in the Wild: An Insider's Account of Making Movies in the Animal Kingdom* (San Francisco: Sierra Club Books, 2010).

16  Some 170 million: "U.S. Pet Ownership Statistics," December 30, 2009, http://www.humanesociety.org/issues/pet_overpopulation/facts/pet_ownership_statistics.html.

16  The pet product and services industry: American Pet Products Association, "Industry Statistics and Trends," accessed May 6, 2010, http://www.americanpetproducts.org/press_industrytrends.asp.

## Chapter One—The Ties That Bond

22  a conditioned fear of snakes: Öhman, Arne, and Susan Mineka, "The Malicious Serpent: Snakes as a Prototypical Stimulus for an Evolved Module of Fear," *Current Directions in Psychological Science* 12 (2003): 5–9.

22  by age five or so: Olmert, *Made for Each Other,* 95.

22  Primates react much as we do: King, G. E., "The Attentional Basis for the Primate Response to Snakes" (paper presented at the Meeting of the American Society of Primatologists, San Diego, CA, June 1997).

23  Likely the greatest pandemic ever: Anitei, Stefan, "Top 10 Infectious Diseases That Have Killed Millions of People," Softpedia.com, November 13, 2007, http://news.softpedia.com/news/Top-10-Infectious-Diseases-That-Have-Killed-Most-People-70741.shtml.

25  Oxytocin is a neuropeptide: Strathearn, Lane, Peter Fonagy, Janet Amico, and P. Read Montague, "Adult Attachment Predicts Maternal Brain and Oxytocin Response in Infant Cues," *Neuropsychopharmacology* 34 (2009): 2655–2666.

25  Olmert argues that oxytocin: Olmert, *Made for Each Other.*

25  trigger a mother's protective aggression: MacDonald, Kai, and Tina Marie MacDonald, "The Peptide that Binds: A Systematic Review of Oxytocin and Its Prosocial Effects in Humans," *Harvard Review Psychiatry* 18 (2010): 1.

25  it has a broader prosocial: Olmert, *Made for Each Other.*

25  It is a social recognition hormone: MacDonald and MacDonald, "The Peptide That Binds," 16.

25  administering oxytocin to people: Reyes, Teófilo L., and Jill M. Mateo, "Oxytocin and Cooperation: Cooperation with Non-kin Associated with Mechanisms for Affiliation," *Journal of Social, Evolutionary, and Cultural Psychology* 2 (2008): 234–246.

26  much of what she learned: Carter, C. Sue, "Oxytocin and the Prairie Vole: A Love Story," in *Essays in Social Neuroscience,* ed. John T. Cacioppo and Gary G. Berntson (Cambridge, MA: The MIT Press, 2004), 53–64.

26  "mammals to make the connection between . . .": Szalavitz, Maia, and Bruce D. Perry, *Born for Love: Why Empathy Is Essential—and Endangered* (New York: HarperCollins, 2010), 30.

26  the interactions almost doubled: Odendaal, J. S. J., and R. A. Meintjes, "Neurophysiological Correlates of Affiliative Behaviour between Humans and Dogs," *Veterinary Journal* 165 (2003): 299.

26  "one of the most potent triggers": Olmert, *Made for Each Other,* xv; Miller, Suzanne C., Cathy Kennedy, Dale DeVoe, Matthew Hickey, Tracy Nelson, and Lori Kogan, "An Examination of Changes in Oxytocin Levels in Men

and Women Before and After Interaction with a Bonded Dog," *Anthrozoos* 22 (2009): 31–42.

27 In 2009, the U.S. Army began: Lorber, Janie, "For the Battle-Scarred, Comfort at Leash's End," *New York Times,* April 2, 2010, http://www.nytimes.com/2010/04/04/us/04dogs.html.

27 "an innate tendency . . .": Wilson, Edward O., *Biophilia: The Human Bond with Other Species,* 12th ed. (Cambridge: Harvard University Press, 2003), 1.

27 "From infancy . . .": Ibid.

27 To support his conclusions: Ibid.

28 Population geneticists at the University of Utah: Wade, Nicholas, "Genome Study Provides a Census of Early Humans," *New York Times,* January 18, 2010, http://www.nytimes.com/2010/01/19/science/19human.html.

28 There is evidence that hominids: Leakey, Richard, and Roger Lewin, *Origins Reconsidered: In Search of What Makes Us Human* (New York: Anchor Books, 1992).

28 not until the arrival of modern humans: Jurmain, Robert, Lynn Kilgore, Wenda Trevathan, and Russell L. Ciochon, *Introduction to Physical Anthropology* (New York: Wadsworth Cengage Learning, 2010), 392.

28 not until about twelve thousand years ago: Kottak, Conrad Phillip, *Physical Anthropology and Archaeology,* 2nd ed. (New York: McGraw-Hill, 2006), 210.

28 about eleven thousand years ago: Ibid.

28 This "overkill" hypothesis: Burney, D. A., and T. F. Flannery, "Fifty Millennia of Catastrophic Extinctions After Human Contact," *Trends in Ecology and Evolution* 20 (2005): 395–401.

29 climate change may explain: Ibid.

29 "on an exact learned knowledge . . .": Wilson, Edward O., "Biophilia and the Conservation Ethic," in *The Biophilia Hypothesis,* ed. Stephen R. Kellert and Edward O. Wilson (Washington, DC: Island Press, 1993), 32.

29 "walking encyclopedias of natural history . . .": Diamond, Jared, *Guns, Germs, and Steel: The Fates of Human Societies* (New York: W.W. Norton & Company, 1997), 143.

29 "For every one of Kulambangra's . . .": Diamond, Jared, "New Guineans and Their Natural World," in *Biophilia Hypothesis,* ed. Kellert and Wilson, 251–271.

29 "people can turn into animals, . . .": Personal communication with James Serpell, February 3, 2010.

30 Social anthropologist Tim Ingold adds: Ingold, Tim, "Human-Animal Relations," in *What Is an Animal?,* ed. Tim Ingold (New York: Routledge, 1994), 15.

30 Stephen Kellert of the Yale School: Kellert, Stephen R., "The Biological Basis for Human Values of Nature," in *Biophilia Hypothesis,* ed. Kellert and Wilson, 42–69.

30 the dog at least fifteen thousand years ago: Clutton-Brock, Juliet, "Origins of the Dog: Domestication and Early History," in *The Domestic Dog: Its Evolution, Behaviour and Interactions with People,* ed. James A. Serpell (Cambridge, MA: Cambridge University Press, 1995), 245–256.

30 the cat some four thousand years ago: Serpell, James A., "Domestication and History of the Cat," in *The Domestic Cat: The Biology of Its Behaviour,* ed. Dennis

C. Turner and Paul Patrick Gordon Bateson (Cambridge, MA: Cambridge University Press, 2000), 177–192.

31 "The various expressions of biophilia . . .": Kellert, Stephen R., *Kinship to Mastery: Biophilia in Human Evolution and Development* (Washington, DC: Island Press, 1997), 4.

31 thirteen million Americans go hunting: U.S. Fish and Wildlife Service, "Hunting Statistics and Economics," 2006, http://www.fws.gov/hunting/huntstat .html.

31 hunting behavior is still encoded: Shepherd, Paul, *The Tender Carnivore and the Sacred Game* (Athens: University of Georgia Press, 1998).

31 Many hunters speak of "buck fever": The Buck Hunters Blog, "5 Tips for Coping with Buck Fever," 2007, http://www.buckhuntersblog.com/5-tips-for -coping-with-buck-fever.

32 "Humans are not hardwired to hunt . . .": Personal communication with James Serpell, February 3, 2010.

32 "the ancestors of the modern Pekingese . . .": Serpell, James A., "Pet-Keeping in Non-Western Societies: Some Popular Misconceptions," in *Animals and People Sharing the World,* ed. Andrew N. Rowan (Hanover, NH: University Press of New England, 1988), 37.

32 "The practice of capturing and taming . . .": Serpell, James, "Pet-Keeping in Non-Western Societies: Some Popular Misconceptions," *Anthrozoos: A Multidisciplinary Journal of the Interactions of People & Animals* 1 (1987): 166–174.

33 "with greater care than they bestow on . . .": Serpell, "Pet-Keeping," in *Animals and People Sharing,* ed. Rowan, 45.

33 "In the New Guinea villages where I work": Diamond, *Guns, Germs, and Steel,* 165.

33 According to Serpell: Serpell, "Pet-Keeping," in *Animals and People Sharing,* ed. Rowan, 42.

33 There is also evidence that native peoples: Lorenz, Konrad, *Studies in Animal and Human Behavior,* vol. 2 (London: Methuen & Co. Ltd., 1971).

34 has an estimated seven hundred million pigs: "China—Slaughtering 14 Million Pigs per Week," *Meat Trade News,* May 2, 2010, http://www.meattradenewsdaily .co.uk/news/040510/china___slaughtering__million_pigs_per_week_.aspx.

34 India has nearly three hundred million cattle: U.S. Department of Agriculture, "Live Cattle Selected Countries Summary," Foreign Agricultural Service, Office of Global Analysis, Circular Series DL&P 2–07, November 2007, www .fas.usda.gov.

35 Christian tradition places the newborn Jesus: Steiner, Gary, "Descartes, Christianity, and Contemporary Speciesism," in *A Communion of Subjects: Animals in Religion, Science, and Ethics,* ed. Paul Waldau and Kimberly Patton (New York: Columbia University Press, 2006), 117–131.

35 Domestication is among: Diamond, *Guns, Germs, and Steel.*

35 Our special regard for dogs: Driscoll, C. A., D. W. Macdonald, and S. J. O'Brien, "From Wild Animals to Domestic Pets, an Evolutionary View of Domestication," *Proceedings of the National Academy of Sciences* 106, Suppl. 1 (2009): 9971– 9978.

36 Sheep, goats, pigs, and cattle . . .: Diamond, *Guns, Germs, and Steel,* 167.

36   Humans domesticated plants: Ibid., 83.

36   These expanded communities: Ibid., 265.

36   progress began with the domestication: Ibid.

36   U.S. cattle industry is worth: USDA, "U.S. Beef and Cattle Industry: Background Statistics and Information," http://www.ers.usda.gov/news/BSECoverage.htm.

37   thirty million to forty million nomadic pastoralists: Wikipedia, "Nomadic Pastoralism," http://en.wikipedia.org/wiki/Nomadic_pastoralism.

37   In India, which has more cows: Wikipedia, "Cattle," http://en.wikipedia.org/wiki/Cattle#Population.

37   venerated in Vedic scriptures: Harris, Marvin, *The Cultural Ecology of India's Sacred Cattle* (New York: Columbia University Press, 1966).

37   The followers of Jainism: Chapple, Christopher Key, "Reverence for All Life: Animals in the Jain Tradition," *Jain Spirit* 2 (1999): 56–58.

37   only fourteen have been successfully domesticated: Diamond, *Guns, Germs, and Steel,* 168.

37   a few of the species that did not submit: Ibid., 398.

37   Humans have had more luck: Ibid., 158.

37   Horses were among the most difficult: Anthony, David W., "Bridling Horse Power: The Domestication of the Horse," in *Horses Through Time,* ed. Sandra Olsen (Lanham, MD: Robert Rinehart Paperback, 2003), 57–82.

37   the domestication of horses: Ibid.

38   With the aid of horses: Diamond, *Guns, Germs, and Steel,* 67.

38   "Horses changed the way people hunted . . .": Anthony, "Bridling Horse Power," 59.

38   "From being independent coequals . . .": Serpell, James, "Animals and Religion: Towards a Unifying Theory," in *The Human-Animal Relationship: Forever and a Day,* ed. Francien Henriëtte de Jonge and Ruud van den Bos (The Netherlands: Royal Van Gorcum, 2005), 15.

39   Historian Lewis Mumford argues that: Mumford, Lewis, *Technics and Human Development: The Myth of the Machine,* vol. 1 (New York: Harvest/HBJ Books, 1967), 146.

39   Rifkin believes the practice: Rifkin, Jeremy, *The Empathetic Civilization: The Race to Global Consciousness in a World in Crisis* (New York: Jeremy P. Tarcher/Penguin, 2009).

40   "drank of his own cup . . .": King James Bible, 2 Samuel 12:3.

40   Even the animal sacrifices: Klawans, Jonathan, "Sacrifice in Ancient Israel: Pure Bodies, Domesticated Animals, and the Divine Shepherd," in *A Communion of Subjects: Animals in Religion, Science, and Ethics,* ed. Paul Waldau and Kimberly Patton (New York: Columbia University Press, 2006), 65–80.

40   For thousands of years: Patton, Kimberley, "Animal Sacrifice: Metaphysics of the Sublimated Victim," in *Communion of Subjects,* ed. Waldau and Patton, 391–405.

40   Laura Hobgood-Oster says: Hobgood-Oster, Laura, *Holy Dogs and Asses: Animals in the Christian Tradition* (Champaign: University of Illinois Press, 2008).

40   Jonathan Klawans writes of sacrifice: Klawans, "Sacrifice in Ancient Israel," 70.

41   act of mercy to a stray lamb: King James Bible, Exodus Rabbah 2:2.

41   Historian Kimberly Patton notes: Patton, "Animal Sacrifice," 391–405.

41   Patton also argues: Ibid.

41   famed ethnographer Bronislaw Malinowski: Quoted in Hubert, Henri, and Marcel Mauss, *Sacrifice: Its Nature and Functions* (Champaign: University of Chicago Press, 1964), 33.

42   in the Christian tradition: Patton, "Animal Sacrifice," 391–405.

42   animal sacrifice continues: Wise, Steven, "Animal Law and Animal Sacrifice: Analysis of the U.S. Supreme Court Ruling on Santería Animal Sacrifice in Hialeah," in *Communion of Subjects,* ed. Waldau and Patton, 585–587.

42   Gadhimai Festival in southern Nepal: Humane Society of the United States, "Mass Animal Sacrifice Planned in Nepal," November 20, 2009, http://www. hsus.org/hsi/confronting_cruelty/animal_cruelty_around_the_world/gadhi mai_festival_112009.html.

43   Hialeah, Florida, outlawed animal sacrifice: Wise, "Animal Law and Animal Sacrifice," 585–587.

44   "If to be feeling alive . . .": William Wilberforce, quoted in Carey, Brycchan, "William Wilberforce's Sentimental Rhetoric: Parliamentary Reportage and the Abolition Speech of 1789," *The Age of Johnson: A Scholarly Annual* 14 (2003): 281–305.

44   "his solicitous care . . .": John Paul II, Message of Reconciliation, March 12, 1982, as quoted in Scully, Matthew, *Dominion: The Power of Man, the Suffering of Animals, and the Call to Mercy* (New York: St. Martin's Griffin, 2002), 23–24.

45   "tyranny or cruelty towards any brute . . .": "Charters and General Laws of the Colony and Province of Massachusetts Bay," 1641, 95.

45   John Locke advised parents: Locke, John, *Some Thoughts Concerning Education and of the Conduct of the Understanding,* ed. Ruth W. Grant and Nathan Tarcov (Indianapolis, IN: Hackett Publishing Co., Inc., 1996).

45   Bentham famously argued: Bentham, Jeremy, *An Introduction to the Principles of Morals and Legislation* (London: Clarendon Press, 1823), 311.

45   "Can there be one kind of justice . . .": Lawrence, John, *A Philosophical Treatise on Horses, and on the Moral Duties of Man towards the Brute Creation* (London: Longman, 1796–1798), quoted in Henry Salt, *Animals' Rights Considered in Relation to Social Progress* (London: Macmillan, 1894), 149.

45   "All things are void of terror": Shelley, Percy Bysshe, *Queen Mab, A Philosophical Poem with Notes* (New York: Online at Google Books, 1831), http://books .google.com/books?id=6NcIAAAAQAAJ&printsec=frontcover&source=gbs _ge_summary_r&cad=0#v=onepage&q&f=false, 59.

45   "The assumption that animals are without rights . . .": Schopenhauer, Arthur, *On the Basis of Morality,* trans. E. F. J. Payne (Providence, RI: Berghahn Books, 1995).

46   the reformer Henry Salt: Salt, *Animals' Rights,* 10, 114.

46   America's first anticruelty laws: Favre, David, and Vivien Tsang, "The Development of Anti-Cruelty Laws During the 1800s," *Detroit College of Law Review* 1 (1993): 8; and Stewart Leavitt, Emily, and Diane Halverson, "The Evolution of Anti-Cruelty Laws in the United States," in *Animal Welfare Institute, Animals and Their Legal Rights: A Survey of American Laws from 1641 to 1990,* 4th ed. (Washington, DC : Animal Welfare Institute, 1990), 4.

46   New York socialite Henry Bergh founded: Lane, Marion S., and Stephen L. Zawistowski, *Heritage of Care: The American Society for the Prevention of Cruelty to Animals* (Westport, CT: Praeger Publishers, 2008).

47   Historian Bernard Unti calls: Unti, Bernard, *The Quality of Mercy: Organized Animal Protection in the United States 1866–1930* (Ann Arbor, MI: Proquest/UMI Dissertation Services, 2002).

47   Katherine Grier notes: Grier, Katherine C., *Pets in America: A History* (Chapel Hill: University of North Carolina Press, 2006), 131.

47   Congress's first animal-protection law: Unti, *Quality of Mercy*.

48   Bernard Unti notes: Ibid., 380.

49   In the nineteenth century, market hunters: Cronon, William, *Nature's Metropolis: Chicago and the Great West* (New York: W.W. Norton & Company, 1991).

49   More than twelve million horses: Sinclair, Upton, *The Jungle* (New York: Sharp Press, 2003).

49   Henry Bergh and others: Unti, *Quality of Mercy*.

49   Katherine Grier argues: Grier, *Pets in America*, 134.

50   "It may enlarge our hearts toward . . .": Unti, *Quality of Mercy;* and Scully, *Dominion,* 14.

51   the very first humane organization: Lane and Zawistowski, *Heritage of Care*.

51   "The care of the defenseless . . .": Scully, *Dominion*

51   "I was convinced," he wrote: Ibid., 15.

## Chapter Two—*The Mismeasure of Animals*

54   visitors would leave with a story: Associated Press, "Gorilla at an Illinois Zoo Rescues a 3-Year-Old Boy," August 16, 1996, http://www.nytimes.com/1996/08/17/us/gorilla-at-an-illinois-zoo-rescues-a-3-year-old-boy.html.

54   behavioral and psychological needs: Fiby, Monika, "Trends in Zoo Design—Changing Needs in Keeping Wild Animals for a Visiting Audience," *International Review of Landscape Architecture and Urban Design,* Topos 62 /2008, http://www.zoolex.org/publication/fiby/zootrends08/fiby_topos62.html.

54   Even if the zoo architects: Associated Press, "Beer Kegs and Christmas Trees Keep Animals Healthy in New 'Enrichment' Programs," March 22, 2009, http://www.foxnews.com/story/0,2933,510060,00.html.

56   in the waters off the Farallon Islands: Fimrite, Peter, "Daring Rescue of Whale Off Farallones/Humpback Nuzzled Her Saviors in Thanks After They Untangled Her from Crab Lines, Diver Says," *San Francisco Chronicle,* December 14, 2005, http://articles.sfgate.com/2005–12–14/news/17403910_1_humpback-crab-pots-whale.

57   "When I was cutting the line . . .": Ibid.

57   German field researchers studying: Boesch, Christophe, Camille Bolé, Nadin Eckhardt, and Hedwig Boesch, "Altruism in Forest Chimpanzees: The Case of Adoption," *PLoS ONE* 5, no. 1 (2010): e8901, doi:10.1371/journal.pone.000890.

57   when Jane Goodall first began: The Jane Goodall Institute, "Study Corner: Biography," 2010, http://www.janegoodall.org/study-corner-biography.

58   Goodall named the chimpanzees: Goodall, Jane, *In the Shadow of Man* (Boston: Houghton Mifflin, 1971), 32–33.

58  No animal other than man: Goodall, Jane, *Through a Window* (Boston: Mariner Books, 1990), 15.

58  It was Leakey who urged Goodall: Goodall, *In the Shadow of Man,* 6.

59  "Now we must redefine 'tool,' . . .": Goodall, Jane, "Learning from the Chimpanzees: A Message Humans Can Understand," *Science* 282 (1998): 2184–2185.

59  make and use rudimentary tools: Choi, Charles Q., "10 Animals That Use Tools," LiveScience, December 14, 2009, http://www.livescience.com/animals/091214–10-tool-users.html.

59  animals can do much more: Moussaieff Masson, Jeffrey, and Susan McCarthy, *When Elephants Weep: The Emotional Lives of Animals* (New York: Delta, 1996).

60  the so-called science of "biological determinists": Gould, Stephen Jay, *The Mismeasure of Man* (New York: W. W. Norton, 1981).

61  "If my cup won't hold but a pint . . .": Mabee, Carleton, and Susan Mabee Newhouse, *Sojourner Truth: Slave, Prophet, Legend* (New York: NYU Press, 1995), 67–82.

61  animals were "mere automatons": Descartes, René, *A Discourse on Method* (New York: Book Jungle, 2008).

61  "Answer me, machinist, . . .": Voltaire, "The Philosophical Dictionary," http://history.hanover.edu/texts/voltaire/volanima.html.

61  mammals "experience (to greater or lesser degrees) . . .": Bekoff, Marc, *The Emotional Lives of Animals* (Novato, CA: New World Library, 2007), 32.

62  mimicking consciousness or intelligence: Griffin, Donald R., *Animal Minds: Beyond Cognition to Consciousness* (Chicago: University of Chicago Press, 1992).

62  most scientists rejected the Cartesian notion: Harriman, Philip Lawrence, ed., *Twentieth Century Psychology: Recent Developments in Psychology* (New York: Philosophical Library, 1946).

62  Morgan called his law of parsimony: Greenberg, Gary, and Maury M. Haraway, *Comparative Psychology: A Handbook* (New York: Taylor and Francis, 1998).

62  In 1913, Watson wrote: Watson, John B., "Psychology as the Behaviorist Views It," *Psychological Review* 20 (1913): 158–177.

63  The most prominent behaviorist of the era: Greenberg and Haraway, *Comparative Psychology.*

63  Masson wrote: Masson, Jeffrey and Susan McCarthy, *When Elephants Weep* (New York: Dell Publishing, 1995), 34.

64  Animal behaviorist Jonathan Balcombe: Balcombe, Jonathan, *Second Nature: The Inner Lives of Animals* (New York: Macmillan, 2010).

65  animals were conscious and emotional: Lorenz, Konrad, *King Solomon's Ring: New Light on Animal Ways* (New York: Routledge, 2002).

65  new field of cognitive ethology: Bekoff, Marc, and Colin Allen, "Cognitive Ethology: Slayers, Skeptics and Proponents," in *Anthropomorphism, Anecdotes and Animals: The Emperor's New Clothes?,* ed. R. W. Mitchessl, N. Thompson, and L. Miles (New York: State University Press of New York State, 1997), 313–334.

65  Harvard zoologist Donald Griffin: Griffin, Donald, *The Question of Animal Awareness* (New York: The Rockefeller University Press, 1976).

65  "Nature," Griffin wrote: Griffin, Donald, *A Communion of Subjects: Animals in Religion, Science, and Ethics* (New York: Columbia University Press, 2006).

66  Animal behaviorist Marc Bekoff told me: Personal communication with Marc Bekoff, December 5, 2009.

66  Marc Bekoff notes that in 2000: Ibid.

66  "You be good, I love you": "'Alex & Me': The Parrot Who Said 'I Love You,'" NPR; August 31, 2009, http://www.npr.org/templates/story/story.php?storyId=112405883.

67  devoted it instead to Alex: "Alex the African Grey," *The Economist*, September 20, 2007, http://www.economist.com/node/9828615?story_id=9828615.

67  When Dr. Pepperberg bought Alex: Pepperberg, Irene, *Alex & Me: How a Scientist and a Parrot Uncovered a Hidden World of Animal Intelligence—and Formed a Deep Bond in the Process* (New York: Harper, 2008).

67  British scientist Nicholas Humphrey: Humphrey, Nicholas, "The Social Function of Intellect," in *Growing Points in Ethology,* ed. P. P. G. Bateson and R. A. Hinde (Cambridge, MA: Cambridge University Press, 1976), 303–317.

67  Alex's education began slowly: Pepperberg, *Alex & Me.*

68  "Scientifically speaking, the greatest lesson . . .": Ibid.

68  In 2003, Cambridge University researchers: Rutledge, Robb, and Gavin Hunt, "Lateralized Tool Use in New Caledonian Crows," *Animal Behaviour* 67 (2004): 327–332, homepages.nyu.edu/~rbr242/RutledgeHuntAB04.pdf.

68  scrub jays have gone even further: Clayton, N. S., and A. D. Dickinson, "What, Where and When: Evidence for Episodic-Like Memory During Cache Recovery by Scrub Jays," *Nature* 395 (1998): 272–274.

69  Rico, a border collie, astounded scientists: Kaminski, Juliane, Josep Call, and Julia Fischer, "Word Learning in a Domestic Dog: Evidence for 'Fast Mapping,'" *Science* 304 (2004): 1682–1683.

69  Stanley Coren, a psychology professor: Coren, Stanley, *The Intelligence of Dogs: A Guide to the Thoughts, Emotions, and Inner Lives of Our Canine Companions* (New York: Bantam, 1995).

69  At Kyoto University, Ayuma: Macrae, Fiona, "I'm the Chimpion! Ape Trounces the Best of the Human World in Memory Competition," Mail Online, January 26, 2008, http://www.dailymail.co.uk/news/article–510260/Im-chimpion—Ape-trounces-best-human-world-memory-competition.html.

70  As Jonathan Balcombe relates: Balcombe, *Second Nature,* 32–33.

70  Balcombe frames the matter: Ibid., 48–49.

70  Dolphins off Western Australia: Mann, Janet, Brooke L. Sargeant, Jana J. Watson-Capps, Quincy A. Gibson, Michael R. Heithaus, Richard C. Connor, and Eric Patterson, "Why Do Dolphins Carry Sponges?," *PLoS One* 3, no. 12 (2008): e3868, doi:10.1371/journal.pone.0003868, http://www.plosone.org/article/info:doi/10.1371/journal.pone.0003868.

71  captive dolphin named Akeakamai: Pack, Adam, and Louis M. Herman, "The Dolphins' Understanding of Human Gazing and Pointing: Knowing What and Where, The Dolphin Institute, University of Hawaii at Monoa and Kewalo Basin Marine Mammal Laboratory, 121, no. 1 (2007): 34–45.

71  In 2009, Australian wildlife rehabilitators: Edwards, Lin, "Scientists Say Dolphins Should Be Treated As Non-Human Persons," PhysOrg.com, January 6, 2010, http://www.physorg.com/news181981904.html.

72  Lifeguard Rob Howes said: "Dolphins Save Swimmers from Shark," CBC News, November 24, 2004, http://www.cbc.ca/world/story/2004/11/24/dolphin_newzealand041124.html.

73  "The shark—a monster great white": Celizic, Mike, "Dolphins Save Surfer from Becoming Shark's Bait," NBC *Today* show, November 8, 2007.

73  The 2010 winner was Kenai: Humane Society of the U.S., "Winners of the Third Annual Dogs of Valor Awards," March 10, 2010, http://www.humane society.org/news/news/2010/03/third_annual_dogs_of_valor_winners.html.

73  The runner-up in the HSUS contest: Ibid.

74  Jack, a terrier mix rescued: "Dogs of Valor Awards," June 25, 2008, http://www.hsus.org/pets/pets_related_news_and_events/dog_of_valor/first_annual/dogs_of_valor_winners.html.

75  Hungarian researchers trained a dog: Kershaw, Sarah, "Good Dog, Smart Dog," *New York Times,* November 1, 2009, http://www.thedogfiles.com/?s=hungarian&x=0&y=0.

75  When Marc Bekoff first began: Bekoff, Marc, *Coyotes: Biology, Behavior and Management* (New York: The Blackburn Press, 2001).

75  coyotes and other canids begin their play fights: Bekoff, Marc, and Colin Allen, "The Evolution of Social Play," in *The Cognitive Animal: Empirical and Theoretical Perspectives on Animal Cognition,* ed. Marc Bekoff, Colin Allen, and Gordon M. Burghardt (Boston: Massachusetts Institute of Technology, 2002).

76  Bekoff tells other stories: Bekoff, Marc, and Jessica Pierce, *Wild Justice: The Moral Lives of Animals* (Chicago: University of Chicago Press, 2009).

76  vampire bats even appear to exhibit reciprocal altruism: Bekoff, Marc, and Jonathan Balcombe, "Q & A: Minds of Their Own, " *All Animals*, March/April 2010, 33.

76  "they seem to care for . . .": Schmid, Randolph, "Monkeys Like to Both Give and Receive," Msnbc.com, http://www.msnbc.msn.com/id/26394973/.

76  Bekoff notes that such grief: "Grief in Animals: It's Arrogant to Think We're the Only Animals Who Mourn," Psychology Today Blog, October 29, 2009, http://www.psychologytoday.com/blog/animal-emotions/200910/grief-in-animals-its-arrogant-think-were-the-only-animals-who-mourn?page=2.

77  procession to mourn the loss of an elder: Heussner, Ki Mae, "Chimps Mourn Passing of One of Their Own," ABC News, October 28, 2009, http://abcnews.go.com/Technology/AmazingAnimals/chimps-mourn-passing/story?id=8937053.

77  rats show increased levels of dopamine: Bozarth, M. A., "Pleasure Systems in the Brain," in *Pleasure: The Politics and the Reality,* ed. D. M. Warburton (New York: John Wiley & Sons, 1994), 5–14.

77  Neuroscientist Jaak Panksepp notes: Panksepp, Jaak, *Affective Neuroscience: The Foundations of Human and Animal Emotions* (New York: Oxford University Press, 1998).

77  researchers at the University of Sussex: Thornhill, Cher, "Gorillas and Humans Use Similar Body Language to Communicate," Mail Online, October 17, 2008, http://www.dailymail.co.uk/sciencetech/article–1078543/Gorillas-humans-use-similar-body-language-communicate.html.

77 Scientists in Uganda: Shaikh, Thair, "Elephants Never Forget . . . and Cannot Forgive," *The Times,* February 16, 2006, http://www.timesonline.co.uk/tol/news/world/article731367.ece.

78 delinquent elephants were responsible: Slotow, R., and G. Van Dyk, "Role of Delinquent Young 'Orphan' Male Elephants in High Mortality of White Rhinoceros in Pilanesberg National Park, South Africa," *Koedoe—African Protected Area Conservation and Science* 44, no. 1 (2001).

78 "They are certainly intelligent enough . . .": Shaikh, "Elephants Never Forget."

79 send the elephants living at the Detroit Zoo: Farinato, Richard, "Detroit Zoo Sends Its Elephants Packing. Should Others Follow Suit?," May 27, 2004, http://www.hsus.org/wildlife/wildlife_news/detroit_zoo_sends_its_elephants_packing_should_others_follow_suit.html.

79 On his first encounter: Personal communication with Ron Kagan, April 17, 2010.

80 Ron went to Kenya: Ibid.

80 "In the most polite way, she told me . . .": Ibid.

81 Kagan and his colleagues decided: Farinato, "Detroit Zoo Sends Its Elephants Packing."

82 Matt told me people rarely know: Personal communication with Matt Smith, April 19, 2010.

83 Mira Tweti provides a similar estimate: Tweti, Mira, *Of Parrots and People: The Sometimes Funny, Always Fascinating, and Often Catastrophic Collision of Two Intelligent Species* (New York: Penguin Books, 2009).

## Chapter Three—A Message from Hallmark: Exposing Factory Farming

92 the state's nearly eighteen million registered voters: Bowen, Debra, *California Secretary of State Voter Registration Report,* October 31, 2008, http://yubanet.com/california/Bowen-Reports-Record-Number-of-Registered-Voters-in-California.php.

92 both sides would spend the bulk of their money: Lawrence, Steve, "California Initiative Campaigns Cost 227 Million," The Associated Press, February 3, 2009.

93 the extreme confinement of twenty million animals: U.S. Department of Agriculture, National Agricultural Statistics Service, *2007 Census of Agriculture,* Vol. 1, U.S. Summary and State Reports, Table 13. Poultry—Inventory and Sales: 2007 and 2002, p. 411, http://www.agcensus.usda.gov/Publications/2007/Full_Report/usv1.pdf.

94 From its founding in the 1950s: Unti, Bernard, *Protecting All Animals: A Fifty-Year History of the Humane Society of the United States* (Washington, DC: Humane Society Press, 2004.)

94 we help animals of every kind: *Annual Report—Rescue, Reform, Results,* The Humane Society of the United States, 2009 Annual Report, http://www.humanesociety.org/about/overview/financials/.

94 I pledged that America's largest animal-protection group: Oldenburg, Don, "Vegan in the Henhouse: Wayne Pacelle, Putting Animals on (and off) the Table," *Washington Post,* August 9, 2004.

94 Nearly ten billion animals: U.S. Department of Agriculture, National Agricultural Statistics Service, Quick Stats, "U.S. & State—Slaughter" and "U.S. & State—Poultry Slaughter," http://www.nass.usda.gov/Data_and_Statistics/Quick_Stats_1.0/index.asp.

95 the nation's top farm state in revenues generated: U.S. Department of Agriculture, Economic Research Service, *Farm Income: Data Files, Income and Production Expenses,* U.S. and State income and production expenses by category, 1949–2009, Value of Agricultural Production, http://www.ers.usda.gov/Data/FarmIncome/FinfidmuXls.htm.

95 It is the biggest dairy state in the nation: U.S. Department of Agriculture, National Agricultural Statistics Service. *2007 Census of Agriculture,* Vol. 1, U.S. Summary and State Reports, Table 11. Cattle and Calves—Inventory and Sales: 2007 and 2002, p. 381, http://www.agcensus.usda.gov/Publications/2007/Full_Report/usv1.pdf; U.S. Department of Agriculture, Economics, Statistics, and Market Information System, 2009, *Milk Production, Disposition and Income: 2008 Summary,* http://usda.mannlib.cornell.edu/usda/nass/MilkProdDi//2000s/2009/MilkProdDi–05–29–2009.pdf.

95 midwestern corn-growing states have surpassed California: U.S. Department of Agriculture, Economics, Statistics, and Market Information System, Chicken and Eggs, http://usda.mannlib.cornell.edu/MannUsda/viewDocumentInfo.do?documentID=1028.

96 an egg factory farm in Riverside County: Blume, Howard, "Footage of Mistreated Hens Released in Support of Proposition 2," *Los Angeles Times,* October 14, 2008.

98 up to a 100 percent annual turnover rate in personnel: Schlosser, Eric, *Fast Food Nation* (New York: Harper Perennial, 2005).

98 Accidents are common: Ibid.

98 Through selective breeding, we've engineered Holsteins: Hansen, L. B. "Consequences of Selection for Milk Yield from a Geneticist's Viewpoint," *Journal of Dairy Science* 83, no. 5 (2000): 1145–1150; and Tsuruta, S., I. Misztal, and T. J. Lawlor, "Changing Definition of Productive Life in U.S. Holsteins: Effect on Genetic Correlations," *Journal of Dairy Science* 88, no. 3 (2005): 1156–1165.

98 Upton Sinclair wrote a hundred years ago: Sinclair, *The Jungle.*

98 17 percent of U.S.-produced beef: Troutt, H. F., and B. I. Osburn, "Meat from Dairy Cows: Possible Microbiological Hazards and Risks," *Revue Scientifique et Technique de l'Office International des Epizooties* 16, no. 2 (1997): 405–414.

98 The average cow produces: U.S. Department of Agriculture, Economics, Statistics, and Market Information System, Milk Production, http://usda.mannlib.cornell.edu/MannUsda/viewDocumentInfo.do?documentID=1103.

99 More than 90 percent of America's dairy herds: The NAHMS survey shows more than 90 percent but could just be one cow in a herd. U.S. Department of Agriculture, APHIS-NAHMS, 2007, *Dairy 2007. Part I: Reference of Dairy Cattle Health and Management Practices in the United States,* http://nahms.aphis.usda.gov/dairy/dairy07/Dairy07_dr_PartI.pdf.

99 Many also have foot rot: Manure is only cited as a factor in foot rot—there can be others. Singh, G. R., Aithal H. P. Amarpal, and P. Kinjavdekar, "Lameness in Cattle—A Review," *Indian Journal of Animal Sciences* 75, no. 6 (2005): 723–740.

99   Though not strictly forbidding the practice: 9 C.F.R. § 313.2(b).

100  U.S. Immigration and Customs Enforcement raided: Shu, Spencer, "Immigration Raid Jars Small Town," *Washington Post,* May 18, 2008.

100  Hallmark's daily total was: U.S. Department of Agriculture, Office of the Inspector General, *Evaluation of FSIS Management Controls Over Pre-Slaughter Activities,* Report No: 24601–07-KC, November 2008, p. 9; *see also* 13 C.F.R. § 121.201 and 61 Fed. Reg. 38,806 (July 25, 1996).

100  Smithfield Foods runs: *See* Plaintiffs' Complaint for Damages and Equitable Relief at p. 3, ¶9 in *Smithfield Foods, Inc. and Smithfield Packing Co. v. United Food and Commercial Workers International Union, et al.,* Civil Action No. 3:07CV641 (U.S. Dist. Ct., E.D. Va.), http://www.prconversations.com/wp-content/uploads/2008/02/rico_1017072.pdf.

101  33 million cows: U.S. Department of Agriculture, National Agricultural Statistics Service, Quick Stats, "U.S. & State—Slaughter," http://www.nass.usda.gov/Data_and_Statistics/Quick_Stats_1.0/index.asp.

101  And the USDA doesn't apply: Humane Methods of Livestock Slaughter, 7 U.S.C. §§ 1901 *et seq.* U.S. Department of Agriculture, National Agricultural Statistics Service, Quick Stats, "U.S. & State—Slaughter" and "U.S. & State—Poultry Slaughter." http://www.nass.usda.gov/Data_and_Statistics/Quick_Stats_1.0/index.asp.

101  After he completed his morning inspection: U.S. Department of Agriculture, Office of the Inspector General, *Evaluation of FSIS Management Controls Over Pre-Slaughter Activities,* Report No: 24601–07-KC, November 2008, at p. 18 (stating "The PHV [public health veterinarian] stated that he took 'shortcuts' in the ante-mortem process to save time, and stated that he had complained in the past about lack of staffing"). *See also* "Contaminated Food: Private Sector Accountability," Hearing before the Subcommittee on Oversight and Investigations, of the House Committee on Energy and Commerce, written testimony of Dr. Michael Greger on behalf of the Humane Society of the United States, February 26, 2008, at pp. 3–4, http://energycommerce.house.gov/images/stories/Documents/Hearings/PDF/110-oi-hrg.022608.Greger-testimony.pdf.

102  "If I relied on my mail, . . .": Unti, Bernard, *Protecting All Animals* (Washington, DC: Humane Society Press, 2004), 45.

102  *Slaughterhouse* about the USDA's failures: Eisnitz, Gail, *Slaughterhouse: The Shocking Story of Greed, Neglect, and Inhumane Treatment Inside the U.S. Meat Industry* (New York: Prometheus Books, 1997).

102  In 2001, the *Washington Post* reported: Warrick, Joby, "'They Die Piece by Piece,' in Overtaxed Plants, Humane Treatment of Cattle Is Often a Battle Lost," *Washington Post,* April 10, 2001.

103  seventeen third-party audits: Steve Mendell, "Westland/Hallmark Meat Co. Statement on Beef Meat Recall," February 3, 2008 (stating, "During 2007, we had 17 third party audits of our operation to confirm that we meet the statutorily mandated humane handling and food safety standards").

106  the average American consumes: U.S. Department of Agriculture, Economic Research Service, Food Availability (Per Capita) Data System, http://www.ers.usda.gov/Data/FoodConsumption/FoodAvailSpreadsheets.htm.

106 Michael Pollan has pointed out: Pollan, Michael, *In Defense of Food* (New York: Penguin Press, 2008), and Pollan, Michael, *The Omnivore's Dilemma* (New York: Penguin Press, 2006).

107 Men, on average, weigh: Schlosser, Eric, *Fast Food Nation* (New York: Houghton Mifflin, 2001).

109 The *Post* story was powerful: Weiss, Rick, "Video Reveals Violations of Laws, Abuse of Cows at Slaughterhouse," *Washington Post,* January 30, 2008.

110 fifty million pounds of ground beef: Geoffrey S. Becker, *Nonambulatory Livestock and the Humane Methods of Slaughter Act,* Congressional Research Service, Report no. RS22819, March 24, 2009 at p. 1 (stating "About 50 million pounds were distributed to the school lunch and other federal nutrition programs in at least 45 states"); *see also* Matthew L. Wald, "Meat Packer Admits Slaughter of Sick Cows," *New York Times,* March 13, 2008 (stating, "Of the 143 million pounds of beef that were recalled, about 50 million pounds went to school lunch programs or federal programs for the poor or elderly, Mr. Stupak said").

110 Daniel Ugarte Navarro, a plant worker: Kim, Victoria, "Cruelty Charges Filed Against Slaughterhouse Boss," *Los Angeles Times,* February 16, 2008.

112 the largest meat recall in American history: Brown, David, "USDA Orders Largest Meat Recall in U.S. History," *Washington Post,* February 18, 2008.

112 downer cows are much more likely: Doherr, M. G., D. Heim, R. Fatzer, C. H. Cohen, M. Vandevelde, and A. Zurbriggen, "Targeted Screening of High-Risk Cattle Populations for BSE to Augment Mandatory Reporting of Clinical Suspects," *Preventive Veterinary Medicine* 51, no. 1–2 (2001): 3–16.

113 He assured the committee: Zhang, Jane, "Meatpacker Admits Ailing Cattle Used at Slaughterhouse," *Wall Street Journal,* March 13, 2008.

113 seventy thousand people took to the streets of Seoul: "South Korean Government Ready to Resign of Resumption of U.S. Beef Imports," October 3, 2010; McNeil, Donald, "Questions on U.S. Beef Remain," *New York Times,* June 10, 2008.

114 eventually announced nearly four months after: The USDA Office of Communications, "Agriculture Secretary Schafer Announces Plan to End Exceptions to Animal Handling Rule," May 20, 2008, press release, Release Number 0131.08.

114 a federal ban on processing: Food Safety Inspection Service, "Requirements for the Disposition of Cattle That Become Non-Ambulatory Disabled Following Ante-Mortem Inspection," *Federal Register* 74, no. 51 (March 18, 2009).

114 Steve Mendell . . . retreated: Goad, Ben, "Beef Recall Costs Reach $67.2 Million and Rising," *Press Enterprise,* April 9, 2008 (stating "Mendell lives in a house in the Corona del Mar area of Newport Beach that has been valued at more than $4 million").

116 conference committee led by leaders: Cong. Rec., 30 September 2003: H8956.

116 Congressman Ackerman argued: U.S. Department of Agriculture Office of Inspector General, Animal and Plant Health Inspection Service and Food Safety and Inspection Service: Bovine Spongiform Encephalopathy (BSE) Surveillance Program—Phase I, August 18, 2004, www.oig.usda.gov/webdocs/50601-9-final.pdf; National Renderers, *Livestock Mortalities: Methods of Disposal and Their*

*Potential Cost,* http://nationalrenderers.org/Economic_Impact/MortalitiesFinal .pdf; and Stull, C. L., M. A. Payne, S. L. Berry, and J. P. Reynolds, "A Review of the Causes, Prevention, and Welfare of Nonambulatory Cattle," *Journal of the American Veterinary Medical Association* 231, no. 2 (2007): 227–234.

116 "That sick animal [in Rep. Ackerman's photo] . . .":: Cong. Rec., 14 July 2003: H6653. Print.

117 more than one hundred deaths due to mad cow disease: United Kingdom National CJD Surveillance Unit, 2010, CJD Statistics, October 4, http://www .cjd.ed.ac.uk/figures.htm; and Collee, J. G., R. Bradley, and P. P. Liberski, "Variant CJD (vCJD) and Bovine Spongiform Encephalopathy (BSE): 10 and 20 Years on: Part 2," *Folia Neuropathologica* 44, no. 2 (2006): 102–110.

117 Nervous consumers started eating: R-CALF USA, 2006, "Cattle and Beef Trade, BSE, and the U.S. Cattle Industry," July, http://www.r-calfusa.com/ BSE/060701-CALFBackgroundPaperOnBSE.pdf.

117 nonambulatory cattle are forty-nine to fifty-eight times: Doherr, M. G., D. Heim, R. Fatzer, C. H. Cohen, M. Vandevelde, and A. Zurbriggen, "Targeted Screening of High-Risk Cattle Populations for BSE to Augment Mandatory Reporting of Clinical Suspects," *Preventive Veterinary Medicine* 51, no. 1–2 (2001): 3–16.

117 It required all of this to compel: "Requirements for the Disposition of Cattle That Become Non-Ambulatory Disabled Following Ante-Mortem Inspection," 74 Fed. Reg. 11,463, 11,464 (March 18, 2009) (to be codified at 9 C.F.R. pt. 309). (Industry opposition continued even after the events at Hallmark/ Westland as documented in the *Federal Register* when USDA published the final rule closing the downer loophole in 2009: "*Comment:* Several industry commenters who opposed the proposed amendment stated that the current regulatory provision should remain unchanged, allowing cattle that become non-ambulatory disabled after passing ante-mortem inspection to be reevaluated by an FSIS PHV. *Response:* FSIS disagrees with the comment. The events at the Hallmark/Westland establishment demonstrate that FSIS inspection personnel are not always notified when cattle become non-ambulatory disabled after they pass ante-mortem inspection. Thus, under the former regulations, and specifically at Hallmark/Westland, non-ambulatory disabled cattle that had not consistently received proper and adequate ante-mortem inspection were slaughtered for human food. In addition, the events at Hallmark demonstrate that requiring re-inspection of cattle that become non-ambulatory disabled after ante-mortem inspection may have created an incentive for establishments to inhumanely attempt to force these animals to rise. Therefore, FSIS has determined that a change in the regulation is needed to ensure more effective and efficient implementation of inspection procedures and compliance with humane handling requirements at official establishments.")

118 It was July 2007 when the agency published: U.S. Department of Agriculture, Food Safety and Inspection Service Notice 05–06, "Re-examination of Bovine That Become Non-ambulatory After Passing Ante-mortem Inspection," January 18, 2006.

119 A 2008 report revealed: Doering, Christopher, "Mad-Cow Ban Cost U.S. $11 Billion in Beef Exports," Reuters, October 7, 2008.

119 more than a dozen BSE-positive: Campbell, D., "Killer Mad Cow Disease Strikes in Alberta," *Calgary Herald* (Alberta, Canada), December 9, 1993, p. D1; Canadian Food Inspection Agency, "Summary of the Report of the Investigation of Bovine Spongiform Encephalopathy (BSE) in Alberta, Canada," July 2, 2003, http://www.inspection.gc.ca/english/anima/heasan/disemala/bseesb/ab2003/evalsume.shtml; USDA press release, "USDA BSE Update," December 27, 2003, http://www.usda.gov/wps/portal/!ut/p/_s.7_0_A/7_0_1OB/.cmd/ad/.ar/sa.retrievecontent/.c/6_2_1UH/.ce/7_2_5JM/.p/5_2_4TQ/.d/7/_th/J_2_9D/_s.7_0_A/7_0_1OB?PC_7_2_5JM_contentid=2003/12/0445.html&PC_7_2_5JM_navtype=RT&PC_7_2_5JM_parentnav=TRANSCRIPTS_SPEEC; Canadian Press, "BSE Confirmed in Alberta Dairy Cow," *Ottawa Sun,* January 3, 2005, p. 20; Johnsrude, L., and G. Richards, "Feed Bought After Ban Fed to Latest Mad Cow: 104 Other Calves Had Access to Same Feed in Spring of 1998, Innisfail-Area Farmer Says," *Edmonton Journal* (Alberta), January 14, 2005, p. A1; FDA, "Commonly Asked Questions About BSE in Products Regulated by FDA's Center for Food Safety and Applied Nutrition (CFSAN)," September 14, 2005, http://www.cfsan.fda.gov/~comm/bsefaq.html; Canadian Food Inspection Agency, *Report on the Investigation of the Fifth Case of Bovine Spongiform Encephalopathy (BSE) in Canada.* June 16, 2006, http://www.inspection.gc.ca/english/anima/heasan/disemala/bseesb/bccb2006/5investe.shtml; Canadian Food Inspection Agency, *Report on the Investigation of the Sixth Case of Bovine Spongiform Encephalopathy (BSE) in Canada,* August 8, 2006, http://www.inspection.gc.ca/english/anima/heasan/disemala/bseesb/mb2006/6investe.shtml; Canadian Food Inspection Agency, *Report on the Investigation of the Seventh Case of Bovine Spongiform Encephalopathy (BSE) in Canada,* August 24, 2006, http://www.inspection.gc.ca/english/anima/heasan/disemala/bseesb/ab2006/7investe.shtml; Canadian Food Inspection Agency, *Report on the Investigation of the Eighth Case of Bovine Spongiform Encephalopathy (BSE) in Canada,* December 18, 2006, http://www.inspection.gc.ca/english/anima/heasan/disemala/bseesb/ab2006/8investe.shtml; and Canadian Food Inspection Agency, *Report on the Investigation of the Tenth Case of Bovine Spongiform Encephalopathy (BSE) in Canada,* July 25. 2007, http://www.inspection.gc.ca/english/anima/heasan/disemala/bseesb/bccb2007/10investe.shtml.

123 The average hog excretes: Seigley, L. S., and D. J. Quade, "An Introduction to Hogs in Iowa," The Iowa Department of Natural Resources, http://www.igsb.uiowa.edu/inforsch/iahogs/iahogs.htm (hog excretes about 10.5 lb/day); and U.S. Environmental Protection Agency, "National Pollutant Discharge Elimination System Permit Regulation and Effluent Limitation Guidelines and Standards for Concentrated Animal Feeding Operations (CAFOs)—Final Rule," *Federal Register* 68, no. 29 (February 12, 2003): 7176, 7180, http://www.epa.gov/npdes/regulations/cafo_fedrgstr.pdf (person excretes about 0.518 tons/yr = 2.8 lb/day—so hog excretes at least three times more than a person per day—or per year).

123 These curious, social animals: Jensen, P., *The Ethology of Domestic Animals: An Introductory Text* (Wallingford, UK: CABI Publishing, 2002), 159–172.

123 For nearly their entire four-month pregnancies: Marchant, J. N., and D. M. Broom, "Effects of Dry Sow Housing Conditions on Muscle Weight and Bone

Strength," *Animal Science* 62 (1996): 105–113; and Anil, L., S. S. Anil, and J. Deen, "Evaluation of the Relationship Between Injuries and Size of Gestation Stalls Relative to Size of Sows," *Journal of the American Veterinary Medical Association* 221, no. 6 (2002): 834–836.

125    the world's largest factory farmer of pigs: Collins, Kristin, "Pork Producer Uncages Some Pigs," *Charlotte News & Observer,* February 13, 2007.

125    American Veal Association, a trade group: Smith, Rod, "Veal Association Policy Urges Group Housing," *Feedstuffs,* August 2, 2007.

127    Wild turkeys can run twenty-five miles an hour: Handwerk, B., "Wild Turkeys Invading Suburban U.S.," National Geographic News, 2007, http://news.nationalgeographic.com/news/2007/11/071119-wild-turkeys.html; and Michigan Department of Natural Resources, "Wild Turkey (*Meleagris gallopavo*)," www.michigan.gov/dnr/0,1607,7–153–10370_12145_12202–52511—,00.html.

127    toms have difficulty mating: Scanes, C. G., G. Brant, and M. E. Ensminger, *Poultry Science,* 4th ed. (Upper Saddle River, NJ: Pearson Prentice Hall, 2004), 282–283.

127    They suffer joint and hip problems: Julian, R., and P. Gazdzinsky, "Lameness and Leg Problems: Turkeys," *World Poultry–Elsevier Special* (2000): 24–31; Julian, R. J., "Tendon Avulsion as a Cause of Lameness in Turkeys," *Avian Diseases* 28, no. 1 (1984): 244–249; Duff, S. R. I., "The Morphology of Degenerative Hip Disease in Male Breeding Turkeys," *Journal of Comparative Pathology* 94, no. 1 (1984): 127–139.

127    Even if they were not confined in buildings: Voris, J. C. "California Turkey Production," University of California Cooperative Extension, Poultry Fact Sheet No. 16c, accessed October 5, 2010, http://animalscience.ucdavis.edu/Avian/pfs16C.htm; Scanes et al., *Poultry Science,* 270; and Healy, W. M., "Behavior," in *The Wild Turkey: Biology and Management,* ed. J. G. Dickson (Harrisburg, PA: Stackpole Books, 1992), 46–65.

128    Factory-farmed turkeys now weigh: There isn't one source that says this. This reference says that wild turkeys weigh 7.7 pounds in four months: Healy, "Behavior." This one says domestic birds weigh twenty-five pounds when they reach four months: Hulet, R. M., P. J. Clauer, G. L. Greaser, J. K. Harper, and L. F. Kime, "Small-Flock Turkey Production," Pennsylvania State University, Agricultural Research and Cooperative Extension, 2004, accessed August 4, 2008, http://agalternatives.aers.psu.edu/Publications/SmallflockTurkeys.pdf.

128    The welfare of farmed turkeys: Voris, "California Turkey Production"; and Austic, R. E., and M. C. Nesheim, *Poultry Production,* 13th ed. (Philadelphia, PA: Lea and Febiger, 1990), 231.

128    actually eighteen to twenty-six million turkeys: Ibid. *See also* Austic and Nesheim, *Poultry Production,* 231. And applying the 7–10 percent to the turkey slaughter numbers from: U.S. Department of Agriculture, National Agricultural Statistics Service, Quick Stats, "U.S. & State—Poultry Slaughter," http://www.nass.usda.gov/Data_and_Statistics/Quick_Stats_1.0/index.asp.

128    Dr. Ian Duncan notes: Duncan, Ian, "Welfare Problems of Poultry." In *The Well-Being of Farm Animals,* ed. J. B. Benson and B. E. Rollin (Ames, IA: Blackwell, 2004), 310.

129   As many as 20 to 30 percent of chickens: Weber, R.M., M. Nogossek, L. Sander, B. Wandt, U. Neumann, and G. Glunder, "Investigations of laying hen health in enriched cages as compared to conventional cages and a floor pen system." *Wiener Tierarzliche Monatsschrift* 90, no. 10 (2003): 257–266.

129   Dead-on-arrival estimates: Nijdam, E., P. Arens, E. Lambooij, E. Decuypere, and J.A. Stegeman, "Factors Influencing Bruises and Mortality of Broilers During Catching, Transport, and Lairage." *Poultry Science* 83 (2004): 1610–1615; Warriss, P.D., E.A. Bevis, S.N. Brown, and J.E. Edwards, "Longer Journeys to Processing Plants Are Associated with Higher Mortality in Broiler Chickens." *British Poultry Science* 33 (1992): 201–206.

132   "in a way, win or lose": Jones, Maggie, "The Barnyard Strategist," *New York Times Magazine*, October 26, 2008.

## *Chapter Four—A Culture of Cruelty: Animal Fighting in America*

135   eighteen-month sentence for dogfighting: Battista, Judy, "Vick Finishes His Sentence; Future Is Cloudy," *New York Times,* July 20, 2009, http://www .nytimes.com/2009/07/21/sports/football/21vick.html.

137   Police had raided Vick's home: Haaser, Brian R., "Bad Newz Kennels, Smithfield, Virginia—Animal Fighting," USDA HY 3330–0018, August 28, 2008, p. 2.

137   selected first in the NFL: "Falcons Quarterback Michael Vick Signs Richest NFL Deal in History," *Jet,* January 17, 2005, http://findarticles.com/p/articles/ mi_m1355/is_3_107/ai_n9771537/.

137   Police had been tipped off: John Goodwin, personal communication with Virginia State Police officer at the Virginia Animal Fighting Task Force meeting in Hanover, VA, 2006.

137   she told a friend, "We got him. . . .": Dohrmann, George, "The House on Moonlight Road," May 29, 2007, sportsillustrated.com, http://sportsillustrated .cnn.com/2007/football/nfl/05/29/vick0604/index.html.

137   they ran a dogfighting enterprise: U.S. District Court for the Eastern District of Virginia. Indictment: *U.S. v. Purnell A. Peace, Quanis L. Phillips, Tony Taylor and Michael Vick,* July 17, 2007, p. 5.

137   "sheds and kennels associated with . . .": Ibid.

138   "It's unfortunate I have to take the heat": Associated Press, "Vick Blames Family for Neglected Dogs at Virginia Home," April 27, 2007, http://sports.espn.go .com/nfl/news/story?id=2851640.

138   Poindexter's investigation was lackluster: WAVY-TV, "Surry County Commonwealth Attorney Denies Evidence in Michael Vick Investigation," May 12, 2007, http://www.wavy.com/Global/story.asp?S=6498964; and Smith, Michael D., "Prosecutor Suggests Vick Dogfighting Case Is a 'Witch Hunt,'" aolsportsblog.com, May 16, 2007.

139   we had fought hard in Congress: Humane Society of the U.S., "Taking Down Dog-fighting," July 27, 2009, http://www.hsus.org/acf/fighting/dogfight/ taking_down_dogfighting.html.

139   And we were just completing: Ibid.

139  In testimony before a congressional committee: Humane Society of the U.S., "Congress Urged to Crack Down on Animal Fighting," February 6, 2007, http://www.hsus.org/press_and_publications/press_releases/congress_urged_to_crack_down.html.

139  obstructionist tactics of Oklahoma senator Tom Coburn: Project Votesmart; "Humane Society of the U.S. Ratings," 2005–2006, http://www.votesmart.org/issue_rating_detail.php?r_id=3489.

139  animal fighting is often tied in: Humane Society of the U.S., "Congress Urged."

140  The county sheriff's office had asked: Haaser, "Bad Newz Kennels," 1.

140  obtained their warrant: Ibid., 2–3.

140  District Attorney Poindexter objected: Ibid.; WBZ-TV, "Police Suspect Dogs Buried on Vick's Former Property," May 29, 2007, http://www.wsbtv.com/news/13407157/detail.html; and Associated Press, "Feds May Take Over Vick Dogfighting Probe," June 7, 2007.

140  obtained a fresh warrant: Haaser, "Bad Newz Kennels," 8.

140  extending from New York: Indictment filed in the U.S. District Court for the Eastern District of Virginia, Criminal No. 3:07 CR 274, July 17, 2007.

140  announced indictments: U.S. District Court for the Eastern District of Virginia, Indictment: *U.S. v. Purnell A. Peace, Quanis L. Phillips, Tony Taylor and Michael Vick,* July 17, 2007.

141  five smaller structures—all painted jet black: Haaser, "Bad Newz Kennels," 13.

141  One cooperating witness: Ibid., 4.

141  Tony Taylor had designed the complex: Strouse, Kathy, *Bad Newz: The Untold Story of the Michael Vick Dog Fighting Case* (Charleston, SC: BookSurge Publishing, 2009), 89, 90.

141  The Bad Newz gang: USDA, Office of the Inspector General, *Report of Investigation, Bad Newz Kennels, Smithfield, VA,* August 28, 2008.

141  watched them "roll" or "test": U.S. District Court for the Eastern District of Virginia, Indictment: *U.S. v. Purnell A. Peace, Quanis L. Phillips, Tony Taylor and Michael Vick,* July 17, 2007, item 12, p. 6.

141  Peace shot one of the weak dogs: U.S. District Court for the Eastern District of Virginia, Indictment: *U.S. v. Purnell A. Peace, Quanis L. Phillips, Tony Taylor and Michael Vick,* July 17, 2007, p. 6.

141  ". . . hanging, drowning, and slamming . . .": Ibid., 17.

141  When federal authorities put all the evidence together: Ibid., 3.

142  First, Taylor pleaded: Haaser, "Bad Newz Kennels," 8.

142  Then Peace and Phillips confessed: Ibid., 1.

142  In August 2007, Vick pleaded guilty: Ibid., 8.

142  Blank had no choice: ESPN, "Vick Suspended Indefinitely from the NFL," October 24, 2007, http://sports.espn.go.com/nfl/news/story?id=2990157.

142  U.S. attorney indicted a fifth member: Haaser, "Bad Newz Kennels," 9.

142  Hudson sentenced Vick: Ibid., 15.

143  We shared dogfighting footage: HSUS, "Taking Action to Stop Dogfighting," November 2, 2009, http://www.humanesociety.org/issues/dogfighting/tips/dogfighting_action.html.

143  as many as forty thousand people: Meyers, Jessica, "Dallas Animal Cruelty Task Force in Works," December 26, 2008, http://www.dallasnews.com/

sharedcontent/dws/news/localnews/stories/1227dnmetdogfighting.38299fc
.html.

143 Vick had started off as a street fighter: U.S. Dept of Agriculture, "Memoran-
dum of Interview," USDA HY 3330–0018, June 28, 2007, p. 1.

144 had called and written protest letters: Garofoli, Joe, "NFL Star's Woes Go to
League Nike/Activists Press Them to Decry Dogfighting" *San Francisco Chronicle,* July 21, 2007.

148 The U.S. Fish and Wildlife Service runs: Fish and Wildlife Service, U.S. Duck
Stamp Office, http://www.fws.gov/duckstamps/stampdesign.htm.

149 ". . . We've started programs . . .": HSUS, "End Dogfighting Program," November 2, 2009, http://www.humanesociety.org/issues/dogfighting/end_dog
fighting.html.

152 watched a segment of the *Oprah Winfrey Show*: *The Oprah Winfrey Show,* "Investigating Puppy Mills," January 4, 2006, http://www.oprah.com/oprahshow/
Investigating-Puppy-Mills.

152 which also included cockfighting: HSUS, "Campaigns," http://www.humane
society.org/issues/campaigns/.

153 to upgrade the federal animal-fighting law: HSUS, "Animal Cruelty and Fighting Victories," May 2009, http://www.hsus.org/acf/campaign/victories.html.

153 the third one since 2002: Ibid.

153 upgrade more than twenty-five state laws: ASPCA, "Dog Fighting FAQ," http://
www.aspca.org/fight-animal-cruelty/dog-fighting/dog-fighting-faq.html.

153 had launched community-based programs: HSUS, "Michael Vick and End
Dogfighting," August 24, 2009, http://www.aspca.org/fight-animal-cruelty/
dog-fighting/dog-fighting-faq.html.

153 anticruelty crusader named Henry Bergh: Lane, Marion S., and Stephen L.
Zawistowski, *Heritage of Care: The American Society for the Prevention of Cruelty to
Animals* (Westport, CT: Praeger Publishers, 2008).

153 By 1880, the ASPCA: Ibid.

154 semiwild jungle fowl were pitted against: Sherman, David M., *Tending Animals
in the Global Village: A Guide to International Veterinary Medicine* (Baltimore: Williams and Williams, 2002), 46.

154 have migrated eastward: Encyclopedia Britannica Online, "Cockfighting,"
2009, http://www.britannica.com/EBchecked/topic/123691/cockfighting.

154 attaching sharpened blades: Curnutt, J., *Animals and the Law: A Sourcebook*
(Santa Barbara, CA: ABC-CLIO, 2001), 276.

154 Themistocles even staged cockfights: Ibid., 277.

154 elephants, bison, lions: Ibid., 231.

154 hundreds of thousands of animals: Ibid.

154 In A.D. 1050, Edward the Confessor: Ibid., 283.

154 During her reign, Queen Elizabeth I: Birley, D., *Land of Sport and Glory: Sport
and British Society 1887–1910* (Manchester, UK: Manchester University Press,
1995), 62–64.

154 but both times to no avail: Gardiner, S., *Sports Law* (London: Routledge Cav-
endish, 2006), 120.

155 Three pits stood in the city: Shevelow, K., *For the Love of Animals: The Rise of the
Animal Protection Movement* (New York: Henry Holt and Co., 2008), 43–46.

155 what one author has called "Bulldog Nation": Russell, Edmund, "The Michael Vicks of Yore," *Washington Post,* September 2, 2007, http://www.washington post.com/wpdyn/content/article/2007/08/31/AR2007083101466.html.

155 to create the modern bull terrier: Encyclopedia Britannica Online, "Pit Bull Terrier," 2009, http://www.britannica.com/EBchecked/topic/1309178/pit -bull-terrier.

155 a monkey named Jacco Macacco: Russell, "The Michael Vicks of Yore."

155 Britain's Parliament considered eleven bills: Malcolmson, R. W., *Popular Recreations in English Society 1700–1850* (Cambridge: Cambridge University Press Archive, 1979), 124.

155 secured fifty-two petitions of support: Shevelow, *For the Love of Animals,* 266–272.

156 1835 and 1849 passed bills: Malcolmson, *Popular Recreations,* 124.

156 brought cockfighting, dogfighting, bull baiting, gander pulling: Fischer, David H., *Albion's Seed: Four British Folkways in America* (New York: Oxford University Press, 1989), 148, 360–364, 552–555; and Isaac, Rhys, *The Transformation of Virginia* (Chapel Hill: University of North Carolina Press, 1982), 101–103.

156 The 1641 Massachusetts "Body of Liberties": Unti, Bernard, "Colonial Era Blood Sports," in *The Quality of Mercy: Organized Animal Protection in the United States 1865–1930* (Ann Arbor, MI: Proquest/UMI Dissertation Services, 2002).

156 "great inhumanity, and a scandalous . . .": Mather, Increase, *A Testimony Against Several Prophane and Superstitious Customs Now Practised by Some in New-England* (London: n.p., 1687), http://www.covenanter.org/IMather/increasemather testimony.htm.

156 "rude and riotous sports": Unti, "Colonial Era Blood Sports."

156 temporarily banned blood sports: Withington, Ann, *Toward a More Perfect Union: Virtue and the Formation of American Republics* (New York: Oxford University Press, 1991), 185–216, 246–248.

156 even before the founding: Lane and Zawistowski, *Heritage of Care.*

156 Cockfighting had even been an issue: Curnutt, *Animals and the Law,* 278; and Sharon MacPherson, letter to Eric Sakach, May 12, 1993. In the author's possession.

156 The first state prohibitions: Unti, *Quality of Mercy.*

156 New York prohibited all forms: Ibid.

157 pit bull debuted in America: Bulldogbreeds.com, http://www.bulldogbreeds.com/americanpitbullterrier.html.

157 the Ohio and Mississippi railroads: Gibson, Hanna, "Dog Fighting Detailed Discussion," Animal Legal and Historical Center, Michigan State University, 2005, http://www.animallaw.info/articles/ddusdogfighting.htm#s2.

157 United Kennel Club (which had previously sanctioned dogfighting): Ibid.

158 no felony cockfighting laws: Curnutt, *Animals and the Law,* 281.

159 among Amazon's 150 top-selling periodicals: HSUS, "Amazon.com Faces Lawsuit for Illegal Cockfighting Magazines," July 18, 2006, http://www.hsus .org/acf/news/amazon_cockfighting_magazines.html.

160 thousands of gamefowl breeders: HSUS, "New Federal Law Will Help Crack Down on Illegal Cockfighting," May 13, 2004, http://www.hsus.org/acf/news/new_federal_law_will_help.html.

160   David Mitchell of Rattlesnake Game Farm: *The Feathered Warrior,* November 2001, p. 3.

160   "Pure Aggression" was advertised: Kilborn, Peter T., "In Enclaves of Rural America, a Cockfighting Industry Thrives," June 5, 2000, http://www.mangos subic.com/cock_fighting.htm; *The Feathered Warrior,* November 2001, p. 3.

161   Florida banned cockfighting in 1986: Animal Protection of New Mexico, "Cockfighting Ban Dates by State," 2003, http://www.apnm.org/campaigns/ cockfighting/ban_dates.html.

161   Kentucky Court of Appeals in 1994: *Munn v. Com.,* 889 S.W.2d 49 (Ky. App. 1994) (defendant could be prosecuted for second-degree cruelty to animals for engaging in cockfighting).

162   Traditionally, the people involved: Gibson, D., "Vick Dog Fighting Case Gives Ugly Glimpse into the Black Community," American Daily, August 24, 2007, http://www.americandaily.com/article/2004.

162   international markets for the dogs: Campisi, G., "Pit Bulls Are New Export: Dogfighting Gains Popularity in Europe," *Philadelphia Daily News,* July 10, 2000.

162   street fighting that emerged in the 1980s: Malanga, S., "The Sick Hipness of Dog Fighting," *Chicago Sun-Times,* June 17, 2007.

162   John Goodwin called the "cockfighting corridor": John Goodwin, quoted in Stabley, Matthew, "Suspected Cockfighting Operation Busted in Virginia," April 20, 2010, http://www.nbcwashington.com/news/local-beat/Suspected -Cockfighting-Operation-Busted-in-Virginia–91653074.html.

162   In Alabama, the maximum penalty: HSUS, "Ranking of State Cockfighting Laws," September 31, 2009, http://www.hsus.org/acf/fighting/cockfight/ state_cockfighting_laws_ranked.html.

162   In Ohio, it was $250: Ibid.

162   the nation's largest cockfighting pit: HSUS, "More Than 140 Arrested in Tennessee as FBI Raids One of the Nation's Largest Illegal Cockfighting Pits," June 12, 2005, http://www.hsus.org/press_and_publications/press_releases/ fbi_raids_cockfighting_pit_in_cocke_county_tennessee.html.

163   State authorities raided Del Rio: Avent, Jan, "Hundreds Are Cited at Cockfight," *Knoxville News Sentinel,* June 12, 1988.

163   led a successful campaign to reduce the penalty: Humphrey, Tom, "Agent Claims Payoff Taken," *Knoxville News Sentinel,* April 10, 2008.

163   When the FBI raided: HSUS, "More Than 140 Arrested."

163   they arrested not only the pit's owners : HSUS, "Congress Urged to Crack Down on Animal Fighting," February 6, 2007, http://www.hsus.org/press_ and_publications/press_releases/congress_urged_to_crack_down.html.

163   crowds of six hundred to seven hundred people: U.S. Government Printing Office, "Native American Methamphetamine Enforcement and Treatment Act of 2007, the Animal Fighting Prohibition Enforcement Act of 2007, and the Preventing Harassment through Outbound Number Enforcement (Phone) Act of 2007," Hearing before the Subcommittee on Crime, Terrorism, and Homeland Security of the Committee on the Judiciary House of Representatives. 110 Congress, First Session. H.R. 545, H.R. 137, and H.R. 740. Serial No. 110–5, February 6, 2007, p. 51.

163  "approximately 182 cock fights at the Del Rio cockfight pit": Ibid.

163  "cooperating witness observed a girl . . .": Ibid.

163  cockfighting contributed to its spread: HSUS, "Cockfighting and the Spread of Bird Flu," August 17, 2006, http://www.hsus.org/farm/news/ournews/cock fighting_bird_flu.html.

163  thirty million fighting cocks: Chunsuttiwat, S., "Response to Avian Influenza and Preparedness for Pandemic," *Respirology* 13, Suppl. 1 (2008): S36.

164  Oklahoma alone had: Kilborn, P. T., "In Enclaves of Rural America, A Cockfighting Industry Thrives," June 5, 2000, http://www.mangossubic.com/cock_fighting.htm.

164  estimated the state had 2.8 million: Associated Press, "Senator Shurden Announces Plans to Lower Cockfighting Penalty," *Okmulgee Daily Times,* November 8, 2002; and Campbell, Jim, "Cockfighting Battle Predictions Aired," Oklahoma Press Association—Capitol Newsbureau, November 2002.

164  In 1976, Congress had passed a law: USDA, "Public Law 94–279, Animal Welfare Act Amendments of 1976," April 22, 1976, http://awic.nal.usda.gov/nal_display/index.php?info_center=3%20&tax_level=4&tax_subject=182&topic_id=1118&level3_id=6735&level4_id=11094&level5_id=0&placement_default=0.

164  Senator Ford succeeded in amending: Ibid.

164  authorities hadn't prosecuted: Hearing Before the Committee on Agriculture, *Prohibition of Interstate Movement of Live Birds for Animal Fighting,* September 13, 2000, Serial No. 106–59.

166  Allard and Peterson agreed: HSUS, "Victory against Cockfighting in New Mexico," March 12, 2007, http://www.hsus.org/acf/news/victory_new_mexico_cockfighting.html.

166  Symms of Idaho and J. Bennett Johnston of Louisiana: Simon, Rich, "Settling the Dog Fight Over Chickens," *Los Angeles Times,* National section, May 5, 2002.

167  "The purpose of this amendment, . . .": C-SPAN Video Library, E. Blumenauer Farm Security Act of 2001, House Session, http://www.cspanvideo.org/video Library/clip.php?appid=596276456.

167  Combest and Stenholm took to the floor: Ibid.

167  Blumenauer's amendment succeeded: Ibid.

169  John had been the cockfighters' point man: Markarian, M., "Soft on Crime, Soft on Cruelty," Huffington Post, October 3, 2009, http://www.huffington post.com/michael-markarian/soft-on-crime-soft-on-cru_b_302186.html.

169  "I strongly support the cockfighting industry in Louisiana": Congressman Chris John, Hearing Before the Committee on Agriculture, *Prohibition of Interstate Movement of Live Birds for Animal Fighting*, September 13, 2000, Serial No. 106–59.

169  "cockfighting is a cultural, family-type thing": *Baton Rouge Advocate,* December 12, 2001.

169  Vitter, on the other hand, had been an opponent: HSUS, "The Humane Society of the United States and U.S. Senator David Vitter Urge Swift End to Cockfighting in Louisiana," June 21, 2007, http://www.hsus.org/press_and_publications/press_releases/hsus_senator_vitter_louisiana_cockfighting.html.

170 Humane USA, a political action committee I founded: Humane USA, http:// www.humaneusa.org/.

170 the first Republican to win: McGill, K., "David Vitter Faces Untainted GOP Challenger," July 18, 2010, http://www.salon.com/news/feature/2010/07/18/ us_louisiana_senate.

170 A postelection poll showed: Markarian, M., "Soft on Crime, Soft on Cruelty," Huffington Post, October 3, 2009, http://www.huffingtonpost.com/michael -markarian/soft-on-crime-soft-on-cru_b_302186.html.

170 Governor Bill Richardson finally: Associated Press, "Gov. Bill Richardson Signs Bill Outlawing Cockfighting in New Mexico," *USA Today,* March 12, 2007, http://www.usatoday.com/news/nation/2007–03–12-cockfighting -nm_N.htm.

171 proposed phasing out cockfighting: HSUS, "Louisiana Considers Immediate Ban on Cockfighting," May 3, 2007, http://www.hsus.org/acf/news/louisiana _considers_ban_cockfighting.html.

171 a one-year phaseout was negotiated: Anderson, E., "Cockfighting Ban Awaits Blanco's Signature," June 27, 2007, http://blog.nola.com/updates/2007/06/ cockfighting_ban_awaits_blanco.html.

171 At the same time he introduced: HSUS, "Cockfighting Is History in the United States," July 12, 2007, http://www.hsus.org/acf/news/cockfighting_ louisiana_50th.html.

171 several of the state's major pits shut down: Ibid.

171 Cockfighting was now illegal: HSUS, "State Cockfighting Laws," May 2009, http://www.hsus.org/acf/fighting/cockfight/state_cockfighting_laws.html.

172 His reform, prohibiting the training": July 29, 2009, http://www.johnkerry.com.

172 the House had already approved: House Agriculture Committee, Farm Bill, H.R. 2419, June 18, 2009, http://agriculture.house.gov/inside/2007FarmBill .html.

172 John Conyers of Michigan broadened: HSUS, "Congress Enacts Key Animal Protection Measures in Farm Bill," May 22, 2008, http://www.hsus.org/press_ and_publications/press_releases/congress_enacts_key_animal_protection _measures_in_farm_bill_052208.html.

173 He was serving the sixty-day: Bryant, H., "Vick's Reality: Life as an Ex-Con," ESPN.com, May 19, 2009, http://sports.espn.go.com/nfl/columns/ story?columnist=bryant_howard&id=4180017.

174 "Virginia O"—Oscar Allen—indicted in October of 2007: Associated Press, "Man Who Sold Dog to Vick Is Sentenced," January 25, 2008; CBS News, January 24, 2008, http://www.cbsnews.com/stories/2008/01/25/sports/ main3752885.shtml.

174 Vick had paid $1 million: *Orange County Register,* "No Duh: Vick Voted Most-Hated Athlete," http://ocpets.ocregister.com/2010/09/16/no-duh-vick-voted -most-hated-athlete/67030.

175 we did our first event with him: HSUS, "The Road Back from Ruin," September 30, 2009, http://www.humanesociety.org/news/news/2009/09/vick_ dc_093009.html; and CBS News, *60 Minutes,* "Michael Vick Vows to Help End Dogfighting," August 16, 2009, http://www.cbsnews.com/video/watch/?id= 5245553n&tag=mncol;lst;5.

## Chapter Five—For the Love of Pets

178  America's largest emergency animal shelter: Halligan, Karen, "Hurricane Katrina: The Animals and the Aftermath," 2006, http://www.dochalligan.com/katrina/katrina.shtml.

179  the world saw that helping animals: Anderson, Allen, and Linda Anderson, *Rescued: Saving Animals from Disaster* (Novato, CA: New World Library, 2006), 8.

179  thousands of rescued animals: Merrit, Clifton, "Hurricane Katrina and Rita Rescuers Shift Gears from Rescue & Reunion to Rehoming," *Animal People,* December 2005, 6.

180  many people had fled their homes: Anderson and Anderson, *Rescued,* 83–86.

181  rising waters overwhelmed: CBC News Online, "Hurricane Katrina Timeline," September 4, 2005, http://www.cbc.ca/news/background/katrina/katrina_timeline.html.

181  The Louisiana SPCA: Louisiana SPCA, "Hurricane Katrina," 2010, http://la-spca.org/Page.aspx?pid=297.

181  twenty-three dogs and cats: Manning, Anita, "Animal Welfare Groups Rescue Abandoned Pets," September 3, 2005, http://www.usatoday.com/news/nation/2005–09–03-katrinapetrescues_x.htm.

181  the largest animal-rescue mission: "Katrina's Animal Rescue," PBS, November 20, 2005, http://www.pbs.org/wnet/nature/episodes/katrinas-animal-rescue/introduction/2561/.

182  There were veterinary teams—VMATs: American Veterinary Medical Association, "AVMA Veterinary Medical Assistance Teams (VMAT)," 2010, http://www.avma.org/vmat/default.asp.

182  billions given for the response: ABC News, "Billions of Dollars in Donations Post-Katrina, Yet Very Little Relief," August 3, 2006, http://blogs.abcnews.com/theblotter/2006/08/billions_of_dol.html.

182  seven thousand phone calls: 2005 HSUS Annual Report, "Responding to Katrina: Meeting the Challenge," 2005, www.humanesociety.org/assets/pdfs/2005_annual_report.pdf, pp. 12–13.

182  National Guard troops had walled it off: Goldblatt, Jeff, Steve Harrigan, Rick Leventhal, Liza Porteus, and the Associated Press, "Official: Astrodome Can't Take More Refugees," September 2, 2005, http://www.foxnews.com/story/0,2933,168112,00.html.

183  large-lettered markings on the front: Shiley, Mike, "Dark Water Rising: Survival Stories of Hurricane Katrina Animal Rescues," Smithsonian Permanent Archive Material, 2006–07–01; 75 minutes.

183  some 70 percent of households: U.S. Pet Ownership Statistics 2005, Compiled from the American Pet Products Manufacturers Association (APPMA) 2003–2004 National Pet Owners Survey, accessed March 1, 2005, http://www.hsus.org/pets/issues_affecting_our_pets/pet_overpopulation_and_ownership_statistics/us_pet_ownership_statistics.html.

184  One was reserved for horses: Anderson and Anderson, *Rescued,* 113.

184  cap of two thousand dogs: Ibid.

184  Among them was John Wallman: Chad Sisneros, Gonzales, Louisiana, interviews from video shot by HSUS at the Lamar-Dixon Center, September 2005.

185  Jeremy Campbell, a man in his late twenties: Ibid.

185  That's how it was for Richard Colar: Personal communication with Corey Smith, HSUS staffer; and "After the Storm," *All Animals* 8, no. 3 (Summer 2006).

187  rescue the shivering, growling terrier: Law, Steve, "Oregon Soldiers Face Adversity in Bid to Help New Orleans," *Salem Statesman Journal*, September 8, 2005.

187  a Japanese professor and his faithful Akita: Tremain, Ruthven, *The Animals' Who's Who: 1,146 Celebrated Animals in History, Popular Culture, Literature, & Lore* (New York: Scribner, 1984), 105.

187  a little boy and his dog Snowball: Associated Press, "Has Snowball Finally Been Found?," MSNBC, September 9, 2005, http://www.msnbc.msn.com/id/9255741/.

189  Pet Evacuation and Transportation Standards (PETS) Act: Simmons, Rebecca, "No Pet Left Behind: The PETS Act Calls for Disaster Plans to Include Animals," April 20, 2006, http://www.hsus.org/pets/pets_related_news_and_events/no_pet_left_behind_the_pets.html.

189  Tom Lantos, a Democrat of California: U.S. Congressional Record, House of Representatives H6806, September 20, 2006.

190  had told a reporter that: Associated Press, "President Bush Signs Bill to Leave No Pet Behind in Disaster Planning and Evacuation," October 6, 2006, http://www.hsus.org/press_and_publications/press_releases/president_bush_signs_pets_Act.html.

190  sixteen states also enacted: Allen, Laura, "Are Government Officials Ready to Evacuate and Shelter Animals in Disasters?," August 29, 2008, http://www.animallawcoalition.com/animals-and-politics/article/580.

190  "I promised them . . .": Chad Sisneros, HSUS video interview shot in Miami, Florida, January 2010.

191  As Paul later told a reporter: Bryan, Susannah, "Two Dogs Taking the Long Way Home," *Sun Sentinel*, January 28, 2010, http://articles.sun-sentinel.com/2010–01–28/features/fl-katrina-haiti-dogs–20100128_1_paul-fowler-dogs-long-way-home.

192  Animal Planet in recent years: *Animal Cops*, 2010, http://animal.discovery.com/tv/animal-cops/.

192  highest-rated show on the network: *Whale Wars*, 2010, http://animal.discovery.com/tv/whale-wars/.

192  *American Sportsman*, which aired on ABC: Amory, Cleveland, *Mankind? Our Incredible War on Wildlife* (New York: Harper and Row, 1974).

193  Discovery Channel introduced: Wikipedia, "Animal Planet," 2010, http://en.wikipedia.org/wiki/Animal_Planet.

193  like the popular *Mutual of Omaha's Wild Kingdom*: Wikipedia, "Wild Kingdom," 2010, http://en.wikipedia.org/wiki/Wild_Kingdom.

193  "dog whisperer" Cesar Millan: "Cesar's Way," 2010, http://www.cesarsway.com/.

193  171 million dogs and cats: U.S. Pet Ownership Statistics 2009, Compiled from the American Pet Products Manufacturers Association (APPMA) 2008–2009 National Pet Owners Survey, December 30, 2009, http://www.hsus.org/pets/issues_affecting_our_pets/pet_overpopulation_and_ownership_statistics/us_pet_ownership_statistics.html.

194   products and services industry: American Pet Products Association, 2010, "Industry Statistics & Trends," http://www.americanpetproducts.org/press_industrytrends.asp.

194   PetSmart, a big-box store: Wikipedia, "Petsmart," 2010, http://en.wikipedia.org/wiki/PetSmart.

194   have made PetSmart Charities: Wikipedia, "Petsmart," 2010, http://en.wikipedia.org/wiki/PetSmart.

194   Petco, also has more than a thousand stores: Wikipedia, "Petco," 2010, http://en.wikipedia.org/wiki/Petco.

194   When Petfinder.com went nationwide in 1998: Wikipedia, "Petfinder.org," 2010, http://en.wikipedia.org/wiki/Petfinder.org.

195   Many of the nation's twenty-eight veterinary schools: Association of American Veterinary Medical Colleges, "AAVMC Annual Report 2008–2009," 2009, http://www.aavmc.org/.

195   From 1984 to 1996, total veterinary income: "The Current and Future Market for Veterinarians and Veterinary Medical Services in the United States," May 1999, www.avma.org/reference/mega715c.pdf; and personal communication with Andrew Rowan, October 10, 2010.

195   Banfield pet hospital runs: Wikipedia, "Banfield (pet hospitals)," 2010, http://en.wikipedia.org/wiki/Banfield_(pet_hospitals).

195   services that we used to associate: Martinez, M., and M. J. Rathbone, "Linking Human and Veterinary Health: Trends, Directions and Initiatives," *AAPS PharmSci* 4, no. 4 (2002).

195   Every day across America: National Council on Pet Population Study and Policy, "The Shelter Statistics Survey, 1994–97," 1997, http://www.petpopulation.org/statsurvey.html.

196   Local animal shelters opened their doors: Unti, *Quality of Mercy,* 468–469.

196   roughly thirty-five hundred brick-and-mortar facilities: Rowan, Andrew, "Counting the Contributions," *Animal Sheltering Magazine,* November/December 2006, p. 36.

197   Upwards of $1 billion a year: Becker, Debbie, "Las Vegas Bets on Life," *USA Today,* June 23, 1998, http://www.adoptapet-wa.org/LasVegasLife.htm; Rowan, "Counting the Contributions."

197   American shelters killed thirteen to fifteen million: Rowan, "Counting the Contributions," 35.

197   Today, shelters euthanize: Humane Society of the U.S., "Common Questions about Animal Shelters," October 26, 2009, http://www.humanesociety.org/animal_community/resources/qa/common_questions_on_shelters.html.

197   intellectual spark for the no-kill movement: Crawford, Lynda, "No Kill, the New Goal in Animal Control," *Gotham Gazette,* February 2, 2009, http://www.gothamgazette.com/article/iotw/20040202/200/856.

199   "Whether strays or surrenders . . .": Duvin, Edward, "In the Name of Mercy," Animalines, 1989, www.bestfriends.org/nomorehomelesspets/pdf/mercy.pdf.

200   1989 essay entitled "In the Name of Mercy": Ibid.

200   He added later: Duvin, Edward, "Specieism: Alive and Well," Animal's Voice, 1990, www.bestfriends.org/nomorehomelesspets/pdf/speciesism.pdf.

200   "adoption guarantee": Best Friends Animal Society, "Richard Avanzino," 2010,

http://www.bestfriends.org/nomorehomelesspets/weeklyforum/bioravanzino .cfm.

201 Rich Avanzino now runs: Ibid.

204 slightly less than 25 percent: Humane Society of the U.S., "Common Questions about Animal Shelters."

206 Pine Bluff Kennels—was an Internet seller: Fisher, Victoria, "Websites Masks Cruel Reality of Puppy Mills," July 14, 2008, http://www.hsus.org/pets/pets_ related_news_and_events/pine_bluff_kennels_website_071408.html.

207 hellhole we raided in Pennsylvania: Humane Society of the U.S., "More Than 200 Dogs Rescued from Penn. Puppy Mill," June 25, 2009, http://www.hsus .org/pets/pets_related_news_and_events/nearly_300_dogs_rescued_from.html

207 At one particularly sickening mill: Humane Society of the U.S., "Third Major Quebec Puppy Mill Bust," December 12, 2008, http://www.humanesociety .org/news/press_releases/2008/12/third_major_quebec_puppy_mill_121208 .html.

208 We estimate that: Humane Society of the U.S., "Taking a Bite out of Puppy Mills: Animal Planet Investigates: Petland," May 18, 2010, http://www .humanesociety.org/news/news/2010/animal_planet_investigates_petland _050610.html.

208 we found nearly a thousand: Humane Society of the U.S., "Virginia: The Next Puppy Mill State?," November 1, 2007, http://www.hsus.org/pets/pets_related _news_and_events/virginia_the_next_puppy_mill.html.

208 two to four million puppies: Humane Society of the U.S., "Puppy Mills: Frequently Asked Questions," July 8, 2010, http://www.humanesociety.org/ issues/puppy_mills/qa/puppy_mill_FAQs.html.

208 In 2009, on the morning: Humane Society of the U.S., "The HSUS Rescues More Than 100 Animals from Arkansas Puppy Mill," October 6, 2009, http:// www.humanesociety.org/news/press_releases/2009/10/arkansas_puppy_ mill_rescue_100609.html.

209 he was given a name: Gentle Ben: Pacelle, Wayne, "Gentle Ben: Before and After," July 11, 2009, http://hsus.typepad.com/wayne/2009/06/gentle-ben .html.

210 bill before the Oregon Legislature: Balzar, John, "A Brave Voice Against Puppy Mills," *All Animals,* November 12, 2009, http://www.humanesociety.org/ news/magazines/2009/07–08/brave_voice_against_puppy_270x224.html.

211 The Dog Press and other websites: Bates, Doug, "Breeders Howling at Oregon AKC judge," *The Oregonian,* April 2, 2009, http://www.oregonlive.com/ opinion/index.ssf/2009/04/breeders_howling_at_oregon_akc.html; and Ted Paul, interview by author, Los Angeles, May 30, 2010.

212 Founded in 1884 . . . the AKC pledged: The American Kennel Club, Rules Applying to Registration and Discipline, amended to April 1, 2010, foreword.

212 The group recognizes 167 breeds: The American Kennel Club.

212 AKC Canine Health Foundation: AKC Canine Health Foundation, 2010, http://www.akcchf.org/.

213 That proposed amendment: HCS HJR 86—Right to Raise Animals Sponsor: Loehner Committee, www.house.mo.gov/billtracking/bills101/sumpdf/ HJR0086c.pdf.

214  The registration fees that AKC gets: American Kennel Club, "Dog Registration Fee Increase Questions," 2010, http://www.akc.org/contact/answer_center/faq_dogreg_fee_increase.cfm.

214  "The best use of pedigree papers . . .": Lemonick, Michael, "A Terrible Beauty," *Time* 144, no. 24, December 12, 1994.

216  The inspector general of the USDA: APHIS Animal Care Program Inspections of Problematic Dealers, Audit Report 33002–4-SF, Washington, DC, May 2010.

216  "The investigators visited . . .": Associated Press, "Dogs Suffer Over Lax Kennel Violation Follow-up," Boston.com; May 26, 2010, http://www.boston.com/news/nation/articles/2010/05/26/dogs_suffering_because_of_lax_kennel_violation_enforcement_report_says/..

216  The report describes one case: APHIS Animal Care Program Inspections of Problematic Dealers, Audit Report 33002–4-SF, Washington, DC, May 2010, pp. 10–12.

217  inherited disorders and congenital health problems: Rooney, Nicola, and David Sargan, *Pedigree Dog Breeding in the UK: A Major Welfare Concern?,* Independent Scientific Report Commissioned by the RSPCA.

218  A study in the *American Naturalist*: Careau, Vincent, et al., "The Pace of Life Under Artificial Selection: Personality, Energy Expenditure, and Longevity Are Correlated in Domestic Dogs," *The American Naturalist* 175 (June 2010): 753–758.

218  "We have allowed some breeds": BBC, *Pedigree Dogs Exposed*, Documentary, August 19, 2008.

219  skull malformation called syringomyelia: Cavalier Health.org, "Syringomyelia (SM) and the Cavalier King Charles Spaniel," http://www.cavalierhealth.org/syringomyelia.htm, 2010; and Rooney and Sargan, *Pedigree Dog Breeding.*

219  English bulldogs: The Bulldog Club of America, *The Bulldog: An Illustrated Guide to the Standard,* http://thebca.org/BulldogGuide.pdf.

219  They are susceptible: Rooney and Sargan, *Pedigree Dog Breeding.*

219  The BBC ceased broadcasting: Margolis, Jonathan, and Fiona Macrae, "BBC Could Drop Crufts Over Unhealthy 'Freak Show' Breeds," August 19, 2009, http://www.dailymail.co.uk/news/article–1046614/BBC-drop-Crufts-unhealthy-freak-breeds.html.

219  "When I watch Crufts, . . .": BBC, *Pedigree Dogs Exposed,* Documentary, August 19, 2008.

220  distinctive ridge on their back: Ibid.

220  Breeder Ann Woodrow told the BBC: Ibid.

220  The Kennel Club and the Dog's Trust: Bateson, Patrick, *Independent Inquiry into Dog Breeding* (Halesworth, Suffolk: University of Cambridge, Micropress, Ltd., 2010).

220  King Charles Cavaliers, for instance: Rooney and Sargan, *Pedigree Dog Breeding.*

## Chapter Six—The Cull of the Wild

222  I hiked almost all: DuFresne, Jim, *Isle Royale National Park: Foot Trails and Water Routes* (Seattle: The Mountaineer Books, 2002).

223 The park was established: Ibid., 16.

230 "Is it any wonder then . . .": Oversight Hearing Before Subcommittee on National Parks, Forests and Public Lands, 110th Congress, Serial No. 110–7, March 20, 2007, p. 3.

231 Bruce Babbitt, an ardent environmentalist: McNamee, Thomas, *The Return of the Wolf to Yellowstone* (New York: Henry Holt, 1997), passim.

232 Interior Secretary Dirk Kempthorne: Miller, John, "Rule Delisting Wolves in Idaho, Montana Imminent?," *Casper Star Tribune,* September 3, 2006.

232 In 2009, President Barack Obama's: Delisting, 74 Fed. Reg. 15121 (April 2, 2009); and Lang, Brent, "Salazar Approves Removing Gray Wolves from Endangered Species List," CBS News, March 6, 2009.

232 In 2009, Idaho sold: Russell, Betsy Z., "Thousands Buy Idaho Wolf Hunting Tags," *Spokesman Review,* August 24, 2009.

232 Idaho governor Butch Otter: Ritter, John, "Idahoans Eager to Thin Resurgent Gray Wolf Packs," *USA Today,* March 14, 2007.

232 Montana followed suit: Murphy, Kim, "Montana Wolf Hunt Is Stalked by Controversy," *Los Angeles Times,* October 25, 2009.

232 Fortunately, the same federal judge: Court decision: *Defenders of Wildlife v. Salazar,* No. 09–0077 (D. Mont., Aug. 5, 2010).

234 Just before the Smithsonian requested: Golden, Tim, "Big Game Hunter's Gift Roils the Smithsonian," *New York Times,* March 17, 1999; and Balzar, John, "Smithsonian Museum in Cross-Hairs of Debate," *Los Angeles Times,* March 21, 1999.

234 Using a helicopter, Behring: "Hunters Red-Faced Over Elephant Shoot," *Johannesburg Mail and Guardian,* April 23, 1999, http://www.mg.co.za/article/1999 -04-23-hunters-red-faced-over-elephant-shoot.

234 Behring had tried: Scully, 69–71.

234 the government concluded: Declaration by the Mozambique Ministry of Agriculture and Fisheries National Directorate for Forestry and Wildlife, issued January 11, 1999, by National Director Arlito Cuco.

235 Congress . . . closed this loophole: Goldfarb, Zachary, "Pension Bill Add-on Would Gun Down Safari Tax Write-off," *Seattle Times,* August 6, 2005.

236 "It takes dedication to the animal,": Santa Cruz, Nicole, "Taxidermist Unable to Make an Escape from the Wildlife," *Los Angeles Times,* February 14, 2010.

236 *Taxidermy and Longing,* stuffed with: Ibid.

236 on his storied African safari: Unti, *Quality of Mercy,* 534.

237 an ennobling pursuit that embodied: Nash, Roderick F., *Wilderness and the American Mind,* 3rd ed. (New Haven: Yale University Press, 1982), 149–153.

237 "Mr. Roosevelt, when are you . . .": Johnson, Robert Underwood, *Remembered Yesterdays* (Boston: Little, Brown, and Co., 1923), 388.

237 He created 150 national forests: Brinkley, Douglas, *The Wilderness Warrior: Theodore Roosevelt and the Crusade for America* (New York: HarperCollins, 2009), passim.

238 elk and bison roamed the forests: Matthiessen, Peter, *Wildlife in America,* rev. ed. (New York: Penguin, 1978), passim.

238   the settlement of the West quickened: Cronon, William, *Nature's Metropolis:*
      *Chicago and the Great West* (New York: W.W. Norton and Company, 1992),
      passim.

241   "A thing is right when it tends to preserve . . .": Leopold, Aldo, *A Sand County*
      *Almanac* (New York: Oxford University Press, 1949), 224–225.

241   Leopold opposed this idea: Leopold, A., "Deer Irruptions," *Wisconsin Conserva-*
      *tion Bulletin* 8 (1949): 3–11.

241   This idea influenced the 1937 Pittman-Robertson Act: Dunlap, Thomas R.
      *Saving America's Wildlife: Ecology and the American Mind, 1850–1990* (Princeton,
      NJ: Princeton University Press, 1988), 78–79.

244   "You just can't let nature run wild": Hevesi, Dennis, "Walter Hickel, Nixon
      Interior Secretary, Dies at 90," *New York Times,* May 8, 2010.

244   "I will not be a part of Alaska . . .": Waterman, Jonathan, "Sheep's Clothing—
      Alaskan Wolf Control: Is It Wildlife Biology? Or Plain Old Politics?," *Outside,*
      May 1993, p. 52.

246   Hickel took the press: Williams, Ted, "Alaska's War on Wolves," *Audubon Mag-*
      *azine,* May/June 1993.

246   "We don't manage your cattle, . . .": Ibid.

247   But the most destructive thing: Regelin, Wayne L., "Wolf Management in
      Alaska with an Historic Perspective," Alaska Department of Fish and Game
      website, http://www.wc.adfg.state.ak.us/index.cfm?adfg=wolf.wolf_mgt.

247   They launched aerial-gunning programs: Medred, Craig, "Aerial Hunting Pro-
      gram Kills 124 Wolves," *Anchorage Daily News,* May 19, 2008.

248   "Wolves don't pay for hunting licenses": Williams, Ted, "Who's Managing the
      Wildlife Managers?" *Orion Nature Quarterly,* 5, no. 4 (1986): 20.

248   Only California bans trophy hunting: Pacelle, Wayne, "Bullets, Ballots, and
      Predatory Instincts," in *Shadow Cat,* ed. by Susan Ewing and Elizabeth Gross-
      man (Seattle: Sasquatch Books, 1999), 199–208.

249   Montana trophy hunters kill: Curtis, Sam, and Tom Dickson, "A Close Look
      at Mountain Lions," *Montana Outdoors,* July–August 2008 at http://wildfelid
      .com/Montana%20Mountain%20Lions%20DeSimone.pdf; Toweill, Dale,
      Steve Nadeau, David Smith, eds., *Proceedings of the Ninth Mountain Lion Work-*
      *shop,* Sun Valley, Idaho, Idaho Department of Fish and Game, May 5–8,
      2008.

249   Ted Williams calls it "garbaging for bears": Williams, Ted, "Hunters Close
      Ranks, and Minds," *High Country News,* March 3, 1997.

249   "never feed bears": U.S. Forest Service, "Be Bear Aware," n.d.

250   hunters shoot four thousand bears: Schweitzer, Sarah, "The Bear Necessities
      Hit Maine: Referendum on Certain Hunting Practices Fuels Debate," *Boston*
      *Globe,* October 31, 2004.

250   Add in the 89,710 coyotes: U.S. Department of Agriculture, "Animals Taken
      by Component/Method Type and Fate by the Wildlife Services Program—FY
      2008,"   http://www.aphis.usda.gov/wildlife_damage/prog_data/2008_pdr/
      PDR_G/TableG_long/Table_G_FY2008_by_Species_Alphabetically_All
      States.pdf.

252   more than one thousand captive bird-shooting facilities: Wingshooting USA,
      http://wingshootingusa.org/index.cfm.

252 only 12.5 million Americans went hunting in the United States in 2006: U.S. Fish and Wildlife Service, *National Survey of Hunting, Fishing and Wildlife Associated Recreation*, 2006. http://wsfrprograms.fws.gov/Subpages/NationalSurvey/reports2006.html.

254 six million seals migrate: International Union for Conservation of Nature, Red List of Threatened Species—*Pagophilus groenlandicus*. April 2010. Available at http://www.iucnredlist.org/apps/redlist/details/41671/0.

255 the annual toll taken by the sealers: Mowat, Farley, *Sea of Slaughter* (Boston: Atlantic Monthly Press, 1984).

255 "A very colorful page": *Canadian Scene*, "A Dying Industry," *Canadian Broadcasting Corporation*, March 14, 1958. Archives.

255 the sealers tinkers with their calendar: Aldworth, Rebecca, and Stephen Harris, "Canada's Commercial Seal Hunt" in *The State of the Animals* 2007, eds. Andrew N. Rowan and Deborah J. Salem, 93–109.

256 the cod had finally reached: Kurlansky, Mark, *Cod: A Biography of the Fish That Changed the World* (New York: Penguin Book, 1998).

256 Harp seals do eat cod: Ibid.

256 "Mr. Speaker, I would like to see": John Efford, Minister of Newfoundland Fisheries and Aquaculture; from Newfoundland House of Assembly Proceedings, Vol. XLIII, No. 18, May 4, 1998; Kaimet, Kate, "Minister Flaunts Sealskin Coat," *Ottawa Citizen*, December 14, 2003.

257 "It is not study that we want": Lawrence O'Brien, from the House of Commons Debates, Vol. 138, No. 091, 2nd Session, 37th Parliament, April 29, 2003.

257 the national government ramped up: Department of Fisheries and Oceans, "Thibault announces multi-year Atlantic seal hunt management measures," News release. February 3, 2003. http://www.dfo-mpo.gc.ca/media/newsrel/2003/hq-ac01_e.htm.

262 a team of veterinarians: Burdon, R. L., J. Gripper, J. A. Longair, I. Robinson, and D. Ruehlmann, Veterinary report, "Canadian commercial seal hunt," Prince Edward Island, March 2001. http://www.scandinavianantisealingcoalition.org/Reports/Vet%20report%20march%202001.pdf.

262 Their report noted: Ibid.

263 In 2009, HSUS helped convince: Philip Sherwell, "Canada's Annual Seal Hunt Draws Sealers and Sceptics," *The Telegraph*, April 10, 2010.

263 described by Vladamir Putin: Halpin, Tony, "Slaughter of the Seals in Russia Is Stopped by Vladmir Putin," *The Sunday Times*, March 20, 2009.

263 In 2010, the allowable kill: Fisheries and Oceans Canada, "Minister Shea increases quota for Atlantic Seal Harvest." Press Release, March 15, 2010. http://www.dfo-mpo.gc.ca/media/npress-communique/2010/hq-ac11-eng.htm.

263 there are only about fourteen thousand licensed sealers: Report of the Seal Seminar, "Seals in the Marine EcoSystem," March 20–21, 2001, Nuuk, Greenland, 57; Fisheries and Oceans Canada, "Frequently Asked Questions about Canada's Seal Harvest." http://www.dfo-mpo.gc.ca/fm-gp/seal-phoque/faq-eng.htm#_13.

263 the industry generated about $1.2 million: Fisheries and Oceans Canada, "Landings and Landed Value by Species," Newfoundland and Labrador region, 2010.

264   There are direct subsidies for seal products: Gallon, G., The Economics of the
       Canadian Sealing Industry. Canadian Institute for Business and Environment.
       2001. Available at http://www.ifaw.org/Publications/Program_Publications/
       Seals/asset_upload_file414_12091.pdf.
265   the value of snow crab exports: Statistics Canada. Canadian international trade
       data April 2004–October 2010.
266   As one Canadian writer has put it: Teitel, Murray, "The Millions Ottawa
       Spends Subsidizing the Seal Hunt," *Financial Post*, April 17, 2008.

### Chapter Seven—Cruelty and Its Defenders

272   "the righteous man regardeth . . .": King James Bible, Proverbs 12:10.
273   "Beef: It's What's for Dinner": Cattlemen's Beef Board and the Federation of
       State Beef Councils, http://www.beef.org; and the California Milk Advisory
       Board, http://www.realcaliforniamilk.com/happycows.
273   guidelines were formulated: Nestle, Marion, *Food Politics: How the Food Industry
       Influences Nutrition & Habits* (Berkeley: University of California Press, 2002);
       and Black, Jane, "Advocates Worry That Dietary Advice Will Be Lost in Trans-
       lation," *Washington Post,* October 3, 2010.
273   the American Dietetic Association tells us: "Position of the American Dietetic
       Association and Dietitians of Canada: Vegetarian Diets," *Journal of the American
       Dietetic Association* 103, no. 6 (2003): 748–765, doi: 10.153/jada.2003.50142.
274   the testing is not only painful: Alttox.org, http://alttox.org/ttrc/tox-test
       -overview/; and Stephens, Martin L., "An Animal Protection Perspective on
       21st Century Toxicology," *Journal of Toxicology and Environmental Health* Part B,
       vol. 13, no. 2 (2010): 291–298.
279   More hunters are using laser sights: Barnard, Jeff, "Technology Pushes Enve-
       lope of Hunting," Associated Press, June 6, 2005, http://sports.espn.go.com/
       outdoors/hunting/news/story?page=c_fea_fading_hunt_AP4_technology;
       Benedetti, Winda, "Electronic Decoys Put Duck Hunting Ethics Under Fire,"
       *Seattle Post-Intelligencer,* August 28, 2001; Barker, Eric, "The Long Rangers,"
       *Lewiston Tribune,* January 17, 2010; "CA Bear Season Won't be Expanded," *San
       Jose Mercury News,* April 21, 2010; and "YO Ranch, A Texas Brand Since 1880,"
       http://www.yoranch.com/history.html.
280   factory farmers suddenly became advocates: *United Egg Producers, The Egg In-
       dustry and Animal Welfare: A Science-Based Approached,* www.unitedegg.org/
       information/pdf/Egg_Industry_Animal_Welfare_Brochure.pdf; Sundberg,
       Paul, The National Pork Board, in the *Proceedings of the 2002 Future Trends in
       Animal Agriculture Symposium,* p. 12; and The Animal Ag Alliance, http://www
       .animalagalliance.org/current/home.cfm?Category=Animal_Care&Section=
       Main.
280   "Do you know what . . .": Smith, Wesley, *A Rat Is a Pig Is a Dog Is a Boy* (New
       York: Encounter Books, 2010).
281   animal rights groups have "hijacked": Ibid., 18.
281   The people who supply dogs: American Anti-Vivisection Society, "Dying
       to Learn: Exposing the Supply and Use of Dogs and Cats in Higher Educa-
       tion," April 2009, accessed September 29, 2010, http://www.dyingtolearn.org/

dyingToLearn.pdf; and The Humane Society of the United States, "Pets Used in Experiments," http://www.humanesociety.org/issues/pets_experiments/.

282 who happen to make up 95 percent: Carbone, Larry, *What Animals Want: Expertise and Advocacy in Laboratory Animal Welfare Policy* (New York: Oxford University Press, 2004).

283 "infinite capacity to rationalize his cruelty": Amory, Cleveland, *Mankind? Our Incredible War on Wildlife* (New York: Harper and Row, 1974), 14.

284 In their continuing quest for respectability: The United Gamefowl Breeders Association, http://www.ugba.net/; and *Declaration of Sandy Johnson, UGBA v. Veneman*, U.S. District Court for the Western District of Louisiana, May 12, 2003, CV03–0970.

285 almost an unbroken political tradition: Nie, Martin, "The Scope and Bias of Political Conflict, State Wildlife Policy and Management," *Public Administration Review* 64 (2004): 221–233; and Hagood, Susan, *Money and Myth: The Pervasive Influence of Hunters, Hunting, Culture and Money* (Washington, DC: The Humane Society of the U.S., 1997).

285 He circulated 160 pages: Schoch, Deborah, "Commissioner Blames NRA for His Ouster," *Los Angeles Times,* September 25, 2007.

286 A 2007 scientific study by the Peregrine Fund: Parish, Chris, William Heinrich, and W. Grainger Hunt, *Lead Exposure, Diagnosis and Treatment in California Condors Released in Arizona* (Boise, ID: The Peregrine Fund, 2006), 3.

286 twelve of the fifty-one condor deaths: Herdt, Timm, "Governor OK's Lead Bullet Ban," *Ventura County Star,* October 14, 2007.

286 "I'm sure the mortality rate . . .": Ibid.

286 U.S. Fish and Wildlife Service, after years: U.S. Fish and Wildlife Service, Migratory Birds, http://www.fws.gov/migratorybirds/currentbirdissues/non toxic.htm.

286 The NRA had bitterly fought: "NRA Opposes Anti-Hunting Petition to Ban Lead Ammunition," The NRA, August 4, 2010, http://www.nrahuntersrights .org/Article.aspx?id=3736

286 the group rallied its members: Schoch, "Commissioner Blames NRA."

286 The NRA arranged for a letter: Ibid.

287 The attack on the independence of the commission: Skelton, George, "Gov. Ignores Gun Lobby, and Condors Get a Lift," *Los Angeles Times,* October 18, 2007.

287 "The science is irrefutable,": Vick, Karl, "Lead from Carrion Killing-Off California Condors," *Washington Post,* October 12, 2007.

287 In his letter of resignation: Personal correspondence from Hanna to Governor Schwarzenegger.

288 It had been a tradition: Blechman, Andrew, *Pigeons, The Fascinating Saga of the World's Most Revered and Reviled Bird* (New York: Grove Press, 2006), 83.

289 The organizers were quick to claim: Worden, Amy, "Critics Still Take Aim at Pa. Pigeon Shoots," *Philadelphia Inquirer,* December 5, 2007.

289 But given that they purchased: "NYC Pigeons Trapped, Kidnapped, Shot for Sport, Group Says," *Wall Street Journal,* May 24, 2010; Fanelli, James, "They Shoot Pigeons, Don't They?," *New York Post,* July 27, 2008; letter from the New York City Bar, Committee on Legal Issues Pertaining to Animals to Joseph

B. Scarnati, President of the PA Senate, and Keith McCall, Speaker of the PA House, May 4, 2010; and *Seeton v. Pike Township Sportsmen's Association,* Civ. No. 01–11736 (Pa. Ct. Common Pleas Berks County, November 2001).

289 No shooter required a license: Worden, "Critics Still Take Aim"; *Seeton v. Pike Township Sportsmen's Association;* and *Covington Twp. v. Moscow Sportsmen's Club, Inc.,* Civ. No. 06–4848 (Pa. Ct. Common Pleas Lackawanna County, September 21, 2006) (Exh. E)/ PA Game Code Title 34 Chapter 1 Section 102.

289 shooting squirrels just for fun: Sanborn, F. B., ed., *Familiar Letters of Henry David Thoreau* (Boston: Houghton-Mifflin, 1894), cited in *The Extended Circle,* ed. Jon Wynne-Tyson (Fontwell, Sussex: Centaur Press, 1985).

290 "Pigeon shooting is an historic . . .": NRA-Institute for Legislative Action e-mail alert titled "Pennsylvania: Bird Shoot Ban to be Heard Tomorrow!," July 14, 2009.

291 state humane agents could arrest pigeon shooters: *Hulsizer v. Labor Day Committee,* 734 A.2d 848, 853 (Pa. 1999).

291 "National 'animal rights' extremist groups, . . .": NRA-Institute for Legislative Action e-mail alert, titled "Ban on Pigeon Shooting Could be Voted on Soon!," September 19, 2008, http://www.nraila.org/Legislation/Read.aspx?ID=4174.

292 the NRA has warded off efforts: NRA-Institute for Legislative Action e-mail alert titled "Anti-Hunting State Ballot Initiative Being Circulated for Signatures at the North Dakota State Fair," July 30, 2010, http://www.nraila.org/legislation/read.aspx?id=5984.

292 it fought off an effort in Congress: Letter to Congressman Nick Jo Rahall from the NRA and other hunting organizations, March 6, 2008.

292 "In a normal season we will go through . . .": Testimony of Wayne Pacelle, senior vice president of the Humane Society of the United States, on H.R. 1006 and H.R. 1472. Legislative Hearing before the Subcommittee on Fisheries Conservation, Wildlife and Oceans of the Committee on Resources, 108th Congress, June 12, 2003. Serial No. 108–25, p. 84.

293 Others use "walk-in" baits: Ibid, p. 84.

293 Congressman Elton Gallegly, a Republican: Legislative hearing before the subcommittee on fisheries, conservation, wildlife and oceans, 108–25, 108th Congress, June 12, 2003.

293 twenty-six cosponsors withdrew their support: Bill Summary and Status, 108th Congress (2003–2004), H.R. 1472 Cosponsors, http://thomas.loc.gov/cgi-bin/bdquery/z?d108:HR01472:@@@P.

294 "This is not about me. . . .": Schoch, "Commissioner Blames NRA."

295 Ohio is one of the biggest: U.S. Department of Agriculture, National Agricultural Statistics Service, "Chickens and Eggs," 2010, p. 8, http://www.usda.mannlib.cornell.edu.usda/current/ChicEggs/ChicEggs–09–21–2010.pdf; U.S. Department of Agriculture National Agricultural Statistics Service, "Quarterly Hogs and Pigs," 2010, http://www.usda.mannlib.cornell.edu/usda/current/HogsPigs/HogsPigs–09–24–2010.pdf; Barnett, J. L., P. H. Hemsworth, G. M. Cronin, E. C. Jongman, and G. D. Hutson, "A Review of the Welfare Issues for Sows and Piglets in Relation to Housing," *Australian Journal of Agricultural Research* 52 (2001): 1–28; United Egg Producers, 2004, Independent Scientific Advisory Committee, www.unitedegg.org/Scientific/default.cfm;

and United Egg Producers, *Animal Husbandry Guidelines for U.S. Egg Laying Flocks,* 2010, p. 1, www.unitedegg.org/information/pdf/UEP_2010_Animal_Welfare_Guidelines.pdf.

296 issuing a damning analysis: The Pew Charitable Trust, *Putting Meat on the Table: Industrial Farm Production in America—A Report of the Pew Commission on Industrial Farm Animal Production,* April 2008, http://www.pewtrusts.org/uploadedFiles/wwwpewtrustsorg/Reports/Industrial_Agriculture/PCIFAP_FINAL.pdf and www.ncifap.org/bin/e/j/PCIFAPFin.pdf.

296 "the agro-industrial complex . . .": Ibid.

296 But these prices mask: Ibid.; *see* Greger, Michael, "The Human/Animal Interface: Emergence and Resurgence of Zoonotic Infectious Diseases," *Critical Reviews in Microbiology* 33 (2007): 243–299; and Donham, K. J., S. Wing, D. Osterberg, et al., "Community Health and Socioeconomic Issues Surrounding Concentrated Animal Feeding Operations," *Environmental Health Perspectives* 115, no. 2 (2007): 317–320.

296 a network of land-grant colleges: Act of July 2, 1862 (Morrill Act), Public Law 37–108, which established land grant colleges, July 2, 1862; and Enrolled Acts and Resolutions of Congress, 1789–1996, Record Group 11, General Records of the United States Government, National Archives.

297 Since 1960 milk production: Keusch, Gerald, et al., *Sustaining Global Surveillance and Response to Emerging Zoonotic Diseases* (Washington, DC: National Academies Press, 2009), p. 80, http://www.nap.edu/catalog.php?record_id=12625.

297 The broiler chicken of our day: Aho, P. W., "Introduction to the U.S. Chicken Meat Industry," in *Commercial Chicken Meat and Egg Production,* 5th ed., ed. D. D. Bell and W. D. Weaver (Norwell, MA: Kluwer Academic Publishers, 2002), 805.

297 There are now about 2 million farmers: U.S. Department of Agriculture, NASS. 2007 Census of Agriculture. Farm numbers. http://www.agcensus.usda.gov/Publications/2007/Online_Highlights/Fact_Sheets/farm_numbers.pdf; Hallberg, Milton C., *Economic Trends in U.S. Agriculture and Food Systems Since World War II* (Ames: Iowa State University Press, 2001), 13.

298 a budget now approaching $150 billion: USDA, USDA Budget Summary and Annual Performance Plan, FY 2011, p. 6, http://www.obpa.usda.gov/budsum/FY11budsum.pdf.

298 "the federal government paid out . . .": Cook, Ken, "Government's Continued Bailout of Corporate Agriculture," The Environmental Working Group, Farm Subsidy Database, http://farm.ewg.org/summary.php.

298 mainly for the largest and most profitable farms: Gaul, Gilbert M., Sarah Cohen and Dan Morgan, "Federal Subsidies Turn Farms Into Big Business," *Washington Post,* December 21, 2006.

298 federal subsidies account for about 20 percent: Roberts, 135, and Cook, "Government's Continued Bailout," at http://www.ewg.org/agmag/2010/05/farm-income-data-debunks-subsidy-myths/.

298 multibillion-dollar subsidies to the corn industry: Sweet, William, "Corn-O-Copia," *IEEE Spectrum Magazine,* January 2007; and Morgan, Dan, "Harvesting Cash, The Ethanol Factor—Corn Farms Prosper, but Subsidies Still Flow," *Washington Post,* September 28, 2007.

299 Congress and the USDA ask for nothing: Dilanian, Ken, "Bill Includes Billions in Farm Subsidies," *USA Today*, May 15, 2008; Cook, Christopher, "Farm Bill: Making America Fat and Polluted, One Subsidy at a Time," *Christian Science Monitor*, April 23, 2008; Karnowski, Steve, "Election Unlikely to Change US Farm Subsidies," *Boston Globe*, October 15, 2010.

299 They have opposed reform efforts: Reilly, Sean, "Alabama Farmers Federation Members Get Watch List from National Leader," *Birmingham News*, March 25, 2010; Jolly, Cliff, "Jolley: Five Minutes With Bob Stallman & The Defense Of American Agriculture," *Drovers*, November 19, 2010. Available at http://www.cattlenetwork.com/Jolley-Five-Minutes-With-Bob-Stallman-The-Defense-Of-American-Agriculture/2010-11-19/Article.aspx?oid=1284445&fid=CN-LATEST_NEWS; Fahrenthold, David A., "Manure Becomes Pollutant as Its Volume Grows Unmanageable," *Washington Post*, March 1, 2010.

299 The federal government funnels: Edwards, Chris, Agricultural Subsidies, The Cato Institute, June 2009, available at http://www.downsizinggovernment.org/agriculture/subsidies; Environmental Working Group at http://www.ewg.org/agmag/2010/05/governments-continued-bailout-of-corporate-agriculture/.

299 Given that more than 97 percent: Steinfeld, H., P. Gerber, T. Wassenaar, V. Castel, M. Rosales, and C. de Haan, *Livestock's Long Shadow: Environmental Issues and Options,* Food and Agriculture Organization of the United Nations, 2006, pp. 39, 43.

299 According to a 2007 study by Tufts University: Starmer, Elanor, and Timothy Wise, *Feeding at the Trough—Industrial Livestock Frims Saved $35 Billion from Low Feed Prices,* The Global Development and Environmental Institute, Tufts University, Policy Brief No. 07–03, p. 2, December 2007.

299 The same study found: Ibid.

299 USDA doled out hundreds of millions: Congressional Research Service, Farm and Food Support Under USDA's Section 32 Program, January 12, 2010, http://www.nationalaglawcenter.org/assets/crs/RL34081.pdf.

300 the hens have fragile bones: Gregory, N. G., and L. J. Wilkins, "Broken Bones in Domestic Fowl: Handling and Processing Damage in End-of-Lay Battery Hens," *British Poultry Science* 30, no. 3 (1989): 555–562.

300 As many as 24 percent of hens: Ibid.

300 Yet the U.S. government: Gregory, C., "Spent Layers: A Valuable Resource?," Poultry Service Industry Workshop, October 5–7, 2004, http://poultryworkshop.com/uploads/PDFs/PSIW%20proceedings%202004.pdf.

300 USDA purchases about 10 percent: Morrison, B., P. Eisler, and A. DeBarros, "Old-Hen Meat Fed to Pets and Schoolkids," *USA Today,* December 16, 2009, http://www.usatoday.com/news/education/2009-12-08-hen-meat-school-lunch_N.htm; and Gregory, "Spent Layers."

300 these hen carcasses: Roy, Parimal, A. S. Dhillon, Lloyd Lauerman, D. M. Schaberg, Daina Bandli, and Sylvia Johnson, "Results of Salmonella Isolation from Poultry Products, Poultry, Poultry Environment, and Other Characteristics," *Avian Diseases* 46, no. 1 (January 2002): 17–24.

300 It also manipulated egg exports: Reuters, "Land O'Lakes in $25 Mln Egg Price-Fixing Accord," June 8, 2010; Complaint and Demand for Jury Trial, *The*

*Kroger Co. et al. v. United Egg Producers, Inc. et al.,* Case No. 2:10-cv–06705-GP (E.D. Pa. Nov. 16, 2010); Complaint and Demand for Jury Trial, *Publix Super Markets, Inc. v. United Egg Producers, Inc. et al.,* Case No. 2:10-cv–06737-GP (E.D. Pa. Nov. 16, 2010).

300 Hallmark/Westland slaughter plant that closed in 2008: Flynn, Dan, "Suit Seeks Lunch Money Refund for Downer Beef," *Food Safety News,* October 2, 2009; and Durbin, Dick, "USDA Must Act to Ensure Safety of Food in the National School Lunch Program," January 20, 2008, http://durbin.senate.gov/showRelease.cfm?releaseId=291516.

301 the National Pork Producers Council (NPPC) asked Congress: Testimony of the National Pork Producers Council on the U.S. Pork Industry Economic Crisis Before the U.S. House Committee on Agriculture Subcommittee on Livestock, Dairy and Poultry, October 22, 2009, pp. 14–15.

301 Governors of nine leading pork-producing states: Lucht, Gene, "Governors Seek Help for Pork," *Iowa Farmer Today,* August 12, 2009.

301 When the House Agriculture Committee: Hearing to Review the Economic Conditions Facing the Pork Industry, Hearing before the Subcommittee on Livestock, Dairy and Poultry, 111th Congress, First Session, October 22, 2009, Serial Number 111–33.

301 the hog industry has degraded: The Pew Charitable Trust, *Putting Meat on the Table.*

301 the state imposed a moratorium: North Carolina General Assembly, Senate Bill 1465, Swine Farm Environmental Performance Standards, 2007, http://www.ncga.state.nc.us/Sessions/2007/Bills/Senate/PDF/S1465v7.pdf.

301 The NPPC has also led efforts: Eckholm, Erik, "U.S. Meat Farmers Brace for Limits on Antibiotics," *New York Times,* September 14, 2010.

301 a practice that has hastened: Silbergeld, Graham, "Industrial Food Animal Production, Antimicrobial Resistance, and Human Health," *Annual Review of Public Health* 29 (2008): 151–169; Ladely, Scott R., et al., "Development of Macrolide-Resistant *Campylobacter* in Broilers Administered Subtherapeutic or Therapeutic Concentrations of Tylosin," *Journal of Food Protection* 70, no. 8 (August 2007): 1945–1951; and Sapkota, Amy R., et al., "What Do We Feed to Food Production Animals? A Review of Animal Feed Ingredients and Their Potential Impacts on Human Health," *Environmental Health Perspectives* 115 (2007): 663–670 (published online February 8, 2007).

301 factory farms are the most dangerous: Greger, Michael, "The Human/Animal Interface: Emergence and Resurgence of Zoonotic Infectious Diseases," *Critical Reviews in Microbiology* 33, no. 4 (2007): 243–299.

301 A hog factory run: Schmidt, C. W., "Swine CAFOs & Novel H1N1 Flu: Separating Facts from Fears," *Environmental Health Perspectives* 117, no. 9 (2009): A394-A401.

301 American factory farms threaten: Saenz, R. A., H. W. Hethcote, and G. C. Gray, "Confined Animal Feeding Operations as Amplifiers of Influenza," *Vector-Borne and Zoonotic Diseases* 6, no. 4 (Winter 2006): 338–346.

302 the industry has managed: United States Code Annotated Currentness, Title 7, Agriculture, Chapter 48, "Humane Methods of Livestock Slaughter."

302 and birds account for more than 95 percent: U.S. Department of Agriculture, National Agricultural Statistics Service, Quick Stats, "U.S. & State—Slaughter" and "U.S. & State—Poultry Slaughter," http://www.nass.usda.gov/Data_and_Statistics/Quick_Stats_1.0/index.asp.

304 The American Veterinary Medical Association: Executive Summary of the AVA Response to the Final Report of the Pew Commission on Industrial Animal Farm Production, November 5, 2009, http://www.avma.org/advocacy/PEWresponse/.

304 "dangerous and under-informed recommendations . . .": Ibid.

304 Medical assistance for dogs and cats: Personal communication with Andrew Rowan, October 10, 2010.

305 even if that means leaving particular animals to suffer:

305 The American Association of Swine Veterinarians: http://www.aasv.org/aasv/aasvisc.php says "Members of the American Association of Swine Veterinarians Industry Support Council provide financial support for the publication of *The Journal of Swine Health and Production.* The 2010 Council includes: Alpharma Inc., Bayer Animal Health, Boehringer Ingelheim Vetmedica, Inc., Elanco, Harrisvaccines, Inc., Intervet/Schering-Plough Animal Health, MVP Laboratories, Newsham Choice Genetics, Novartis Animal Health U.S., Inc., Pfizer Animal Health, PIC International.

305 The American Society of Laboratory Animal Practitioners: Position statement of the American Society of Laboratory Animal Practitioners, approved by board on July 15, 2007, available at http://www.aslap.org/ClassBDealers2007.pdf.

305 The American Association of Bovine Practitioners: Personal communication with AABP president Roger Saltman, January 3, 2010.

306 the AVMA sat on the sidelines: Stull, Caroline, et al., "A Review of the Causes, Prevention and Welfare of Non-Ambulatory Cattle," *Journal of the American Veterinary Medical Association* 231, no. 2 (July 15, 2007): 227–234.

306 The veterinary association has refused: "AVMA Takes No Position on Foie Gras, Opposes Database Mining," *Online Journal of the American Veterinary Medical Association,* September 1, 2007, http://www.avma.org/onlnews/javma/sep07/070901f.asp.

306 And until just a few years ago: "AVMA Board Opposes Five of Six Resolutions Delegates Will Vote on in July," *Journal of the American Veterinary Medical Association News,* June 15, 2003.

306 AVMA supported confining calves: McPheron, Tom, "AVMA Passes Groundbreaking Animal Welfare Policies," American Veterinary Medical Association, press release, July 21, 2008.

306 Dosing the animals with these drugs: Office of Technology Assessment, *Drugs in Livestock Feed.* Vol. 1: *Technical Report* (Washington, DC: U.S. Government Printing Office, 1979), 41, www.princeton.edu/~ota/disk3/1979/7905/7905.PDF.

307 "now we're seeing increasing numbers . . .": Kristof, Nicholas, "The Spread of Superbugs," *New York Times,* March 6, 2010.

307 "Even back then," he writes: Kennedy, Donald, "Cows on Drugs," *New York Times,* April 17, 2010.

307  The best-known public-health organizations: "Keep Antibiotics Working, The Campaign to End Antibiotic Overuse," http://www.keepantibioticsworking. com; and Kennedy, "Cows on Drugs."

307  Only one major medical group opposes: Ibid.; and AVMA Issue Brief on S619-HR.1549, Preservation of Antibiotics for Medical Treatment Act of 2009.

308  "Federal standards for farm animal welfare,": Pew Charitable Trust, *Putting Meat on the Table.*

308  confinement systems are: Smith, Jeff, "Vets Group Takes Ethical Stance on Ballot Measure," *Modesto Bee,* August 7, 2008.

308  and leaned hard on the CVMA: Personal communication with Jeff Smith, September 2008.

308  vehement opposition to federal legislation: Nolen, R. Scott, "U.S. Horse Slaughter Exports to Mexico Increase 312%," *Journal of the American Medical Association,* January 15, 2008. Available at http://www.avma.org/onlnews/javma/jan08/080115a.asp.

309  Agriculture groups, led by the Farm Bureau: Skrinjar, Janelle, "Farm Bureau Blasts Horse Slaughter Ban," *Farm and Dairy,* March 15, 2007. Available at http://www.farmanddairy.com/news/farm-bureau-blasts-horse-slaughter-ban/427.html.

309  If slaughter plants aren't allowed: American Veterinary Medical Association, "Unwanted Horses and Horse Slaughter: Frequently Asked Questions," September 5, 2008. Available at http://www.avma.org/issues/animal_welfare/unwanted_horses_faq.asp.

309  tens of thousands of abandoned, unwanted horses: Ibid.

310  telling a congressional committee: Testimony of Dr. Douglas Corey, DVM, before a House Congressional Hearing regarding H.R. 503, A Bill to Amend the Horse Protection Act, 109th Congress, 2005–2006. July 5, 2006, p. 115.

310  Dr. Nicholas Dodman, a veterinarian: The testimony of Dr. Nicholas Dodman before the Subcommittee on Crime, Terrorism and Homeland Security, The Prevention of Equine Cruelty Act of 2008 & the Animal Cruelty Statistics Act of 2008, 110th Congress, July 31, 2008, p. 62.

310  "Any group or organization . . .": Ibid.

## Chapter Eight—The Humane Economy

312  Whales and dolphins have learned: "State of the Sanctuaries 2006 Accomplishments Report," National Marine Sanctuaries, http://sanctuaries.noaa.gov/sos2006/stellwagen.html; and Provincetown Center for Coastal Studies, http://www.coastalstudies.org/what-we-do/stellwagen-bank/baleen-whales.htm.

313  producing a kind of bill of particulars: Salt, Henry S., *Animals' Rights Considered in Relation to Social Progress* (London: George Bell, 1892), 10, 114.

314  industrial whaling fleets slaughtered: Adams, Mark, *Against Extinction: The Story of Conservation* (London: Earthscan, 2004); and the United Nations Environmental Data.

314  We have now heard the "songs": Payne, Roger, and Scott McVay, "Songs of Humpback Whales," *Science* 173, no. 3997 (1971): 585–597; and Hildebrand,

John, and Erin Oleson, "Behavioral Context of Call Production by Eastern North Pacific Blue Whales," *Marine Ecology Progress Series Journal* 330 (2007): 269–284.

317    the famed Harvard economist Joseph Schumpeter: Schumpeter, Joseph A., *Capitalism, Socialism, and Democracy,* 5th ed. (New York: Harper and Brothers, 1942; London: George Allen and Unwin, 1976), chapter 7.

319    Matthew Sleeth, who himself grew up: Matthew Sleeth, interview by author, March 25, 2010; and Sleeth, Matthew, *The Gospel According to the Earth* (New York: HarperOne, 2010).

320    The prospect of seeing whales: Tourism New Zealand—The Official Website of the New Zealand Tourism Board, May 27, 2010, http://www.newzealand .com/travel/media/press-releases/2010/5/tourism-news-whale-watch-award_ press-release.cfm/.

320    as a nation generates more than $80 million: International Whaling Commission, Report of the Conservation Committee, 2010, IWC/62/Rep 4.

321    more than thirteen million people: O'Connor, S., R. Campbell, H. Cortez, and T. Knowles, *Whale Watching Worldwide: Tourism Numbers, Expenditures and Expanding Economic Benefits*, a special report from the International Fund for Animal Welfare, Yarmouth, MA, USA, prepared by Economists at Large, 2009; and Garret, Peter, speech given by Australia's Minister for Environmental Protection during the International Whaling Commission meeting in Agadir, Morocco, June 21–25, 2010.

321    a gusher in Titusville, Pennsylvania, in 1859: PBS; "Who Made America?," http://www.pbs.org/wghb/theymadeamerica/whomade/drake_hi.html; and Giddens, Paul H., *The Early Days of Oil* (Princeton, NJ: Princeton University Press, 1948).

322    a thriving global watching industry: O'Connor et al., *Whale Watching Worldwide.*

322    younger generations of Norwegians and Japanese: "Big Three Fisheries Companies Say No to Re-Entering Business of Commercial Whaling," *Asahi Shimbun,* June 13, 2008.

324    rebuffing lobbying by the Safari Club International: Mbaria, John, "Law on Culling Could Promote Game Hunting," *The Nation,* March 15, 2007, http:// www.bushdrums.com/news/index.php?shownews=865; and "Will Kenya Learn from Its Southern Neighbors?," *African Indaba e-Newsletter* 2, no. 6 (November 2004): 8–9, http://bigfivehq.com/no2–6.pdf.

324    The market for ecotourists dwarfs: Leonard, Jerry, "Wildlife Watching in the U.S.: The Economic Impacts on National and State Economies in 2006," http://library.fws.gov.pubs/nat_survey2006_economicvalues.pdf.

325    Whale meat is stockpiled: Japan, Ministry of Agriculture, Forestry and Fisheries monthly statistics on frozen marine product stockpiles, www.maff.go.jp .www/info/bunrui/bun06.html#tsuki4.

326    wildlife watching is officially: Leonard, "Wildlife Watching in the U.S."

329    Recent studies on the concept of "cognitive mapping": Sawa, Kosuke, J. Kenneth Leising, and Aaron P. Blaisdell, "Sensory Preconditioning in Spatial Learning Using a Touch Screen Task in Pigeons," *Journal of Experimental*

*Psychology, Animal Behavior Processes* 31, no. 3 (2005): 368–375; and Tolman, Edward C., "Cognitive Maps in Rats and Men," *Psychological Review* 55, no. 4 (1948): 189–208.

330 These commercial operators: Cea, John, "Alternatives for Nuisance Animal Disposal," *Animal Damage Control* 4, no. 1 (1996): 11–12; and Dr. John Hadidian, interview by author, Gaithersburg, Maryland, July 13, 2010.

332 "We are about at the point . . .": Dr. John Hadidian, interview by author, Gaithersburg, Maryland, July 13, 2010.

332 a co-op board abruptly decided to destroy the nest: Lueck, Thomas J., "New York Celebrities Evicted on Fifth Ave., Feathers and All," *New York Times*, December 8, 2004, pp. B1, B3.

333 Some planned communities, such as Harmony: Florida Sustainable Communities Network, Smart Communities of Avalon Park and Celebration; and Harmony, http://www.harmonyfl.com; Congress for New Urbanism, http://www.cnuflorida.org.

333 first rule of intelligent thinking: Leopold, Aldo, *Round River* (New York: Oxford University Press, 1996), 145–146.

335 their product of choice is OvoControl: Friess, Steve, "Feeling Pooped by Pigeons, Cities Try Bird Birth Control," AOL News, July 2, 2010.

335 but the alternative is far worse: Dr. John Hadidian, interview by author, Gaithersburg, Maryland, July 13, 2010.

335 using the vaccine to control elephant populations: "HSI Announces New Elephant Immunocontraceptive Method," August 31, 2010, http://www.humane society.org/news/press_releases/2010/08/hsi_announces_new_elephant_contra ception_083110.html; and Rutberg, Allen T., and Rick Naugle, "Testing the Effectiveness of One-Shot Immunocontraceptives on White-Tailed Deer at Fripp Island, South Carolina—2008 Progress Report to the South Carolina Department of Natural Resources and the Fripp Island Property Owners Association," February 2, 2009, http://www.fipoa.org/fyi_files/FIdeerReport08.pdf.

335 Researchers Jay Kirkpatrick and John Turner: Kirkpatrick, Jay F., "Management of Wild Horses by Fertility Control: The Assateague Experience," *National Park Service Scientific Monograph*, No. 26 (1995): 60.

336 "fast disappearing from the American scene": The Wild Free-Roaming Horses and Burros Act of 1971—Public Law 92–195, http://www.wildhorsepreserva tion.com/resources/theact.pdf.

336 rescued by Cleveland Amory: Amory, Cleveland, *Ranch of Dreams, The Heartwarming Story of America's Most Unusual Animal Sanctuary* (New York: Penguin Publishing, 1997), 56–91.

336 the agency has swamped the system: Lombardi, Kristen, "BLM Fights to Keep Secret Names of Ranchers with Grazing Permits," Sunlight Foundation Reporting Group, April 21, 2010; and Committee on Natural Resources, Legislative Hearing, The Restore Our American Mustangs Act, H.R. 1018, March 3, 2009.

336 the captive population of wild horses: U.S. Government Accountability Office, Bureau of Land Management, *Effective Long Term Options Needed to Manage Unadoptable Wild Horses*, October 2008, pp. 44–45.

337 Gary Michelson has pledged $75 million: "The Michelson Prize and Grants in Reproductive Biology," http://www.foundanimals.org/index.php/About -Michelson/the-michelson-prize.html.

337 approaches have lowered annual: The Humane Society of the United States, "Common Questions about Animal Shelters," October 26, 2009, http://www .humanesociety.org/animal_community/resources/qa/common_questions_ on_shelters.html.

338 chemicals have been routinely: U.S. Code of Federal Regulations, Title 40, Part 158.500; and International Conference on Harmonization of Technical Requirements for Registration of Pharmaceuticals for Human Use; European Union Regulation No 1907/2006 on the Registration, Evaluation, Authorization and Restriction of Chemicals (REACH).

338 Estimates are that ten to twenty million: Taylor, K., N. Gordon, G. Langley, and W. Higgins, "Estimates for Worldwide Laboratory Animal Use in 2005," *Alternatives to Laboratory Animals* 36 (2008): 327–342.

338 the animals are rarely given: *OECD Guidelines for the Testing of Chemicals* (Paris, France: OECD), http://www.oecd-ilibrary.org/content/serial/20745788; jsessionid=136mnajj661qm.delta; ICH Test Guidelines for the safety testing of human pharmaceuticals, http://www.ich.org/cache/compo/276–254–1.html; and VICA Test Guidelines for the safety testing of veterinary pharmaceuticals, http://www.vichsec.org/en/guidelines2.htm.

338 poisoning animals might tell us: European Parliament and Council, Regulation No. 1907/2006 on the Regulation, Evaluation, Authorization and Restriction of Chemicals (REACH), http://bit.ly/9mYso2; and The Humane Society of the United States, "EU Initiative to 'AXLR8' Move to High-Tech, Animal-Free Methods for Chemical and Drug Testing," April 15, 2010, http:// www.humanesociety.org/news/press_releases/2010/04/axlr8_eu_initiative _041510.html.

339 new drug candidates have only: U.S. Food and Drug Administration, *Challenges and Opportunities Report* (Washington, DC: Author, March 2004).

339 animal testing can be problematic: Fox, Maggie, "Government Labs Try Non-Animal Testing," Reuters, February 14, 2008; and Weise, Elizabeth, "Three U.S. Agencies Claim to End Animal Testing," *USA Today,* February 14, 2008.

339 Nicholson, a brilliant businessman: Bill Nicholson, interview by author, July 14, 2010.

340 Greg put an end to animal testing: "Amway Halts Animal Tests," *New York Times,* August 22, 1989, Business Day section.

340 law does not even cover: Carbone, *What Animals Want.*

341 In 2008, in response to a campaign: European Commission, Letter of Commitment to the Parliament, May 5, 2008, Brussels.

342 Some experts believe: Weise, "Three U.S. Agencies."

342 highly sophisticated computer-based approaches: Berg, Ninna, et al., "Toxicology in the 21st Century—Working Our Way Towards a Visionary Reality," *In Vitro Testing Industrial Platform,* November 26, 2009; and Humane Society International, "EU Initiative."

342 a conventional animal test: Billington, Richard, et al., "The Mouse Carcino-

genicity Study Is No Longer a Scientifically Justifiable Core Data Requirement for the Safety Assessment of Pesticides," *Critical Reviews in Toxicology* 40, no. 1 (2010): 35–49, http://bit.ly/9LX4Ds; NIEHS Fact Sheet No. 3–NTP–9/96, National Toxicology Program.

342 without any use of animals: Personal communication from Dr. Chris Austin, director of the U.S. National Institutes of Health Chemical Genomics Center.

343 with an additional $30–50 million: Personal communication with Andrew Rowan, October 6, 2010.

345 They went to Capitol Hill: Heilprin, John, "Shark Attack Victims Aim to Protect Sharks," Associated Press, September 14, 2010.

345 Every year close to a hundred million sharks: O'Malley, Mary, The Shark Safe Network, 2009; and Ling, Lisa; "Shark Fin Soup Alters Ecosystem," CNN, December 15, 2008, http://www.cnn.com/2008/WORLD/asiapcf/12/10/pip.shark.finning/index.html.

348 Embedded in the flesh: Conroy, Erin, "Netted Whale Hit by Lance a Century Ago," Associated Press, June 12, 2007.

# INDEX